REALMS YOU WILL NEVER SEE, HEAR, TASTE, TOUCH, OR SMELL

Sunlight, meteoroids, tiny particles, highly energetic rays and radiation—these cosmic forces are continually bombarding the earth. THE CYCLES OF HEAVEN is the first giant step toward understanding how these powerful energies affect man and his environment in such areas as laser acupuncture, biorythms, electromagnetic medicine, cybernetics, meteorological phenomena, animal ESP, and much more.

Guy L. Playfair and Scott Hill document surprising examples of how invisible forces influence our behavior and they raise provocative questions: Are flourescent lights shortening our lives? Do TV sets transmit viruses to humans? What really happened to the astronauts of Soyuz XI?

Backed by years of diligent research, THE CYCLES OF HEAVEN is an exploration of scientific forces that are only just beginning to be understood and that could well change man's view of himself and his universe.

THE CYCLES OF HEAVEN

BY
**GUY L. PLAYFAIR
AND SCOTT HILL**

AVON
PUBLISHERS OF BARD, CAMELOT AND DISCUS BOOKS

AVON BOOKS
A division of
The Hearst Corporation
959 Eighth Avenue
New York, New York 10019
Copyright © 1978 by Guy Lyon Playfair and Scott Hill
Published by arrangement with St. Martin's Press, Inc.
Library of Congress Catalog Card Number: 77-15823
ISBN: 0-380-45419-x

Cover illustration by Michael Marks of Art-So-Fine Studios

All rights reserved, which includes the right to reproduce this book or portions thereof in any form whatsoever. For information address: St. Martin's Press, Inc., 175 Fifth Avenue, New York, New York 10010

First Avon Printing, August, 1979

AVON TRADEMARK REG. U.S. PAT. OFF. AND IN
OTHER COUNTRIES, MARCA REGISTRADA, HECHO EN
U.S.A.

Printed in the U.S.A.

CONTENTS

Introduction	ix
Acknowledgments	xvii
PART ONE: BACKGROUND	**1**
1 The Many-Body Cosmos	3
2 The Imperfect Sun	29
PART TWO: FOREGROUND	**67**
3 Homo Electromagneticus	69
4 Electric Weather	102
5 Earthquake	139
PART THREE: RHYTHMS OF LIFE	**167**
6 Cycles	169
7 Biorhythm	188
8 Bioentrainment	201
9 Biofeedback—Self-Entrainment	221
PART FOUR: BEYOND THE FRONTIERS	**243**
10 Baby and Bathwater	245
11 The Electromagnetic Web	277
12 The Energy Body	296
List of References	325
Index	351

The Eagle soars in the summit of Heaven,
The Hunter with his dogs pursues his circuit.
O perpetual revolution of configured stars,
O perpetual recurrence of determined seasons,
O world of spring and autumn, birth and dying!
The endless cycle of idea and action,
Endless invention, endless experiment,
Brings knowledge of motion, but not of stillness;
Knowledge of speech, but not of silence;
Knowledge of words, and ignorance of the Word.
All our knowledge brings us nearer to our ignorance,
All our ignorance brings us nearer to death,
But nearness to death no nearer to GOD.
Where is the Life we have lost in living?
Where is the wisdom we have lost in knowledge?
Where is the knowledge we have lost in information?
The cycles of Heaven in twenty centuries
Bring us farther from GOD and nearer to the Dust.

T. S. ELIOT

INTRODUCTION

Earthly life is bound by invisible threads to the cosmos, both near and far.

To be subjected to cosmic effects, man does not have to be shot into space. He does not even have to leave his room.[1]

These are not occult speculations. They are scientifically established facts. And they are facts of vital importance.

Every living being on Earth is under constant attack, from birth to death, by the natural forces of the cosmos. Most of these forces are invisible, few of them are either fully understood or consciously perceived, and all of them may be influencing our everyday lives in many ways, for better or worse.

We are in control of our own personal destinies to a lesser extent than some of us would like to think. On the other hand, we *could* be in control of them far more than we are if only we understood a lot more about our cosmic environment and our place in it. This book is an attempt to add to such an understanding, one we believe to be long overdue.

Our claims are based on a vast amount of research—most of it very recent and much of it hitherto unpublished in English—the work of several hundred scientists working in many different specialized fields in a dozen countries. Yet very little of this research has yet become familiar even to other scientists, outside one specialist field or another, let alone to the general public. And no one has pieced it all together, as we have, and found that an entirely new model of man has been emerging these past twenty years.

What are these cosmic forces? Where do they originate? To what extent do they affect us? And do they really matter?

Present-day physics knows of only four forces in nature, and for the most part, only two of these—gravity and electromagnetism—concern us here, although we introduce bionuclear forces in our final chapters. These forces, most of them extraterrestrial in origin, affect us in a number of ways, and as to whether they matter, surely anything that influences human behaviour matters? Especially when it does so without our being *consciously* aware that we are being influenced at all.

Force, says the dictionary, is *capacity to cause physical changes or to affect mind and behaviour*. It is not easy to deal with "extraterrestrial forces" or "cosmic influences" without lapsing into occult speculation or mystical obscurity. So let us make it clear from the start that although ours is a book about very mysterious cosmic influences, it is also a book of fact, based firmly on the published work of established scientists. It is not about past catastrophes or future utopias (though we touch on these subjects) but about the present, and what is happening in it.

When we stroll along a beach in the moonlight and watch the tide coming in or going out, we are witnessing a visible phenomenon of extraterrestrial origin at close quarters. Peaceful as the scene may appear, how many of us are aware of the tremendous forces at work? It is not generally known, for instance, that not only are the waters surrounding the Earth disturbed by the tidal pull of the Moon, the Sun and the planets, but so is the solid Earth itself. So indeed, as we shall see, are people.

There would be no life on Earth as we know it without cosmic forces, such as electromagnetic radiation from the Sun. But there are many more cosmic inputs than just sunlight. These range from solid bodies such as meteoroids to tiny particles and many forms of radiation, some of which are very hard to detect. Luckily for us, our planet's natural defences keep out most of the big stuff, but plenty of highly energetic rays or waves, heavy ions, protons and electrons manage to slip through into the atmosphere in which we all live and breathe.

Such invisible forces can do more than give and sustain life. They can deform and even destroy it, and until very recently we have been defenceless against them for the simple reason that we could not even identify most of them. This is not to say that we were totally unaware of them, for as we shall point out, man's instincts have always been a step ahead of his technology.

Technology seemed to be catching up when, on 4 October 1957, the launching of Sputnik 1 inaugurated the Space Age. In the two decades that followed this first great venture outwards, we learnt more about the cosmos at first hand than had ever been known throughout the recorded history of our species.

Soon came the Information Explosion, and with it the fragmentation of many existing scientific disciplines into specialist fields, and fields within fields. Yet Nature, unlike universities, is not divided into faculties, departments and subdepartments. It is not divided into anything. It is one whole, and it affects all of us all the time. Much as we need the specialist to probe the mysteries of the macroscopic quasar or the microscopic quark, we also need the interdisciplinary overview. Every so often, we must collect the fragments, put them together, and see what new picture emerges. We need such a picture of the forest as much as we need detailed analyses of each tree, bird, animal, insect and plant in it. This book is an attempt to present such a coherent picture, not of a forest, but of man in his cosmic environment.

Today's specialist is a busy man. He can hardly keep pace with the published literature on his own subject, let alone anybody else's. However, as the editor of *Nature*, one of the world's leading scientific magazines, has suggested, the time has come when scientists can no longer afford *not* to talk to each other. (We quote him more fully in Chapter 6).

One of man's main concerns has always been to connect one thing with another. In 1820, Ørsted and Ampère found the link between electricity and magnetism, thus discovering *electromagnetism*, without the many practical applications of which our lives would be very much the poorer. When Einstein linked time, space, energy and matter, we had to change our whole concept of the universe. More recently, as *cybernetics* and *ecology* have become everyday words, man has come to realize that he is linked with his environment in a single system. More recently still, the biofeedback revolution has done much to solve one of the oldest of mysteries: that of the links between mind and matter.

Some form of invisible force affects everything on, above, and under the Earth, whether living or not. A host of questions arise that cannot be answered by one specialist alone, which may explain why they are seldom asked in the first place. For instance:

> Why do more people commit suicide in April and May than in any other months?
>
> What might the positions of Uranus or Jupiter have to do with the timing of earthquakes?
>
> How do sound waves make us sick?
>
> What could the level of Lake Huron have in common with the abundance of partridge in Hertfordshire?
>
> Is there a link between the appearance of sunspots and the way people vote?
>
> What are the facts behind the sinister headline in May 1977 newspapers: JUMBO JET PASSENGERS COUGH BLOOD?
>
> What really happened to the astronauts of Soyuz XI?

These questions, all of them quite important, are typical of those that turn up again and again when we set out to discover what really goes on in the natural environment that surrounds us, extending beyond our skins, garden gates, city limits, national and continental boundaries, out into the spaces between the planets of our solar system, and far beyond . . .

Our first two chapters deal with Earth as a planet in its cosmic environment centred on its local star, the Sun. We introduce our two principal invisible forces, gravity and electromagnetism, and prepare the scene for Chapters 3 and 4, in which we get to grips with the main theme of the book: the interactions between man and nature, including those of which he is not consciously aware, and the complex effects of such interactions.

In Chapter 5 we look at earthquakes, as examples of natural and very strange forces at their most extreme and destructive, with all too visible effects. Chapters 6 and 7 introduce the very important subject of cycles, in nature, human affairs, and the natural rhythms of the human body. We see how these also interact. Then, in Chapter 8, which is perhaps the most alarming in the book, we show how invisible external forces can, and do, influence our thoughts as well as our actions. Chapter 9, on the other hand, shows how we can fight back and reassert our control over ourselves and at least some of our implacable environment.

In our last three chapters, we venture far beyond the frontiers of established science, although as in the rest of the book, our source material comes either from the published work of qualified scientists or from our own firsthand experience. We discuss the fascinating new concept of man's "energy body" and its radiations, and how it may be interacting with its energetic surroundings for good or ill. We also include the first eye-witness account to appear in English of current research in the Soviet Union of potentially tremendous significance.

More questions arise, some of which would seem absurd were it not for the facts we present:

> Is there a relation between "planetary heredity" and our ability to realize our full potential?
> Does a baby play a part in determining the precise timing of its own birth?
> Do the fortunes of man—both individually and collectively—oscillate in cycles? And are these cycles driven by those of our all-powerful Sun? And are these in turn driven by those of the planets?
> Have Soviet scientists upstaged the Spiritualists with their discovery of the bioplasma body? Can this body be made visible and be photographed?
> Will the introduction of computerized acupuncture diagnosis alter our whole approach to medicine and healing?
> Is the secret of life to be found in an electromagnetic wave?

We have included as much recent research as possible in this book, although this has meant that many of our early drafts went out of date even as we were writing them. We have also rescued some earlier work that has been swept under the rug, and view it in a new, wider context. This rescue operation has led to many surprises. Again and again we have found a "modern" discovery to be no more than an old one rediscovered, as man's technology limps along behind his boundlessly creative imagination. We have also found that many marvels of modern technology have been known to nature for millions of years.

This book is an interdisciplinary undertaking in itself.

It came about after the authors discovered that each was working independently on very similar books, and decided to join forces rather than compete. It is a collaboration between a research scientist (Hill) trained in biophysics—itself an interdisciplinary science combining biology with physics—who over the past six years has amassed a huge pile of material from many sources and also carried out a good deal of experimental work of his own; and a writer (Playfair) with no formal scientific training of any kind who is mainly interested in the unexplained border areas of human activity, about which he has already published two books of his own.* Hill has contributed several articles to specialist journals, and has recently co-translated an important Russian book, to be mentioned in Chapter 12. Between them, the authors speak and read a number of languages, and much of the German, French, Russian and Danish material in this book appears for the first time in English.

The result of such a collaboration, we hope, is a book that presents considerably more technical information than is usually to be found in popular works, written in a style that makes it accessible to anybody interested in learning more about the human condition.

Almost everything we have to report is controversial, and nothing is true just because we say so, or because anybody, however eminent he may be, says so. Conversely, nothing is untrue just because it seems implausible or because sceptics decide it to be impossible. For reasons of space, we have not included every argument and counter-argument that has followed the publication of much of our source material. However, we are well aware of the need for further research in all the areas we mention.

Although, as we have said, this is a book of facts about the present, we have lightened it here and there with some speculation of our own, and with one or two items that may have no scientific basis at all, such as our account of how to levitate blocks of stone! We have done our best to make it clear what we are offering as fact or speculation.

"The Space Age has begun," Michel Gauquelin wrote in 1966, "and it is therefore vitally important to construct

---

* *The Flying Cow* (1975) and *The Indefinite Boundary* (1976) London: Souvenir Press.

a solidly based science of the relations which exist between men and the Universe."[2] Gauquelin himself has done much to help both construct and popularise this new science, and we include a discussion of his work along with that of his predecessors and contemporaries such as A. L. Chizhevsky from Moscow, William Petersen (Chicago), Maki Takata (Japan), Solco W. Tromp (Leiden, Holland), Viktor Inyushin (Alma-Ata, Kazakhstan) and Giorgio Piccardi (Italy), among many others to be mentioned in the appropriate chapters.

"The new sciences regard man as a being of immense and irreducible complexity participating in continual and intimate interaction with his environment," Stuart Holroyd wrote in 1977. "And by environment they mean everything between his skin and the stars."[3] The discoveries of the first two decades of the Space Age show this statement to be fully justified.

The excitement of space travel has captured the imaginations of millions. We now dream of light-year-long odysseys in deep space, of colonizing planets and creating artificial societies in the sky, of contracting intelligences on other worlds, and of popping in and out of black holes. (Into them, anyway.) We have even sent the voice of the U.S. president plus some Bach and pop music out to entertain the galaxies.

But what we have tended to overlook is the huge mass of data the Space Age has already provided us with about *ourselves*. Our discovery of inner space is as exciting as that of outer space, and is of more immediate importance. There is still as much to be learned about the biomolecule as about the Moon, and even more about ways in which one affects the other.

Man has been described as a "living sundial" and a "living barometer" (by Hakata and Piccardi respectively). We terrestrians, Gauquelin has said, "can regard ourselves as living in the interior of the Sun." Chizhevsky, with a touch of Slavonic fatalism, once asked: "Are we all slaves of the Sun?" Tromp has shown that nearly every organ in the human body responds to invisible stimuli. A Polish scientist has even boldly defined life as "an electromagnetic wave."

We are gradually discovering exactly what the forces of nature are doing to us, thereby relating the macrocosmos to the microcosmic universe beneath our skins. Some of these discoveries are encouraging, while as we shall see, others are alarming. Yet we must face up to them all. A better understanding of them can only bring benefit to the entire human race.

Faced with the explosion of data from the early years of the Space Age, we may repeat the question T. S. Eliot asked twenty years before it all began:

Where is the knowledge we have lost in information?

We are about to begin a search for that lost knowledge.

ACKNOWLEDGMENTS

We are grateful for help received in many forms from the following, none of whom can be held responsible for any views expressed except where specifically attributed:

Dr. S. W. Tromp (Leiden), Dr. Michel Gauquelin (Paris), John H. Nelson (Whiting), William H. Kautz (Palo Alto), Dr. John Gribbin (Brighton), Joe Cooper (Leeds), Professor E. R. Laithwaite (London), Hernani G. Andrade (São Paulo), Dr. Richard D. Mattuck and Dr. P. V. Sharma (Copenhagen), Dr. Aleksei S. Romen and Dr. Viktor M. Inyushin (Alma-Ata), Dr. Geoffrey A. Dean (Perth), Dr. E. Stanton Maxey (Stuart, Florida), Arthur A. Prieditis (Chicago), Dr. J. H. L. Playfair and Edward Playfair (London). We thank the librarians and staff of the Science Museum, Natural History Museum, Institute of Geological Sciences, College of Psychic Studies and Society for Psychical Research libraries (London) and of the H. C. Ørsted Institute of the University of Copenhagen.

For permission to use copyright material, we thank:

Dr. David Davies, John H. Nelson and the American Federation of Astrologers, the Science Museum (London), L. Kjellson, the Soviet Academy of Sciences, the American Journal of Chinese Medicine, Dr. V. M. Inyushin, and Resch Verlag (Innsbruck). The lines from *The Rock* (1934) are reprinted by permission of Faber and Faber Ltd. from *Collected Poems 1909-1962* by T. S. Eliot.

Thanks are also due to Robert Horner Photography Ltd. and the Society of Authors (London) for photographic and photocopying work respectively.

Finally, we thank the authors of the 400-plus works

listed at the end of the book. Without their efforts we would have had no book to write.

Neither the authors nor the publisher can be held in any way responsible for any of the commercially manufactured appliances mentioned in the book or listed in the List of References. Full addresses of suppliers are given for the convenience of readers.

PART ONE

BACKGROUND

Chapter 1

THE MANY-BODY COSMOS

In October 1976, British Rail proudly inaugurated its high-speed train service between London and Bristol, and passengers were soon surprised to find that they could enjoy a cup of coffee at 125 mph without the cup rattling on the saucer. Some years previously, an airline publicized the fact that a child could build a card-house while flying in one of the new jetliners at more than 500 mph. Today, astronauts can relax while moving at tens of thousands of miles per hour.

It is just as surprising, when we come to think of it, that we can relax, build card houses and drink coffee in comfort as we sit quietly in our homes. For nothing in the Universe is ever motionless. Earth rotates on its axis at about 1,800 mph (at the equator) while tearing round the Sun at the same time at 66,000 mph, and the entire solar system is hurtling in the general direction of the constellation Sagittarius at around 48,000 mph. There will not be a collision, though, since every point in the Universe is moving away from every other point, like dots on a balloon being blown up. We cannot help being reminded of the hero in a Stephen Leacock story who "flung himself upon his horse and rode madly off in all directions."

Yet we can enjoy our coffee in peace, because everything on spacecraft Earth moves at the same speed and is held in position by the force of gravity. If the Earth were to change its rotational or orbital revolution velocity even slightly, we would become acutely aware of the tremendous velocities involved, and of our own quite appreciable inertia—assuming we had time to think about such things before being annihilated. This may never happen, although some scientists regard it as possible, reminding us that nowadays it is unwise to take anything in nature for granted.

We are lulled into a false sense of security by the seemingly constant motion of our planet. Earth does not in fact move at the same speed all the time; it slows down in September and speeds up in March, at the autumnal and vernal equinoxes. Moreover, it does not revolve smoothly, but wobbles like a spinning top about to fall over.

Not only are we rushing about in all directions, like Leacock's horseman, but we are also under constant attack. We are trapped in the midst of a non-stop bombardment of particles, waves and rays, plus occasional larger objects such as meteorites, which can weigh several tons. (Not to mention whatever it was that crashed into Siberia on 30 June 1908, doing as much damage as an atomic bomb would have done.)

In addition to all this activity, we are being pushed and pulled around by tidal (gravitational) forces from other bodies in our solar system. We are also caught in a spider's web of electrical and magnetic fields of all shapes and sizes, from huge interplanetary superfields to tiny local fields-within-fields. It is a wonder we survive; not in spite of all this energy and motion, but *because* of it.

Even in an age of science, we still take a very unscientific view of our cosmic surroundings. We talk about *sunrise* and *sunset*, although the Sun neither rises nor sets. (What we mean is that as Earth turns on its axis, there comes a moment when our horizon intersects the solar orb.) We say that spring begins when the Sun "enters the sign of Aries," when it is in fact the Earth, moving in relation to the Sun, that is "entering" the sign not of Aries but of Libra (as seen from the Sun), although neither Sun nor Earth is really entering anything at all.

The stars that make up the constellations of Aries, Libra and the rest are useful to us as a backdrop against which we can measure the positions and movements in our tiny little corner of Creation. But we are never going to "enter" them—they are so far away that we have to measure their distances from us in light years, or the distance covered by an object travelling at the speed of light for a year (some 6 trillion miles), and the nearest star to us after our own Sun, Proxima Centauri, is 4.4 light years away. Our solar system is very much on its own, and although it is so small compared even with the visible universe, we still know surprisingly little about it.

Man has been staring at the stars for thousands of years and wondering what they all mean, yet we still do not know for sure exactly how many planets there are orbiting around our star, the Sun. For centuries, it was assumed that there was only Mercury, Venus, Earth (plus its Moon), Mars, Jupiter and Saturn. Then, long after telescopes had been invented, Uranus was discovered (1781), followed by Neptune (1846) and Pluto (1930), and these between them have more than 30 satellites or moons. In addition, there is that curious collection of rubble known as the asteroid belt, the orbit of which comes in between those of Mars and Jupiter. It consists of about 50,000 "planetoids," some probably less than a mile in diameter, all of which may be the remains of an exploded planet, though a more exciting theory is that this orbiting junkyard is the remains of the original building blocks of the rest of the solar system.

It is by no means easy to define a planet. For convenience, the nine listed above are called the "principal planets," although one of Jupiter's satellites is in fact larger than Mercury. Pluto may once have been a satellite of Neptune, and there is no reason to suppose that it is the outermost planet in our solar system. Indeed, there is much evidence that suggests otherwise, and over the past hundred years there have been many predictions of new planets and telescopic searches for them.

Neptune and Pluto were predicted long before anybody actually found them, being assumed to exist because of irregularities in the orbits of the planets nearest to them. Thus Neptune was discovered after studies of the orbit of Uranus, and Pluto after those of Neptune. Now Pluto is a very odd planet. It is very small and very far away, and it orbits the Sun in a direction quite unlike that of any other known planet. Long before its discovery, Professor George Forbes was telling the Royal Society of Edinburgh that he believed there to be *two* planets beyond Neptune. He based his argument on a study of the orbits of comets, a basis still used today by (among others) J. L. Brady, who has predicted the existence of a huge planet twice as far from the Sun as Neptune.

It must be quite a planet. Brady's study of Halley's comet shows that a planet large enough to account for observed effects in its orbit must be three times the size of Saturn (which is about 95 times more massive than Earth), and about 65 times further from the Sun than Earth is. Moreover, its orbit must be even stranger than that of Pluto, the plane of which is 17° from the ecliptic, or plane of Earth's orbit round the Sun. It is thought to have an inclination to the ecliptic of between 60 and 120 degrees, and to revolve backwards in relation to all the other planets.

Such a "Planet X" is no figment of the imagination. There is as much ground for believing it to exist as there was for predicting Neptune and Pluto, while orbital anomalies of Neptune and Uranus also suggest its existence. Indeed, one astronomer (H. H. Kritzinger) has predicted *two* more planets beyond Neptune in addition to Pluto.

All that remains is to find this strange Planet X. This is like finding the proverbial needle in the haystack, and it may be impossible with today's instruments, but it is worth trying. Here is how:

Take a small area of sky, one degree of arc for example, and search it through a telescope, taking photographs at regular intervals and then studying them in sequence. If anything seems to have moved relative to the fixed-star background, and is not an asteroid, comet or meteor, then it could well be a planet. This method, known as "blinking," was used by Clyde Tombaugh to find Pluto, though Planet X may be too faint to photograph, as Brady himself suggests. (Tombaugh, incidentally, spent 15 years searching for other planets without success, though according to Brady he did not search far enough from the ecliptic.)[1]

With 360 degrees of sky to choose from, the question is where to start? Planet X is thought to be near the constellation Cassiopeia in the sign of Aries at present. And astronomers who abandon the search for Planet X can try their hand at locating yet another predicted planet within the orbit of Mercury, provisionally named Vulcan. Several astronomers of the nineteenth century thought they had spotted it, but in this century it has proved extremely elusive.

Astronomy is one of the few sciences in which amateurs can and do make important discoveries. Sir William

Herschel, who discovered the planet Uranus, was originally a musician.* It was an amateur, S. C. Chandler, who first described the wobbling of Earth's poles (in 1893), now known as the Chandler Wobble. In 1977, a 27-year-old Huddersfield postman named Graham Hosty discovered a new star (Nova Sagitta 1977) with defective equipment worth about £10, while another Englishman, George Alcock, has spotted four new comets and three novae, an outstanding record for an amateur.

If a new planet has been predicted, there is a chance that it really exists and can be spotted, unless of course new evidence suggests that it does not exist, as seems to be the case with Vulcan. Neptune and Pluto were both predicted before anybody ever saw them, as was probably the case with Phobos and Deimos, the mysterious moons of Mars. These were mentioned and described very accurately by Swift in his satire *Gulliver's Travels* (1726) but only observed 150 years later by Asaph Hall. It is not known who actually predicted them, though their existence was suspected by Kepler early in the seventeenth century.

Planet X may or may not exist. Until we know for sure how many planets there are in our solar system, we will never fully understand the effects of each of its components on the others. We are faced not only with a "many-body problem," to borrow a term from physics, but also with a "how-many-body?" problem.

The many-body problem, as defined by physicist Richard Mattuck of the University of Copenhagen, who has written a book about it,** is the study of the effects of interaction between bodies on the behaviour of a system in which there are many bodies. (See Fig. 1.) In the nineteenth century, says Dr. Mattuck, the three-body problem was insoluble, as far as exact solutions are concerned. With general relativity and quantum electrodynamics in the twentieth century, the two- and one-body problems became insoluble as well. Now, with quantum field theory, the zero-body or "vacuum" problem is also insoluble! ("No bodies at all is already too many!" Mattuck comments).

---

* Modern professional astronomers who are also talented musicians include Ernst Öpik, Peter van de Kamp, P. Kulikovsky, Sir Bernard Lovell and Patrick Moore.

** *A Guide to Feynman Diagrams in the Many-Body Problem*. New York: McGraw-Hill, 1976. (2nd ed.).

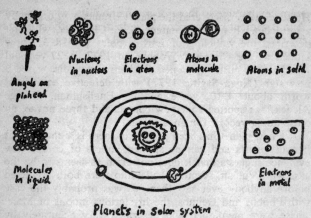

Fig. 1
The solar system is one of many many-body problems, as illustrated here by Dr. R. D. Mattuck of the University of Copenhagen. (Courtesy of the artist).

The solar system is, then, a formidable many-body problem and before the invention of the electronic calculator it could take a lifetime to predict the motions of just one of its bodies, thanks to the host of interactions by gravitational attraction leading to "perturbations" in the motions of the various bodies involved.

The astronomer is unlucky. He cannot study his raw material at first hand; at least, he could not until equipment was landed on the Moon and planets. Anatomists can cut up bodies and study bits of them under microscopes, chemists can drop some of this into some of that and see what happens, while physicists can build machines in their laboratories to test their theories. But until very recently, all the astronomer could do was peer through his telescope and try to make sense of what he saw—or what he thought he saw.

Koestler, in his book on man's changing vision of his Universe, likens the pioneers whose discoveries enabled us to learn a little more about it to sleepwalkers. The methods of Copernicus, Galileo, Kepler and even Newton often seem comparable to those of somnambulists, who get from A to B without knowing how they do it, and without intending consciously to go anywhere at all.[2] Intuition and sleep-walking have served science well in the past, leading to many major discoveries, and we make no apologies for including a number of ideas in this book that will certainly seem somnambulistic to devotees of so-called materialism.

This is not to suggest that the sleepwalkers are always right, even if they are well-trained scientists in their waking states. The history of cosmic theory is one of "collective obsessions and controlled schizophrenias," as Koestler puts it, and not all of these have come true. At various times, the Earth has been thought to be a rectangular box, a flat slab, or a disc surrounded by water. Canals have been spotted on Mars, and suspension bridges on the Moon. Edmund Halley, of comet fame, thought the interior of the Earth was inhabited, while Sir William Herschel thought it quite likely that there were people living on the Sun.

When, on 4 October 1957, the Sputnik 1 satellite orbited the Earth and inaugurated the Space Age, astronomers could at long last look forward to obtaining first-hand evidence from the space beyond Earth's atmosphere, and so much new information now pours in every time a new space probe leaves its launch pad that several new branches of science have had to be founded to deal with it. "Text books of astronomy published before 1965 might just as well be written in Sanskrit, for all their present usefulness," a U.S. scientist wrote in 1974,[3] while the index alone to a single year's publication by the National Aeronautics and Space Administration (NASA) runs to more than 1,000 pages. And yet, as the data pile up and up, we are reminded how little we still know about our neighbouring planets, not to mention the distant stars and galaxies, and how they are affecting us. For, as we shall see, affect us they most surely do.

Long before telescopes were invented, making accurate measurement of the heavens possible, astronomers felt there must be some sort of mathematical order in the universe. As indeed there is, for we live inside a clock with many faces showing different "times"; here is a list of the principal planets' heliocentric orbital periods, or the time they take to make one journey around the Sun (as seen from the Sun) expressed in Earth's days and years:

☿ Mercury	87.97 days	♃ Jupiter	11.86 years
♀ Venus	224.7 days	♄ Saturn	29.46 years
⊕ EARTH	1 year	♅ Uranus	84.02 years
♂ Mars	1.88 years	♆ Neptune	164.78 years
Asteroids	4.6 years (av.)	♇ Pluto	248.4 years

The innermost four are known as the fast planets, because they get round the Sun much more quickly than the others, Mercury's mean orbital velocity of 47.9 km./sec. being ten times that of Pluto.

A few minutes' playing with a pocket calculator is enough to show that there are many intriguing coincidences in the relationships between multiples of planetary orbital periods. These are known as "near commensurabilities," and are not the same as exact resonances, to be mentioned later. For instance, four orbital periods of Mercury are within a couple of weeks of one of Earth, five of Mercury are ten days less than two of Venus, three of Venus are not far from one of Mars, six of Mars are close to one of Jupiter, five of Jupiter are very close to two of Saturn, two of Uranus are just over one of Neptune, while three of Neptune are extremely close to two of Pluto.

To Kepler, who discovered the laws of planetary motion, the universe was both mathematics and music. He was excited to find that the "music of the spheres" could be represented in numbers, and that certain proportions between one planetary motion and another were almost exactly those worked out by Pythagoras for the intervals of the musical scale. Kepler found, for instance, that the ratio of Saturn's velocities at aphelion and perihelion was almost exactly 4:5, the ratio for the major third! Likewise, he found Jupiter to match with a minor third, Mars a perfect fifth, Earth and Venus with semitones; many other musical parallels are to be found when planets are considered in pairs. Coincidence or not, there *is* harmony among the spheres.

As for mathematics, Kepler thought that the five spaces between the six planets known in his day could be made to correspond to the five perfect solids: the pyramid, cube, octahedron, dodecahedron and icosahedron. These are the only solids of which all faces are identical, and Kepler's flash of inspiration came very close indeed to actual proportions as they would be if the planets went round in circles instead of ellipses, which he himself eventually discovered:

hence his discovery mentioned above of the relations between planetary velocities when they are nearest to, and furthest from, the Sun.

In the eighteenth century, a table of numbers was drawn up to account for planetary distances from the Sun. This is sometimes called the Titius-Bode Law, although it is not a law at all and was originally discovered neither by Johann Titius nor Johann Bode, but probably by Christian Wolf in 1724, nearly 50 years before Bode popularized it. To work it out is like a popular game. Think of a number: 3. Add 4 and divide the result by 10. This gives you 0.7, the exact distance of Venus from the Sun in astronomical units, one AU being the mean distance of the Sun from the Earth, about 93 million miles. For Earth you double the number you thought of ($2 \times 3 = 6$), add 4 and divide by 10, giving 1.0. For Mars you add 4 to twice 6, divide by 10 and get 1.6, and so on. (For Mercury you just divide 4 by 10.) Eventually you end up with this:

Planet	Theoretical Distance	Actual Distance
Mercury	0.4	0.4
Venus	0.7	0.7
EARTH	1.0	1.0
Mars	1.6	1.5
Asteroid belt (av.)	2.8	2.8
Jupiter	5.2	5.2
Saturn	10.0	9.5
Uranus	19.6	19.2
Neptune	—	30.0
Pluto	38.8	39.5
X	77.2	—

It can be seen at once that the correspondence is very close as far as Uranus, but then something goes wrong. Neptune should not be where it is, according to the Bode series, while Pluto is sometimes actually closer to the Sun than Neptune, thanks to the inclination of its irregular elliptical orbit which brings it to a distance of 30 AU at perihelion and 50 AU at aphelion.* The X after Pluto represents the planet we hope will be discovered; it is sometimes referred to as Transpluto, or Planet X.

---

* Perihelion and aphelion are the points nearest to and farthest away from the Sun, respectively, in a planet's elliptical orbit. Perigee and apogee are the points closest to and farthest from Earth.

Theodor Landscheidt, a contemporary German who is carrying on his country's tradition of relating the cosmos to numbers, has put forward the fascinating suggestion that the macroscopic ordering of the planets can be related to the microscopic ordering within the atom, by using the fundamental constants (fixed values) of modern physics, echoing the ancient Hermetic doctrine of "as above, so below."[4]

Landscheidt's hypothesis, in which "the pure numbers of physics represent a cosmic principle of order," as he puts it, is complemented by that of the Soviet astronomer A. M. Molchanov, who claimed in 1968 to have discovered a resonant structure in the solar system involving all the planets plus their moons and satellites. Taking the frequency of the largest known planet, Jupiter, as a "reference frequency," and using six "quantum numbers," Molchanov formulated eight simultaneous equations which when solved give values very close to the observed frequencies of each planet. He later did the same for the moons of each planet as an individual quantum system, and claimed that the probability of his overall findings regarding the planets and their satellites being due to chance alone was one in 1,000,000,000.[5]

It may seem strange to speak of planets resonating with each other as if they were strings on a piano vibrating in response to a tuning fork, but the principle is the same. When a piano tuner strikes his A fork, it vibrates at 440 cycles per second (cps. or Hz. for Hertz), and if the piano's A string above middle C is in tune, i.e. at the right combination of length and tension, it too will start resonating, or vibrating in response to the sound waves emitted by the fork. If there is another piano in the room, its A string will start to vibrate as well (if somebody's foot is on the sustaining pedal). The important thing about resonance for our purposes here is that *energy is transferred* in the process, and when two elements are in vibrational harmony, energy can pass from one to the other even if the frequencies are different. For instance, a loud A (440 Hz.) will cause audible vibrations on the A string in the next octave, which vibrates at 880 Hz., or twice as fast. This effect is due to overtones or harmonics to the 440 Hz. "fundamental," and likewise there are undertones and subharmonics at the other end of the scale. We shall see the importance of these in the following chapter, and later in the book we shall see that not only do vibrating piano strings respond to external cyclic stimuli, *but so do people,* which makes the subject of resonance very important to us.

The bottom A on the piano vibrates at only 27.5 Hz., and if there were an A an octave lower we would not be able to hear it, for the average human ear can pick up sound waves only between about 20 and 20,000 Hz.— about two and a half octaves beyond the top note of a piano. But the acoustic waves are still there outside these frequency limits, and later on we shall see what dramatic effects the extremely low (ELF) frequencies, of both acoustic and electromagnetic waves, can have.

If we cannot hear "sounds" of less than 20 cps. we cannot hope to hear the "music of the spheres," with its periods of cycles lasting from 88 days to 248 years, or frequencies of much less than one cycle per day or year rather than of cycles per second. Yet such "music" exists, and like any other music must have its harmonics and resonances.

Molchanov's ideas attracted little attention apart from suggestions that he had simply juggled with numbers to produce a spurious theory, but shortly after he published them, two other scientists announced that they had found an actual *physical* link between two planets, Venus and Earth. It was never possible to measure the rotation rate of Venus, because of the thick clouds that obscure its surface; but after radar beams were bounced off it in 1961, studies revealed that it rotates backwards, or in the opposite direction to all the other planets, taking 243 Earth days to do so. Venus also rotates in such a way that it turns the same face towards Earth at each inferior conjunction (that is, at its closest approach to Earth on the same side of the Sun). It now appears that its spin, or rotation about its axis, is for some reason related to the revolution of Earth in its orbit round the Sun, and it has even been suggested that the two are coupled, or linked, with Earth locking Venus into a resonance. This is an example of links that Molchanov believes to exist between *all* planets. This would mean that energy is being exchanged between them, making the whole question of planetary influence on Earthly events into something far beyond mere superstition.[6]

But are such matters of more than mathematical interest?

Who cares if five orbital periods of Jupiter equal two of Saturn? Can resonances among the planets have any real significance? Let us look at it this way:

A periodic driving force will produce a maximum response from a mechanical system when the frequency of the applied force is equal to the natural vibratory frequency of the system. To illustrate this, place a small child in a swing and push her once or twice to get the motion started. Once she is swinging away happily at a frequency of one cycle every four seconds, all you have to do to keep her swinging for ever and ever is touch her gently on the back every four seconds, at the peak of each cycle. The result is resonance "pumping" of energy from you to the child.

In the three-dimensional many-body problem that is our solar system, we have an apparently infinite number of possible resonances between any of at least nine different orbits and spins, not to mention those of the numerous planetary moons and satellites, and without even considering the position of the whole solar system within its galaxy. (Landscheidt has done this, but we do not want to complicate matters any further for the moment.) It may well be that there is a relation between certain angles or "aspects" between the planets and certain resonances; in later chapters we provide evidence in support of this. An enormous amount of work remains to be done on the interplanetary many-body problem, and if there is no solution yet in sight, there are indications that some have begun to define it correctly.

Next, in our brief study of the energetic music of the spheres, we must take a look at the forces of nature, since any resonant system must have a driving force to make it oscillate.

The Invisible Forces—1

There are generally thought to be only four forces in nature—electromagnetism, gravity, and the strong and weak nuclear forces. The latter operate over such small distances that they need not concern us here. First, then, let us take gravity.

This is the weakest force of them all, yet it seems to be what holds the whole of Creation together, and like so many things that affect all of us all the time, it is an almost total mystery. We think we know how it behaves, thanks to Newton's famous laws, but we have no idea what it really is.

Gravity has some curious properties. It is so weak that every time a baby picks up a toy from the floor, he outpulls the entire Earth. Yet despite this weakness, gravity operates over enormous distances. For instance, the planet Pluto reaches a distance of 4,600,000,000 miles from the Sun, yet it stays put in its orbit without zooming off into outer space thanks entirely to gravity.

Another odd feature of the force of gravity is that it is always positive, as far as we know, so that any two bodies in space must always attract each other. Between Earth and Moon, for example, gravitational forces work both ways; we attract the Moon and it attracts us, and so do all the planets and the Sun likewise attract each other, with a force proportional to the product of their masses and inversely proportional to the square of their distances from each other. This is Newton's law of gravitation, or formulation of how the observed bodies behave, and he decided it must apply to every single particle of matter in the Universe, not just the observable planets.

The most obvious and visible result of all this mutual attraction is the phenomenon of tides. The effect of the Moon's tidal pull on Earth can be seen on any beach. Not only the Moon, however, but also the Sun *and all the planets* cause tides on Earth, although the combined planetary effects are very much less than those of Sun or Moon, the former being so huge and the latter so close to us. Yet planetary tidal effects do exist, though until recently they have been thought to be of no possible biological significance and have been completely ignored. This may have been a mistake.

Tides are not only to be found in the seas and rivers. There are tides in Earth's atmosphere, and even in the solid Earth itself. It is not generally realized that the "solid" surface of our planet actually moves up and down as much as 9 inches in response to the tidal pull of Sun and Moon. This indicates that part of Earth's interior must be liquid, a fact that has interesting implications in connection with earthquakes, as we shall see in Chapter 5.

There are also tides in us. The human body is almost three-quarters water, and like any other body of water it must respond to the influence of the celestial bodies. Again,

16   Background

this fact has some very intriguing implications, which we introduce in Chapter 6.

There are tides on the Sun as well as on Earth, although there is no water there but only superheated ionized gas in a "fourth state of matter" called plasma (the other three being solid, liquid and gas). The plasmic oceans of the Sun must be affected by tides induced by the bodies that orbit around them, just as the atmosphere of Earth is. There are indeed tides everywhere, and their effects keep turning up in the least expected places. (Including "the affairs of man," as Shakespeake thought, and as we show in Chapter 10.)

Here is a table of the theoretical tidal effects of the principal planets, plus Sun and Moon, on the surface of Earth: (Unit = 10^{-13}g).*

BODY	MAXIMUM	MINIMUM	RATIO MAX./MIN.
Sun	500,000	500,000	—
Moon	1,300,000	860,000	1.5
Mercury	0.53	0.023	23
Venus	50.1	0.21	239
Mars	2.8	0.009	311
Jupiter	5.7	2.9	2
Saturn	0.21	0.11	1.9
Uranus	0.0032	0.0024	1.3
Neptune	0.0009	0.0008	1.1
Pluto	0.00004	0.000016	2.5

It is quite plain that the Moon is by far the most important tide-raiser as far as Earth is concerned, although the Sun's tidal force can be more than half as strong as the Moon's at times. By comparison, the tidal forces of the planets are so small that they may seem hardly worth considering. However, as we have said, such forces do exist and planetary tidal effects on Earth are theoretically possible. As to whether they are significant to any observable extent, we shall see again and again in this book that it may be unwise to ignore any force at all that can be measured or calculated, and even some that cannot.

A mathematician who had never seen the sea might argue that the Moon could not have any important effect

---

* Planetary tidal forces on the Sun are mentioned in Chapter 2.

on Earth, since its tidal force is a mere ten millionth of the force that holds us all on Earth's surface, which is weak enough to start with. Yet, if our inland mathematician went to the seaside, he would find that the Moon in fact causes the displacement of colossal masses of water, and if he did some further research he would find that it even pulls the solid mass of the Earth out of shape as if it were an elastic ball, as the astronomer Fred Whipple has put it. The question "how strong is weak?" is one to which we return in a later chapter. It is a question that should not be left unasked any longer.

An important point about the invisible tidal forces around us is that whatever their effects, they cannot be constant. For instance, when both Sun and Moon are aligned with Earth between them, our tides are large, as they also are when the Moon is at perigee. So when both these things happen at once we should expect even bigger tides. But this is not always the case; depending on one's geographic position such a situation can actually lead to minimum tides instead. Thanks to the number of bodies exercising influence on Earth, and the complexity of their orbits now moving closer to us, now receding, and most important of all now and then pairing up to pull together against us, we can see that we must be caught up in a system of *cyclic* influences of horrendous complexity, one which we have only recently begun to examine in any detail.

For an example of such complexity, we need look no further than our Moon. To an astronomer, a month means the period of the Moon's revolution from a given reference point back to that same point. This sounds simple enough, but there are in fact five different "months," as follows:

MONTH	SOLAR DAYS (AV.)	= Days/Hrs./Min./Sec.			
1. Synodic	29.53	29	12	44	2.8
2. Anomalistic	27.55	27	13	18	33.2
3. Sidereal	27.32	27	7	43	11.5
4. Tropical	27.32	27	7	43	4.7
5. Draconic	27.21	27	5	5	35.8

KEY: 1. This is the period between full Moons, or the time it takes for the Moon to go once round Earth with respect to the Sun. 2. This is the time between lunar perigees; the time between positions at which the Moon is closest to Earth. 3. The true mechanical period of the Moon's revolution, this is the time taken for one lunar orbit of Earth with respect to the fixed-star background as seen from Earth. It is shorter than the

*synodic month, because the Earth advances from one new
Moon to the next as it moves along its own orbital path round
the Sun. 4. This is the time between lunar passages across
a given celestial longitude and finally—5. The draconic or
nodical month is the time between passages of the Moon
through the ascending* node, *or point where the lunar orbital
plane crosses that of the Earth/Sun orbit.*

As if all these natural months were not enough, we have
invented yet another, the artificial calendar month of 31,
30 or 28 days. It is worth bearing the differences in the
various months in mind: when in later chapters we come to
influences of the Moon on terrestrial events and human
affairs, it will be as well to realize that each of the lunar
"months" can be expected to show up in them.

We have already mentioned Newton's celebrated law of
gravitation, and in our unsettled age in which everything
and everybody tends to be debunked sooner or later, this
is surely one piece of scientific legislation still to be trusted.
Gravity is supposed to be constant, irreversible, and impossible to shield except perhaps by the designers of flying
saucers. Anybody suggesting that this sacred law may be
due for repeal can only expect to be either laughed at or
politely ignored.

The former is what happened to Professor Eric Laithwaite, of Imperial College, London, when he demonstrated
at the Royal Society in 1974 a machine based on a gyroscope that appeared to defy gravity,* while the latter was
the fate of Maurice Allais, a professor at the School of
Mines in Paris, after he announced in 1959 that he had
observed a series of effects that could not be explained
according to classical gravitation theory. We will not comment on Laithwaite's work until he has written it up himself, but in the case of Allais the experiment has not only
been published but also repeated independently, an essential
step towards acceptance of any new idea in science.

Over a four-year period (1953-1957), Allais carried out
a series of experiments with large pendulums. As visitors to
the Science Museum in London can see, when a large

---

* For a not entirely accurate report of this happening, see *New
Scientist* 14 November 1974, p. 470. See also ibid. 28 November,
pages 679 and 704, and 19 December, p. 895.

pendulum is set in motion, the plane of its swing will appear to rotate slowly, although it is of course the Earth that is rotating and not the pendulum. If the cosmos is behaving as science thinks it should, there cannot be any variation in the pendulum's regular behaviour.

However, after some 220,000 observations, Allais found that there was a definite periodicity in the observed motions of his pendulum that seemed to be related to the motions of the Sun and Moon, being of between 24 and 25 hours. According to The Law, there should be no such thing, but there it was, and Allais, a qualified statistician, decided it was statistically significant.

This discovery was intriguing enough in itself, but there was an even bigger surprise in store for Allais. During his test period there was a total eclipse of the Sun, on 30 June 1954, which lasted in Paris from 11:21 a.m. to 1:55 p.m. It must have been the most fascinating lunch hour Allais ever spent, for at the very moment the eclipse began, his pendulum suddenly shifted its plane of oscillation by five degrees of azimuth, deviating still further until just before the midpoint of the eclipse and falling back to normal just as the eclipse ended. Overall, Allais calculated that the deviations he had observed (and recorded) amounted to *one hundred million* times what could be expected according to classical theory.

Faced with such evidence, which Allais duly published in great detail, with charts and photographs, scientists had the choice of two reactions: to have a good laugh and forget about it, or to wait with an open mind to see if it could be repeated. (There was actually no need to wait for an eclipse, for Allais insists that anybody who observes a pendulum regularly for a few days will find anomalies similar to those he noticed before his eclipse experiment.)

Two American scientists chose the latter course. They were Erwin Saxl of Tenistron Inc., Harvard, and Mildred Allen of Mount Holyoke College in South Hadley, Massachusetts. In 1970, they set up an experiment similar to Allais' for the solar eclipse of 7 March, which was 96.5 percent total in the Boston area. Their pendulum acted in precisely the same way, leading them to conclude that "both our experimental findings and those of Allais cause one to question whether the classical laws of gravitation hold without modification."[7]

Max Planck, who did as much as anybody to revise the laws of physics, once remarked that nothing was more interesting than a fact that ran counter to a theory previously held to be sound. "The real work begins at that point," he said, and it is to be hoped that further repetitions of the work of Allais, Saxl and Allen will be made, leading to new laws. One of us (Hill) is involved in such a project, involving twin pendulums in Sweden and West Germany, which is awaiting a suitable eclipse.

The moral of Allais' story is that nothing in nature can be taken for granted any longer, as we shall see again more than once in this book.

The Invisible Forces—2

If gravity holds the bodies in our solar system in position, it is an entirely different force that governs much of what goes on in it, especially in our immediate environment. This is *electromagnetism* (EM), the name given to magnetism that arises from an electric charge in motion. The EM spectrum ranges from cosmic rays with wavelengths of fractions of billionths of a centimetre, all the way through gamma and x-ray, ultraviolet, visible and infrared light, to radio waves of thousands of kilometres in length. Here is a simplified chart of part of the spectrum:

FREQUENCY (kHz)	WAVELENGTH (m)	DESCRIPTION
1	10^5	Atmospherics,
10	10^4	Alternating current,
10^2	10^3	Radio frequencies
10^3	10^2	from VLF to EHF
10^4	10	
10^5	1	
10^6	10^{-1}	Microwaves (radar)
10^7	10^{-2}	
10^8	10^{-3}	
10^9	10^{-4}	Infrared (heat waves)
10^{10}	10^{-5}	————————
10^{11}	10^{-6}	Visible light
10^{12}	10^{-7}	————————
10^{13}	10^{-8}	Ultraviolet rays
10^{14}	10^{-9}	X-rays
10^{15}	10^{-10}	
10^{16}	10^{-11}	Gamma, cosmic rays
10^{17}	10^{-12}	

It is interesting that while the human ear can pick up about ten octaves of acoustic waves and interpret them as "sound," our eyes can barely register a single octave of EM waves as visible light, the wavelength of which is roughly between four and seven hundred thousandths of a centimetre (3,900 to 7,700 Ångströms). We can consciously perceive a certain amount of radiation to either side of the visible light band—as heat (infrared) or as something that gives us a suntan (ultraviolet). But what of the rest of this huge spectrum?

Over and over again in this book we shall be suggesting that *all living beings, man included, do respond in countless ways to influences of which they are not consciously aware.* Just because our senses are limited, we have no reason at all to suppose that the effects of the EM spectrum are limited to those we can see or feel.

There are two important differences between gravity and EM. The former cannot be shielded, while the latter can, up to a point. Gravity cannot be produced artificially, whereas EM can and is all the time, and since we can produce EM waves to our own specifications in the laboratory, we can study them in detail. This cannot be done with gravity waves as yet (American scientist Joe Weber believes he can, but his findings await general acceptance), nor can we induce an artificial gravitational field, short of gathering enough material together to make a new artificial planet. This means that we are limited to studying gravitational fields as produced by nature.

Escaping from EM waves is normally done by using a contraption called a Faraday cage. This is a cage or a whole room lined with metal netting or plates of a conductor such as copper, and is widely used in physics, psychology and parapsychology laboratories. It does not, as many believe, shield out all EM waves. Very high-energy cosmic rays can penetrate even a lead vault without appreciable attenuation. But few of us spend much of our time in Faraday cages, so we are therefore surrounded at all times and in nearly all places by a wide range of EM fields and waves, whether we like it, or even know it, or not.

Our prescientific ancestors seem to have been more aware of the need to ward off the "evil eye" of EM radiation than we are today. The traditional royal crown, a metallic headband, is a kind of archetypal Faraday cage, while the bishop's crozier is believed to have served originally to ground His Eminence against wicked forces. Even the horseshoe hanging over the door may have more to do with science than with superstition. The Danish physicist Niels Bohr is said to have had one over the door of his country home, and when asked if he believed in that sort of thing, to have replied "No, but they are supposed to bring you luck even if you don't believe in them!" (This story comes from that great scientist-humorist the late George Gamow and has been denied by Niels's son Aage Bohr. However, German scientist H. L. König has suggested that a horseshoe may be an electromagnetic resonator. Who knows?)

New discoveries are made almost daily about ways in which EM waves of all kinds interfere with us. A certain amount is now known about the Earth's magnetic field, or geomagnetic field, and some of its effects (of which more anon), but what do we know about all the fields of unknown origin through which we may be passing as our planet hurtles through space? Thanks to the spiral path of Earth, we are never in the same part of space twice. Intergalactic fields, about which we know next to nothing, may be producing all kinds of effects by interacting with our local fields as we pass through them.

The same question can be asked about both gravity and EM: are not most of the forces concerned so weak that they cannot be significant? They may be of academic interest, but do they matter? The answer is that anything we can detect and measure may well matter, even if the force in question falls below the thresholds of our conscious perception mechanisms.

The Cosmic Blitz

For all the mysteries our solar system still contains, these can be nothing compared with those of the universe as a whole. Our little corner of space is not a closed system; there is no fence around it to keep trespassers out, and much of the energy that sweeps around our planet originates from far beyond its outer limits.

And what energy! Astronomers are already talking about objects one hundred million times the size of our Sun, and such is the mystery surrounding some of the sources of energy we have actually identified that we do not even know what to call them except *pulsars*—things that pulse energy at us—or *quasars*—quasi-stellar radio sources, which were dicovered in 1967 and 1963 respectively.

Pulsars are thought to be neutron stars, so compact and dense that a pulsar with the mass of the Sun would be about ten miles in diameter. More than 100 have been discovered so far, one in the Crab Nebula thought to be the remains of a supernova that exploded in the year 1054. It has a remarkably rapid rotation rate of 30 cycles per second, which is responsible for some of the variations in energy that radiates from its core on all wavelengths from radio to x-rays, amounting to a total radiation output of some 100,000 of our Suns.

Quasars are even more powerful, more distant and more alarming. We are told that some of them emit as much radiation as *a thousand million galaxies* the size of our Milky Way—which contains about 100,000,000,000 stars. Fortunately, quasars are a long way off, as far away as the radius of the known universe, a matter of ten thousand million light years or so.

Faced with statistics of this magnitude, the average mind simply reels, so let us get back to our modest little solar system. Here too, new sources of energy are being identified one after the other. The planets, for instance, were always thought to be electromagnetically "quiet" until 1954, when the startling discovery was made that Jupiter emits powerful EM radiation in the form of radio and microwaves. (It has even been suggested that Jupiter is a Sun in the making.) The following year, a similar discovery was made regarding Venus. Suddenly, the solar system has been found to be an EM system.

We on Earth are in direct contact with this system. We are under constant bombardment from all directions by a host of protons, alpha particles and other nuclear bits and pieces known as cosmic rays. These are most noticeable when they originate from the Sun, but they also come from further afield and when they reach our atmosphere they produce secondary radiation in the from of gamma rays, electrons and mesons.

Luckily, we have our natural defences. The zones of charged particles known as the Van Allen belts, which begin 500 miles from Earth, trap a certain amount of the invading energy, though far from all of it, and our Sun asserts its authority as ruler of its system by sweeping some foreign radiation out of our way with its own powerful "wind" (see below). But the Sun's activity is not constant, as we see in the following chapter. Its energy comes in cycles of just over 11 years, and when the Sun relaxes its output, other cosmic energies pour in unchecked. Therefore, not only the amount of energy bombarding us but also its quality is constantly changing.

Cosmic rays fit into the ultra-short-wave end of the EM spectrum. Their wavelengths are so short, in fact, that these rays might as well be called particles, and the word *wavicle* has been coined to describe such wave-particles. Their frequencies are so high that they represent the frontier of our knowledge at present of SHF (super-high frequencies), and their energy is such that considerable amounts can be counted even at the bottoms of deep mine shafts.

Part of this cosmic bombardment is known as the solar wind. This sounds like a contradiction in terms: how can there be any sort of wind in space, which is a perfect vacuum? The answer is that space is not a perfect vacuum; it is filled with all sorts of waves, fields and particles in addition to the protons and electrons the Sun blows in our direction.

The solar wind, only discovered in the 1940s, is composed of high-energy particles ejected at about a million miles an hour at temperatures of more than a million degrees (Kelvin) from the Sun's outer layer or corona. These particles are lower in energy than cosmic rays from deep space, but they are still able to form a wind that has been described as "extremely gusty," variations in its speed forming velocity waves that sweep far beyond the Earth.[8]

This space-wind has caught the imaginations of scientists and science fiction writers (many of whom are scientists) alike. If one could build a sail only a few atoms thick and several square miles in area, one could catch the solar wind and sail away into deep space. Acceleration would be slow, but like the progress of Aesop's tortoise, very sure. Over the months, thrust would build up and a space "yacht" could reach very high velocities, and if this source of free energy can be harnessed efficiently it will be the answer to the problem of fuel bills on deep space voyages.

Thanks to all the miles of data from the hundreds of satellites launched since *Sputnik*, especially from the IMP 7, *Pioneer 10* and *11* probes of 1972/3, we now know that every planet casts a shadow on the solar wind, just as Earth casts a shadow on the Moon to cause an eclipse. What the effects of sudden variations in the density of this wind may be, we are just beginning to discover.

Scientists now believe that cosmic ray counts and solar flares may have something to do with a measurement we all take for granted: the length of the day. There are indications that the days are getting longer all the time; not much longer, but just enough to measure. This is because the rotation of Earth is slowing down a few thousandths of a second every day, or about one second per year. Not enough to worry about, perhaps, but the stability of our calendar is threatened in other ways . . .

In 1960 a French astronomer, Dr. A. Danjon, made a claim that was so startling that it was simply ignored by most of the scientific community for about ten years. What he suggested was that an unusually large solar flare has caused a *sudden* change in the length of an Earth day. Nobody paid much attention until an enterprising pair of astrophysicists, John Gribbin and Stephen Plagemann (whom we will meet again in Chapter 5), made the same discovery after the great solar storm of 4 August 1972 (the strongest one ever recorded), and published their findings. The idea of an extra millisecond or two on the length of a day may not sound very exciting, until one realizes that the entire Earth, which weighs 6×10^{24} kilograms, has to slow down more or less immediately to make this possible.[9]

When we speak of invisible forces, fields or influences from space, therefore, we are not discussing unimportant little particles that produce insignificant effects, but of forces that can literally shake the foundations of our planet.

And if they can slow down the world, what are they doing to us?

To look for some of the answers to this question, we must turn first to the Sun, but before we do this there is one more mystery to examine.

The Elusive Ether
"Please, Sir, how do all these forces from millions of miles away get here?"

"Propagation, or transmission through space by radiation."

"Yes, but . . ."

"Never mind. They just do. Get on with your work."

Such a dialogue must often take place between student and teacher, and if the student is lucky, the teacher may then launch into a discussion of one of the most controversial topics in science—the ether.

By the end of the nineteenth century, it was widely accepted, without there being any proof, that the whole of creation was filled with an invisible substance, if you could call it a substance, known as the ether, or aether, a word originally referring to the clean air breathed by the gods on top of Mount Olympus. Even if there was no evidence for it, there just had to be something filling what looked like empty space, so that EM and gravity waves could propagate from A to B by "waving." But waving in what?

The concept of an all-pervading substance was nothing new: the *prana* of yoga traditions amounts to much the same thing. But scientists like to measure things, and the ether has persistently refused to be measured, which means either that there isn't any ether or that we have been asking the wrong questions. The old ether concept is often thought to have been debunked once and for all by the Michelson and Morley experiments of the 1880s, and by Einstein's relativity theory of 1905. This is not so.

Albert A. Michelson (first scientist to win a Nobel Prize for the United States) set out with the help of Edward Morley to see if light would travel at slightly different speeds when light beams were placed at right angles to each other, so that one beam would be parallel to the "ether flow" and the other perpendicular to it. The idea was that the ether, if it existed, must be carried along with the rotating Earth, so that a beam sent against the tide, as it were, would be slowed down by the etheric current. It was a good idea, but it didn't work. No such difference in velocity could be found, and to orthodox science that was the end of the ether.

A more elaborate experiment was carried out a few years later by Oliver Lodge, who wanted to see if large steel discs revolving at frighteningly high speeds would cause variations in the speed of light by dragging the ether along with them. They failed to oblige, though Lodge never lost his faith in the ether, which he saw as the essential intermediary substance between individual atoms of matter and even between matter and mind. It welded atoms together by cohesion, planets and stars by gravitation, and it also served to transmit vibrations from one substance to another. It belonged to the physical universe, though it was no ordinary "matter." It was, Lodge concluded, "the living garment of God."[10]

Neither Lodge nor Michelson disproved the existence of the ether. They failed to prove it, which is not the same thing at all. As the saying goes, absence of proof is not proof of absence. Neither experiment took into account the spiral motion of Earth relative to the "galactic ether," only that of Earth's orbit around the Sun, and in this century, Dayton C. Miller claimed he *had* found evidence for "ether drift" after a series of experiments held in all four seasons on top of Mount Wilson, though whether he was right or not is still uncertain.[11]

In more recent years, the idea of an ether has refused to leave physics. In 1951 one of England's most eminent physicists, Paul Dirac, wrote a paper called "Is There an Æther?," and his answer was—yes, we are "rather forced" to admit its existence, although he preferred to call it an "isotropic sea of electrons in random motion."[12] Wolfgang Pauli had no objection to the idea of an ether, except as actual substance, calling it "the totality of those physical qualities which are to be associated with matter-free space," while Louis de Broglie wrote about a "subquantic medium," and the term "neutrino sea" has also come into use. Another recent definition is "the sum of all the unobservable forms of energy that exist in the microcosm beyond our observational limits."[13]

Incidentally, Einstein once told the physicist Horace Dudley that he probably had no knowledge of the Michelson-Morley experiments when he announced his special theory of relativity, and even if he had, he was quite sure that his thinking had not been influenced by them. And by 1933, when he met Lodge for a brief chat, he was prepared to admit that the ether entered into everything.[14]

We are in good company, then, if we postulate the existence of what might as well be called an ether as anything else.

Now it is time to take a closer look at the Sun, and to warn readers that this must *never* be done with the naked eye, especially through a telescope. Galileo is said to have done this, and he probably soon wished he hadn't. Naked-eye observation of the Sun is occasionally possible, but it is best left to the professionals. It is a little ironic that we cannot look at the most important object in our lives without going blind, yet without studying it (indirectly) in detail we must remain blind to the possible cause of countless effects that shape and influence much of our existence. One such effect is that of disturbed radio conditions, and it was a study of these that led to some very surprising conclusions, as we now see.

Chapter 2

THE IMPERFECT SUN

IN APRIL 1946, RCA Communications Inc. set up a small astronomical laboratory on the roof of its main office building in central Manhattan. John Henry Nelson, an engineer who had been with the firm for 25 years and was later to become its chief propagation analyst, was given a six-inch refracting telescope and told to study the Sun. It had been known for 20 years that its behaviour could somehow affect radio communications from time to time, even to the point where the company—the world's largest shortwave radio communications organization—would be temporarily out of business, as magnetic storms and other forms of solar interference made it impossible for their messages to get to their destinations. Company chiefs knew this happened, but they wanted to know more about how it happened— and especially when it would happen next.

Nelson's first job was to find a way of making accurate daily predictions of "radio weather" so that RCA staff around the world could be alerted in advance when reception was likely to be bad. In those days, there was no reliable way of telling when this would happen.

First of all, Nelson turned his attention to the sunspots, those enigmatic blobs near the solar equator that have been known to come and go in fairly regular cycles for centuries. It was known, too, that there was some connection between the appearance of sunspots and magnetic storms, and that the latter were the radio operator's worst enemy. Getting a shortwave signal through the charged particles that clog the

Earth's upper atmosphere during and after a storm was as hopeless a task as shining a weak torch through thick fog.

He studied the spots every day, noting how they varied in shape and size, and on what part of the Sun's face they appeared. He soon found that no two were alike; the appearance of some actually seemed to coincide with improved radio conditions whereas others could be "lethal." After all, it was the ultraviolet radiation from a sunspot group that made shortwave radio communication possible at all, by making the ionosphere more reflective and thus enabling signals to pass around the curve of the Earth. But now and then, it seemed, the appearance of a certain type of spot would wreak havoc. Nelson called them "maverick spots," and their damage meant the loss of several hundred radio-hours every year.

After a couple of years spent observing sunspots and noting what kind of spot proved to be a "maverick," Nelson was predicting radio weather, which is done on a very precisely established scale, with an accuracy of about 70 percent. This was encouraging, but it was not good enough, and by 1949 he had come to the conclusion that something other than the spots had to be involved. He went to the New York Public Library and dug out all he could find on sunspots and their cycles, and very soon it occurred to him what the missing link might be—*the planets*.

Nelson had never been a professional astronomer. Had he been, he would have "known" that planets have no more than insignificant statistical influence on anything, except the brains of astrologers. The Sun is so huge and powerful that no known external force could possibly affect its energy output. Newton established that every body in the solar system attracted every other according to their masses and relative distances from each other, so the Sun, with 99 percent of all known mass in its system, clearly does most of the attracting. It is known to have a colossal gravitational field, and its surface is not solid, but gaseous, which would naturally make it more volatile; but to suggest that a mere planet could interfere with solar behaviour is out of the question. "Utter bilge," (as Britain's Astronomer Royal said about space travel the year before it happened!).

Nor was Nelson an astrologer. He neither believed nor disbelieved in astrology for the simple reason that he had never studied it. He did not even know the symbols for the planets used by both astronomers and astrologers. He merely had a job to do, and wondered if a knowledge of planetary positions might help him do it better. What he did next now seems so obvious that one is led to wonder why nobody seems to have thought of doing it earlier.

On Easter Sunday, 23 March 1940, the worst radio blackout on record had taken place, a severe magnetic storm lasting four days. Nelson collected data on this and subsequent storms, then went back to the library to consult the *American Ephemeris and Nautical Almanac*, the U.S. Navy annual publication that lists the exact positions of the planets for every day of the year. Then he took an ordinary protractor compass, and on a sheet of plain paper he drew up a rudimentary horoscope for the 1940 short-wave radio blackout, which involved nothing more than noting where each planet was at the time, in relation to the Sun.

Horoscopes can be either heliocentric or geocentric; that is, based on the Sun or the Earth as centre point. Nelson made a heliocentric chart, since he was interested in the Sun, and then he measured the angles between the planets. Mercury and Jupiter were lined up with the Sun right between them, in *opposition*, as astronomers say, meaning 180° apart. Mercury was just moving into opposition with Saturn, while Saturn was at right angles to Venus. So there were two angles of 180° and one of 90° (half 180) involved in the four-day period under study. Very interesting. Moving on to later solar disturbance dates, Nelson found the same angles turning up again and again, together with zero degrees, or *conjunction*—that is, when two planets are on the same side of the Sun and in line with it and each other. It was beginning to look like more than coincidence.

On 12 April 1951, RCA Communications issued a press release announcing that evidence had been found pointing to a *direct relationship* between magnetic storms on Earth (i.e. bad radio conditions) and the position of the planets with respect both to each other and to the Sun. Thanks to its new system of prediction, RCA stated, it was now achieving 85 percent accuracy in its daily forecast service. Moreover, Nelson could tell when good days were coming as well as bad; when two or more planets were spaced 120° apart, conditions were likely to be good, while angles of 0°, 90° or 180° meant trouble. These were the "maverick" angles.

The sextile (60°) and the trine (120°) have been among

the favourable angles or "aspects" of astrology since time immemorial, just as the conjunction, square, and opposition (0°, 90°, 180°) have been the unfavourable ones. No astrologer seems to know why, and before Nelson nobody had ever produced scientifically acceptable evidence for such notions. For that matter, it had never been shown beyond any reasonable doubt that there was *any* connection between planetary positions and *any* terrestrial phenomenon. We cannot help wondering why it was left to a non-astrologer and non-professional astronomer to make this discovery.[1]

We will pick up Nelson's story later in this chapter, when we mention some of his other discoveries regarding Sun, planet and Earth. But first, we must try to summarize some of the Sun's vital statistics, features and habits, so that non-scientist readers can get some idea how it works. The next two or three pages will be a bit technical, and if some readers get a little confused they can take comfort in a recent remark by Dr. John Eddy, an astronomer with a special interest in the Sun:

"There is really only one thing that the world needs to know from astronomy, and that is—how does the Sun change?" Then he added: "We've been so poor about answering that question."[2]

"The Sun is God. Everybody can see that."

So the psychologist C. G. Jung was assured by a Hopi Indian chief, and many have shared Aristotle's view that the Sun must be constant, perfect and unblemished, as befits a deity. But it is none of these things, and Homer in his *Odyssey* foresaw something of its true nature when he described it as having a violent temper, liable when angry to flare up and emit thunderbolts. It does indeed flare up frequently and emit electromagnetic discharges, and it also provides us with a plainly visible clue, which has literally been staring us in the face since the dawn of creation, to the mystery of the invisible forces it generates, or perhaps transforms. The clue is the sunspot cycle, of which more anon. But let us forget the Sun's Godlike behaviour—and the legend of the Egyptian goddess Nut who swallows it every evening—and take a more objective look at our nearest star.

The Sun is fairly insignificant in the cosmic context. It is just one of about 10^{20} other stars in the observable universe, though it is important to us because of its proximity;

The Imperfect Sun

no other star is less than 300,000 times farther from Earth. There must be millions of other stars quite similar to it, although by Earth standards its essential statistics are staggering. It has a mass of 5×10^{33}g., or about 335,000 times that of Earth. Its density is extremely high at its core (110g. per cc.) falling to only 10^{-6}g./cm.^{3} in its outer layers. Earth's average density, in comparison, is only 5.52 g./cm.^{3} The Sun's diameter, about 865,000 miles, is more than 100 times that of Earth. The whole of the rest of the solar system would fit comfortably into less than one percent of the Sun. It is thought to be a perpetually exploding hydrogen bomb, fuelled by nuclear fusion, so powerful that every square metre of it emits about 70,000 horsepower of energy into space. To generate energy on this scale, we would have to burn about 11 billion times the world's annual coal output *every second*.

Since early in the seventeenth century, we have generally regarded the Sun as being at the exact centre of its solar system. But this is not strictly correct. The Sun does not remain still, nor does the centre of mass (COM) of the solar system coincide with the exact COM of the Sun. Sometimes, it happens that the true COM of the solar system moves outside the Sun altogether, as it was predicted to do, for example, in the years 1977 to 1984.* Both Sun and planets, therefore, revolve around a common COM, and to add yet one more complex motion to those we mentioned in the previous chapter, the entire solar system itself is revolving around the centre of our spiral galaxy (which is 3×10^{16}km. from the Sun) at a speed of 200 km. per second.

Nobody has ever seen the interior of the Sun, and we may never be certain of what lies below its turbulent photosphere, or surface. It is generally thought to have an inner core or nuclear chain-reaction region, surrounded by a convection region which in turn is surrounded by the photosphere. This is where the sunspots become visible to us. Finally, there is the superhot gaseous layer of plasma called the chromosphere, and surrounding the whole fiery orb is the solar corona. This was observed in ancient times during eclipses, as it still is today, for by fortunate coincidence (?) the Moon is just the right size to block out

---

* We refer to COM motions again in Chapter 5.

all of the Sun except its corona. Kepler made detailed studies of the solar corona in 1605, but it was not until the nineteenth century, when it was first photographed, that it could be seen to extend at least two million kilometres from the Sun itself.

Rays, or "streamers," shoot out through this corona up to distances of many times the solar radius, and photographs reveal clear differences in these streamers at different times in the solar cycle. At times of maximum energy discharge, they are arranged symmetrically around the Sun, whereas at intermediate and minimum periods they seem to elongate at the solar equator, streaming out from the poles like iron filings placed on a sheet of paper above a bar magnet, revealing what must surely be the lines of force of the solar magnetic field. From this observation alone, it can be seen that the solar output is far from constant. (The Sun's magnetic field, which was not measured until this century, plays a dominant role in the solar atmosphere, especially its outer layers, and has been closely linked to the 11-year periodicity of the solar cycle and to the latitude variation within each cycle.)

Solar Flares

Solar flares were first observed in 1859 by two Englishmen independently. Neither knew what he had seen: R. C. Carrington called it a "Singular Appearance" and R. Hodgson a "Curious Appearance."[3] Singular and curious they are: these gaseous outbursts average some 10^{19} sq.cm. in area, about a thousandth of the Sun's total area, but since even a medium-sized flare leaps 10,000 km. into space, its overall area can be more than enough to swallow an object the size of Earth.

By looking at the Sun with a coronagraph, which blocks out the light from its disc rather like an artificial eclipse, we can see that some special flares called "surge prominences" shoot out at speeds of up to 500 km./sec. to distances of 100,000 km. During a solar flare, there is an increase in both emitted solar x-ray intensity and output of ultra-violet and visible light, while extra-strong disturbances can cause havoc in Earth's ionosphere and prevent the propagation of radio signals, as Nelson discovered. During a "solar storm," the solar proton count also goes up, leading to an actual decrease, however, in the amount of cosmic ray protons that reach us, a mechanism called a Forbush Decrease after the man who discovered it.

Most cosmic particles are far more energetic than solar protons, reaching one billion electron volts, which is important when determining how many of them penetrate Earth's atmosphere and reach its surface. During quiet magnetic periods, the natural geomagnetic field at the equator turns away particles with less than a certain momentum, the value varying according to latitude, meaning that solar protons are normally kept out of our atmosphere at latitudes below 60°N, the approximate latitude of Oslo, Stockholm, Helsinki, Leningrad and Anchorage, Alaska. Likewise, cosmic ray proton intensity increases with latitude from the equator, being about four times higher around 60°N. At times of magnetic storms, minimum latitudes at which particles can enter are reduced because of the ring current of particles formed around the Earth. So it is plain that not everybody on Earth gets the same ration of solar and cosmic inputs, the people of the cities mentioned above getting more than their share of proton radiation.

Not all parts of a solar flare reach Earth together. Light takes eight minutes, but solar protons travel slower than the speed of light, and moreover they come not in a straight line but in a spiral path, so it may be more than an hour after the light from a solar flare reaches us that the protons start pouring in. Other forms of energy take even longer, and because more particles from the Sun reach Earth's surface at some latitudes than at others, their effects must also be widely varied.

It is only recently that we have begun to discover in how many ways life on Earth is affected by the Sun. Many, many more things than radio signal quality are influenced by its activity, as we shall see, and to understand the nature of such effects we need a better knowledge of the possible cause; so here is a brief outline of some of the other types of energy that the Sun churns out.

A small fraction of its total flare energy is made up of radio waves. In 1942 a chain of British radar-stations discovered very strong noise signals from the Sun at wavelengths of four to six metres. They lasted a couple of days, and when investigated were found to have been caused by a large solar flare. Such radio bursts are sensitive indicators of solar eruptions, making the latter sometimes easier to forecast than local showers of rain, which seems odd when we think of the relative distances involved.

Magnetic "storms," mentioned above, are sharp and brief fluctuations in the strength of the Earth's magnetic field (or "geomagnetic field") caused by charged particles from the Sun entering our atmosphere. At such times, Earth and Sun are in effect linked by a magnetic umbilical cord, and spectacular visible evidence for this link can be seen in the *aurora borealis* or Northern Lights visible in Canada and Alaska which, like the southern *aurora australis*, is caused by high speed protons and electrons from the solar wind and serves as a clear indication of an alteration in the Earth's magnetic field.

The lines of the Sun's magnetic field curve away from it in an Archimedes spiral, like the curve followed by the spray from a rotating lawn sprinkler, though one end of the field is linked to the Sun and rotates with it. It is along these lines that charged solar particles travel from Sun to Earth: in 1963 the IMP-3 satellite's measurements showed that such particles are indeed "tied" to the solar magnetic field lines as they sweep across the face of our planet.

It was only discovered that the Sun emits electrons as well as protons when NASA's first planetary probe, Pioneer 5, was launched just before the strong geomagnetic storm of 31 March 1960. This caused major disturbances in Earth currents and a total communications blackout over the North Atlantic. Studies of data from this probe indicated that these electrons in space must be the extreme energetic part of the plasma stream associated with the March 31 event.[4,5]

We now know that at least 50 percent more electrons come from the direction of the Sun than from the opposite direction. This is thought to mean that electrons, like protons, follow the Archimedes spiral and reach Earth in a tight column. It is also possible that the Sun emits neutrons, and since these have no charge, they would travel in a straight line, thus arriving before the protons and providing us with another means of forecasting proton storms.

It is clear, therefore, that a good deal more than just light and heat reaches us from the Sun. Before we had the instruments to measure all this activity, it was natural that we, or our ancestors, found the Sun somewhat terrifying and, taking the line of least resistance, preferred to worship rather than study it. Now that we know at least some of the nature of the solar thunderbolts, we can begin to understand many previously inexplicable terrestrial phenomena, and there is no doubt that much of this progress is due to our study of the sunspots.

Sunspots and the Solar Cycles

Sunspots were discovered by European scientists in the year 1611 or thereabouts; though once again the history of western science coincides with the rediscovery of Chinese science, for sunspot records in Chinese literature date back at least to 28 B.C. The earliest known reference to a sunspot has been attributed to Theophrastus, a pupil of Aristotle (who cannot have been too pleased!), and they have been observed over the centuries by several individuals, most of whom had no idea what they were; suggestions ranging from high-flying birds, planets in transit, new planets (Vulcan?), or defects in telescopes.

The Incas, however, knew there were blemishes on the face of Inti, their sun-god and chief deity. Russian chronicles of the fourteenth century give vivid descriptions of sunspots as seen through the haze of forest fires. Kepler saw one in 1607, but thought it was the planet Mercury, and a few years later at least four other Europeans with their newly-invented telescopes had observed them independently. They were Thomas Harriot in Oxford, Johann Fabricius in Wittenberg, Father C. Scheiner, a Jesuit priest-astronomer (one of whose assistants actually saw one before he did) in Bavaria, and Galileo in Italy, who is usually given credit for discovering them all by himself, a claim for which there is no evidence except his own.[6] Scheiner deserves the credit for making the first regular records of sunspots in modern times.

Then a very strange thing happened, the implications of which have still to be explained. Only a few decades after the sunspots were discovered, they disappeared altogether for seventy years, with very rare and isolated exceptions. This fact has been on record at least since 1887, though astronomer John Eddy surprised many scientists when he researched and published the full story of the spotless period in 1976.[7] The period lasted from 1645 to 1715, or the equivalent of what we now reckon to be more than six normal solar cycles. It coincided almost to the day with the reign of Louis XIV, the Sun King, also with the "little ice age" of the seventeenth cenutry, and, more intriguingly, with one of the more peaceful periods in recorded history. The study of sunspot cycles leads to many surprises, as we shall see in a later chapter.

Even when the spots returned, it was fifty years before anybody made much effort to find out what they were. In 1769, Alexander Wilson of Glasgow studied them long enough to note that there were variations in their penumbrae, and at the end of the eighteenth century the musician-astronomer Sir William Herschel studied the Sun in more detail for five years, starting a controversy that continues today with his speculation that seasonal differences in the Earth's climate could be due to "a more or less copious emission of the solar beams." This was one of the earliest suggestions that a study of disturbances on the Sun could be of any particular value.[8]

In 1833, his son Sir John Herschel made a very important discovery: that the spots were electromagnetic in nature. He was also the first to suggest that they had something to do with magnetic storms. Finally, more than 230 years after man first studied a sunspot through a telescope, the German astronomer Schwabe made the momentous discovery that they came and went in *cycles*, which he reckoned to be about ten years in length.[9]

Still nobody showed much interest, except Rudolf Wolf in Switzerland and R. C. Carrington in England. Carrington, one of the first men to spot a solar flare, was the son of a brewer who gave up astronomy at the early age of 36, going back to making beer after discovering the fact that the average latitude of sunspots decreases towards the end of each cycle. Wolf went over all the records he could find, and announced that the average length of a cycle was not ten, but more like 11.1 years, a figure still regarded today as accurate. (Strictly speaking, the cycle is of 22.2 years, since the polarity of the spots is usually reversed after each 11.1-year period.) It was also Wolf who introduced the formula still in use today for calculating daily spot numbers, which are known as Wolf numbers.[10]

The modern age of sunspot observation began after a team of mules had clambered up the 5,670-foot Mount Wilson in California with bits of telescope on their backs— eventually to be replaced by the famous 150-foot solar tower telescope which began operating in 1912 under the direction of George Ellery Hale. It was Hale's photographs that revealed sunspots to be the centres of huge hurricane-like vortices, and also powerful magnets. Lines of force around them resembled those of an ordinary EM field; not that there was anything ordinary about the strength of the solar fields, which could be up to a million times that of the average geomagnetic field.

What exactly is a sunspot?

As seen from Earth, it is a dark blob of *umbra*, dark because slightly cooler than its hazy surrounding *penumbra* area, though some incipient spots are only visible as tiny pores that fade away after a day or two. Those surviving this stage form themselves into groups, with two spots in each group normally dominating the rest, their penumbrae growing until the group reaches maximum size after about ten days. Then the lesser spots disappear, followed by the second of the two main ones, which break up into smaller spots. The leading spot starts to shrink about a month after its formation, and gradually fades away, though it does not break up in the way its former fellow-spot has done. The two main spots in each group have opposing polarities, and the magnetic field associated with them grows in strength until the group starts to break up.

It is while the leading spot in each group develops its penumbra that the solar flares first become visible. Flares can last for up to 25 days, and the decrease in flare activity coincides with the fading of all spots in a group except the leader. There is, therefore, an unmistakably definite connection between sunspot and flare, which is where the spots begin to be of more than academic interest to us. When a spot appears, it means a flare is on the way.

Also firmly established is the link between sunspot appearance and magnetic conditions on Earth. It has been suggested that the size of each solar maximum, or peak of the sunspot cycle, can be determined at the beginning of each cycle, because of the close correlation between magnetic data at sunspot *minimum* and the magnitude of the ensuing maximum. This effect has been noticed throughout eight whole solar cycles, and must be more than coincidence.[11] If so, it makes one aspect of the sunspot mystery clear: both the solar cycle and its apparent effects on Earth must in fact be effects of an unknown third factor. We cannot have cause following effect. Therefore we cannot say for certain that the solar cycle actually *causes* anything. It may coincide with, or precede, or follow a terrestrial cycle, but with a few possible exceptions it cannot cause them. We shall find this theme recurring again and again.

Nevertheless, the sunspot cycle is very useful. It is a daily news bulletin written in a code we are just beginning to decipher. It is a distant early warning system, advising us of variations to come in our quota of solar and cosmic energies. Above all, it is a valuable reference point for numerous other cycles to be mentioned in due course.

More than any other single feature, it is the sunspot cycle that offers us a clue to the workings of the great solar system machine. Spot numbers have been recorded regularly and accurately since 1848, though data on monthly averages goes back to 1749 and some information on their peaks and troughs is available from 1611. Thus the modern cycle-watcher has plenty of good data to work with.

Fig. 2
The sunspot cycle from 1750 to 1976. Each dot indicates the average annual Wolf number (R) as measured on vertical line. Note that the two highest peaks of solar activity yet recorded are 179 years apart. The sunspot cycle has baffled scientists for more than a century, though possible solutions to the problem of its origin are now in sight. (Data from M. Waldmeier, Zürich).

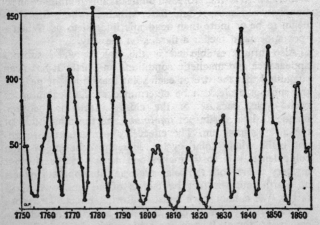

The Imperfect Sun 41

There is still much to be learned about both sunspots and the Sun itself. The first comprehensive textbooks on the spots appeared only in 1964,[12] in the early years of the Space Age, and we live at a time when a single space probe can render a textbook obsolete overnight. For instance, the first manned Skylab mission of May/June 1973 lasted just 28 days, but in that time more hours of observation of the outer solar corona were logged than the total of all observations from Earth by all men in all recorded history! Skylab crews had many surprises; film taken on June 10th revealed "a great transient blob" the size of the Sun itself which moved outwards through the corona at about 400 km. a second, while a couple of months later a coronal spectrograph showed an eruption of helium rising half a million miles above the Sun's surface and then stopping as if blocked by an invisible wall. This remarkable sight was described at the time as "a total mystery."[13]

The origin of sunspots and their cycles is equally mysterious. It is generally believed that the appearance of spots has something to do with the solar magnetic field popping out in a sort of bubble. But what causes this to happen? Here the real arguments begin.

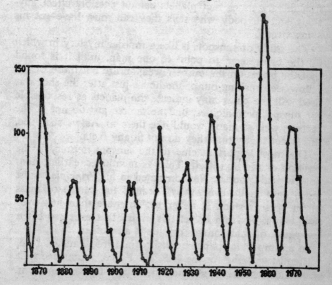

In the 1940s, the Swedish scientist Hannes Alfvén put forward the theory that sunspot fields result from the behaviour of a general internal field. His ideas gave a boost to the new science of magnetohydrodynamics, the study of electrically conducting fluids (such as the solar plasma) moving through electric and magnetic fields, though more recent theories prefer to see the Sun's interior as a dynamo fed by the interaction of its rotation and its convection.[14] This can be made to make sense mathematically, but it does not explain why sunspot cycles are irregular in length, nor can it explain the period from 1645 to 1715 (known as the Maunder Minimum after one of the first astronomers to draw attention to it) when there were hardly any spots at all.

Solar scientists divide into two often hostile camps. One says that all solar events originate in the Sun and nowhere else, while the other insists that the planets have something to do with at least some of them, including the periodicity of the spots' appearance. Nobody is suggesting that external forces *cause* them to form, but that planetary positions *might* be influencing the precise timing of their appearance. We give more space to the second school here, since the first merely states that planets cannot possibly affect anything and anybody who says they can must have got his sums wrong.

The study of sunspots is like a murder mystery in which all the clues seem to point to one man, until it is found that the prints on the murder weapon are not those of the fellow seen rushing out of the house just after the shot was fired. All the clues may indicate the planets as responsible for sunspot periodicities, but the figures just do not fit the facts as well as many would like them to. As we shall see, this is not to say that they do not fit any facts.

The most important clue to the sunspot mystery is the position of the Sun itself. This, as mentioned earlier, does not keep still, but as Newton noted in his *Principia* "must move continually every way" around the centre of mass of the solar system. It never moves more than just over two solar radii from it, yet this means that every so often the centre of mass moves outside it altogether.

In an important paper published in 1965, Paul D. Jose, a scientist working for the U.S. Air Force, announced that after studying the Sun's motion relative to the centre of mass, he had found a period of 178.7 years. This just happens to be the exact interval, plus or minus a few months, between the two highest peaks of solar activity yet recorded—1778 and 1957—which could of course be no more than coincidence, though later in this chapter we shall present more evidence that suggests otherwise.

Based on his findings, Jose made predictions for future maxima and minima of the solar cycle, which had proved extremely accurate up to 1977. His work has stimulated several other studies, and offers a possible solution to the problem of the irregularity of the solar cycle, which averages just over 11 years although individual cycles have ranged in the past from 7.3 to 17.1 years: a fact that critics of planet-Sun theories find hard to explain away.[15]

Now, if it can be shown that the motion of the Sun around the centre of mass of our solar system is caused by planetary influences, and if in turn such influences can be found to have the same periodicity, then surely we have solved the problem?

This may be the case, and several researchers think it is, though many others disagree. But we are running ahead of the story, and before returning to sunspot research in the Space Age, let us pick out a few of the highlights of more than a century of clue-hunting, in order to show that modern theories are based on plenty of good evidence.

The second half of the nineteenth century saw much interest in sunspots from astronomers. This was largely thanks to the work of Carrington and Wolf, whom we mentioned earlier, and theories as to the origins of the solar cycle came thick and fast, as if making up for 250 years of lost time.

The first serious attempt to link planet motion and sunspot appearance was probably that of Wolf, who in 1859 published a formula according to which the mass, distance and angular velocity of the planets could be used to plot a curve broadly similar to that of the growth of the spots.[16] A contemporary of his, H. Fritz, fully supported his theory, while as for Carrington, before he abandoned his telescope for the family brewery, he had time to notice that Jupiter's orbital period of 11.86 years was close to that of the average sunspot cycle. He wondered if there might be a link.

This proved to be a red herring, for the cycles of Jupiter

and the spots were completely out of phase early in the twentieth century. Yet Carrington's idea was a very useful one, because it encouraged others to look harder at the planets and their relative positions, angles and periods. It could easily be calculated, for instance, that the greatest tidal influences on the Sun must be those exerted by Jupiter and Venus, and in 1867 Professor W. A. Norton of Yale suggested that there must be a "physical agency" exerted by these two planets on the solar photosphere.

Other astronomers noticed that conjunctions and oppositions of Jupiter and Saturn came at intervals close to that of the spots' maxima. But again, not quite close enough. And again, this was a red herring that led to useful research, such as that of Professor E. W. Brown, also of Yale. He looked into the tide-raising forces of these two planets not only at conjunction and opposition, but also when they were 90° apart, or "square" to each other, plotting a curve for the previous 300 years which showed considerable agreement with the solar cycle. For once, this was no red herring, and in 1965 W. H. Portig of National Engineering Science Co. reckoned he had actually proved that Jupiter's and Saturn's relative positions were linked to spot formation. It so happens that 15 orbits of Jupiter take almost exactly 178 years, as do nine Jupiter-Saturn synodic periods, and after extracting these two components from the sunspot cycle, Portig found an unidentified period remaining of 11.1 years![17]

One of the earliest breakthroughs in solar cycle study came in 1902, when a British statistician named William Digby tried a new approach. He formed a theory, made a prediction based on it, and got it right. Going back to Newton's (1687) delineation of the Moon's tangential pull as a tide-raising factor, he applied this to the six planets nearest the Sun (Mercury to Saturn), predicting maximum effects when two or more were in conjunction at an angle of 45° to the solar equator, and correctly naming 1905 as the year of the next solar activity maximum. This was an impressive hit, for the 1905 peak was an unusually low one that could not have been reasonably expected by chance.[18]

In 1907, Mrs. Annie Maunder complicated matters somewhat by suggesting that certain planets might be inhibiting sunspot formation rather than helping cause it. She based this idea on her discovery that some 22 percent more sunspots came into view round the east limb of the Sun, or formed close to it, than passed round the west limb or disappeared near it. The same approach was taken later by Fernando Sanford, who thought that the motions of Venus and Earth might have such an inhibiting effect, since when they are in opposition there are 60 percent more visible spots than there are when they are in conjunction.[19]*

Theories of planet-Sun interactions rely heavily on the supposition that the gravitational and electromagnetic influences of the planets set up tides on the Sun in much the same way as the Moon does in the waters of Earth. (And in the solid Earth itself, as we shall see in Chapter 5.) The most thorough study of this complex and controversial subject is that of the Norwegian astronomer K. G. Meldahl, who goes so far as to suggest that sunspots are effects caused by solar flares and prominences, which in turn are started in the corona by external forces, i.e., planets.[20]

Meldahl is the exception to the rule that nobody claims the planets to be the actual cause of solar effects, and in view of the fact that his extensive work has been almost entirely ignored, it is worth summarizing in brief:

It is known that during sunspot maxima periods, the Sun's corona shows ejected matter streaming outwards in many directions around its circumference; whereas at minima such streamers emerge from a belt about 60° wide around the solar equator. (Sunspots seldom appear more than 35° to either side of the equator.) Within this belt, the great centrifugal force of the rotating Sun must surely overcome any other forces at work. Meldahl looked at possible interactions of all the forces involved: gravitational (downwards), radiational (upwards, giving rise to the solar wind), centrifugal (outwards and perpendicular to the solar axis), and finally the tidal action of the planets, consisting of positive and negative radial forces combined with tangential forces. Disturbances on the Sun often appear in opposing polarities, and no force other than tidal, Meldahl thinks, should show such symmetry.

Moreover, he argues, radiation pressure on certain small particles in the corona should balance gravitational forces

---

* It must be remembered that we cannot see the spots on the far side of the Sun.

so nearly as to create a sensitive layer which, when disturbed, would be liable to violent motion of ejection away from the solar surface. Tidal action, he says, would be too weak to create noticeable motion in the heavier photosphere. However, hydrogen prominences are known to appear at levels comparable in height to that of such a sensitive layer. They set in suddenly, becoming more numerous as solar maximum approaches, and the problem is to find the link between them and the tidal forces. Meldahl reckons he has done this: by calculating the effects of the five nearest planets on the solar tides, he has shown that they interact in a way that reveals several different periods of up to 243 years in length. When unfolded, his charts fill a room, which makes reproducing them difficult.*

Coming now to the Space Age, we might expect that the sunspot problem is virtually solved. Before the invention of the high-speed electronic calculator, major studies of sunspot periodicities, cycles, harmonics and planetary resonances just could not be done in a single lifetime, because of the horrendously complicated calculations required. Yet the computer raises as many problems as it solves, at least in this field, and cycles in solar activity resemble UFOs in that some see them and others do not, and that the latter tend either to ignore the former or attack them with much fury.

One who can see them is Dr. E. K. Bigg, who published a study in 1967 on the influence of Mercury on the solar surface based on a hundred years of data. He identified a periodic variation in sunspot activity of 87.976 days, the exact period of Mercury, and also found that when either Venus, Earth or Jupiter is on the same side of the Sun as Mercury during its perihelion approach, the "Mercury Effect" is double what it is when the other planet in question is on the other side of the Sun. "Even though they may be small," he insists, "planetary effects *do* exist." (His italics.)[21]

Dr. Bigg has not been ignored, and though his findings

---

* Meldahl calculates the relative vertical tidal forces on the Sun as follows: Mercury 16.3—57.3 (newtons), Venus 63.4—66.1, Earth 28.9—31.9 and Jupiter 58.9—78.6. By comparison, forces of the other planets are extremely small, e.g., about 3.3 for Saturn and 0.9 for Mars.

have been much challenged, several other scientists have unearthed definite relationships between planetary orbits and the sunspot cycle, at least on paper. Dr. Robert M. Wood of McDonnell Douglas Astronautics has found good correlation between solar activity periods and the synodic period resonances of planets when taken in pairs, also between the variations of the resonance period resulting from orbit eccentricity and variations in the sunspot period.[22] Incidentally, there may be countless other cycles in solar activity in addition to the most obvious 11.1-year one; good evidence has been produced for a short 29-day period, and there are numerous theories of long cycles of up to several centuries in duration.[23]

Dr. Wood's father, Professor Karl D. Wood of the University of Colorado, has studied the effects of the nearer planets on the Sun and found that when Earth, Venus and Jupiter are in conjunction or opposition they can raise tides on the Sun up to 50 percent greater than the largest tide raised by Jupiter alone. As to how significant such planetary tides could be, the astronomer Ernst Öpik has calculated that gravitational effects could raise a tidal flow on the Sun with a velocity of 93 km./sec., compared with 300 km./sec. for lunar tides on Earth. This, Öpik says, is "not at all negligible," as has been suggested, especially in view of the fact that the Sun has 27.6 times the gravity acceleration of Earth. He believes a link between sunspot activity and planetary tides to be "highly probable."[24]

The outer planets are often omitted from calculations of tidal effects on Sun (or Earth). It is hard to imagine how an object as far away as Uranus, let alone Neptune or Pluto, can possibly have an effect at all. Yet this seems to be a mistake. It has been noted, for instance, that Uranus and Neptune seem to be square to each other at times of solar maximum and in either conjunction or opposition at minimum more often than they should be by chance, and it has been suggested that an 86-year periodicity is involved here.[25] As for Pluto, at least one scientist, as we shall see later in the chapter, has managed to bring it into the picture.

Not all planet-Sun theories are confined to studies of tides and gravity. Professor W. A. Luby suggested in 1940 that the action of the planets was by precessional pull on the solar equator, while in 1946 Maxwell O. Johnson went so far as to state that the influences were not gravitational at all, but electrical and magnetic.[26]

The debate continues, with some theories holding up better than others. One of these is that of Jose involving the solar system centre of mass. British astrophysicist and author John Gribbin speculates that sunspots may appear as a result of the "stirring effect" as the centre of mass moves around with respect to the centre of the Sun. The gravitational pull of all the planets together, he points out, swings the Sun around in a loop-like orbit around the centre of mass, and this could produce forces that influence the Sun's internal workings as well as its visible spots.[27]

If the ideas of Jose, Gribbin and other pro-planet researchers hold up and can be proved, which is certainly not impossible, it may lead to a happy reunion between the two warring schools of solar physicists. Proving them will not be easy, and there is always somebody to dismiss a claim correlation as an "artefact of the calculation."[28] What will help more than anything is to be able to make predictions about solar activity and see them come true. This has already been done by Jose, whose predictions to the year 2009 were published in 1965. His first three were right on target, and according to him we are due for maxima in 1984, 1995 and 2009, and minima in 1990 and 2002. Wood (K. D.)'s predictions are entirely different, though he too scored with the 1977 minimum. In fact, he has predicted a maximum for 2003, only one year after Jose's predicted minimum![29]

The next step in the unravelling of the sunspot mystery is to forecast specific solar events, and this too has been done by several independent scientists, as well as by Nelson (to whom we will return in a moment). Indeed, Dr. Jane Blizard of the University of Denver has stated that "long-range prediction of solar activity has now become possible." Proton events on the Sun, she said, had been shown to be related to the positions of Mercury, Venus, Earth and Jupiter. She could not explain the mechanisms involved, but stressed that this did not detract from the value of this method of prediction.

As long ago as 1963, Dr. Richard Head of NASA announced a technique of predicting solar flares by computer, based on a study of planetary gravitational effects. He later claimed to have forecast a week in advance the major solar storm at the end of August 1966, even getting the solar latitude exactly right, the longitude within three degrees, and the time within three hours. We would like to be able to say that the fact that the U.S. has yet to lose a man in space is partly due to NASA's backroom astrologers, but we have insufficient evidence. (We return to this subject later.)[30]

In Germany, Dr. Theodor Landscheidt has claimed a bull's eye at least once, with his prediction over a month in advance of the proton storm of 1 September 1971. This was an interesting one, for the very day before the event the probability of such a storm was rated at zero percent by the Solar Disturbance Forecast Center, which bases its work on information from a worldwide observer network. This, incidentally, was its lowest probability forecast for the whole year. Since Dr. Landscheidt is not a professional scientist, he is seldom listened to, which may be unfortunate.[31]

H. Prescott Sleeper Jr. of Northrop Services Inc. is a professional scientist, however, and in 1972 he wrote a 56-page report for NASA on the subject of Sun-planet interactions which contained a number of very specific predictions, and included a new model of the Sun (about the workings of which very little is really known for certain) for good measure. Like Jose and Wood, he correctly named 1977 as a minimum year, though his ensuing maximum is predicted earlier than by either of them, for 1981. (At least a year either way must be allowed for such predictions, since the actual minimum point of a solar cycle may not come in the year with the lowest average Wolf index.)

Sleeper offered strong support for Jose's earlier work. He reckons that the 11.1 (actually 11.08) -year cycle is due to resonances of the periods of the three inner planets, while the longer 179-year cycle is driven by those of the four biggest planets' synodic periods: that is, the time between the points at which two planets are in line with each other as seen from the Sun. Consider the following list of "coincidences," as first noted, we believe, by Jane Blizard, upon which Sleeper bases his predictions and hypotheses:

Planet	Orbits	Synodic Periods*	Years (Tropical)
Mercury	46		11.079
Venus	18		11.074
Earth	11		11.000
(Moon)		137†	11.077
Mars	6		11.286
Saturn	6		176.746
Jupiter	15		177.933
Jupiter/Saturn		9	178.734
Jupiter/Neptune		14	178.923
Jupiter/Uranus		13	179.562
Saturn/Neptune		5	179.385
Saturn/Uranus		4	181.455

* That is, intervals between planetary conjunctions as seen from Earth.

† = synodic revolutions.

Thus it can be seen that *all* the planets with the exception of Pluto have a harmonic relationship with the two most easily identifiable periods of the Sun. But this is not all. In Chapter 1 we mentioned the discovery that the spin of Venus is related to the orbit of Earth, and Molchanov's claim that all celestial bodies are harmonically related. Sleeper also reckons that there is spin-orbit coupling between the supposed rigid core of the Sun and the orbital period of Earth, the former's synodic rotation of 27.04 days being almost exactly one twenty-seventh of two Earth years. "The Sun's spin," Sleeper says, "may be trapped in an Earth-orbital resonance." Reading his provocative report, one cannot escape the feeling that our fellow-planets are very closely involved with us and our Sun, to an extent perhaps even astrologers never imagined. Arthur C. Clarke's dream of the United Planets may already be here.[32]

Do such coincidences as those listed above have real significance, or are they of interest merely to numerologists? We return to this tricky question in a later chapter. It is easy to become carried away by numbers.

Even so, some of the "near commensurabilities" to be found among the various planetary orbital periods and their multiples are quite surprising, especially some involving Pluto. One of us (Playfair, who is unable to do the simplest sum without computerized help) discovered after pressing the wrong button on his mini-calculator that if you divide the orbital period of Saturn by that of Jupiter, you get exactly one hundredth of the period of Pluto. Likewise:

Saturn × Uranus = 2476   years = 10 × Pluto (Error 3.2%)
Jupiter × Uranus = 997.07 years = 4 × Pluto (Error 1.4%)

and, to the list of commensurabilities quoted by Sleeper, we can also add:

Mars × Saturn ÷ 5 = 11.08 years
Mars × Jupiter ÷ 2 = 11.15 years
Mars × Jupiter × 8 = 178.37 years

This would make a good game for astronomers stranded at airports.

But let us get back to the real world, and pick up the story of John Nelson where we left it at the beginning of this chapter. It should be emphasized that his early forecasting work, based to some extent on planetary positions, was carried out and published long before much of the other research we have mentioned above was even begun. It should also be mentioned that no professional scientist, as far as we know, has yet come close to equalling Nelson's overall record as a forecaster of solar events.

Meanwhile, in Manhattan . . .

By 1953, after six years of study, Nelson was able to tell his professional colleagues at a symposium that *all* his radio-weather forecasting was now being done on a basis of planetary interrelationships together with signal analysis. He had already learned a good deal about sunspots; it was their type and position that seemed to affect radio conditions. Large spots consisting mostly of penumbra might have no effect at all, whereas small spots could be, as he put it, "quite deadly." The larger the umbra, the more potent a spot's effect on radio circuits would be, and rapidly changing spots were the most lethal of all. Nelson had also found that spots on a certain part of the Sun, which he called a "critical centre," did most damage. (It was almost as if he had come to know each spot individually!) During the 405 hours lost to radio communication from July to December 1947, for instance, 89 percent of sunspots were inside this critical centre.

Yet despite his early success, it was only when Nelson

began to study planets as well as the Sun that his accuracy rate began to soar. At first, he took only the six planets nearest the Sun. Like others we have mentioned, he simply could not imagine how Uranus, Neptune or Pluto could affect it, so he left them aside. However, soon after he had published his 1951 report, he was contacted by a colleague, J. H. Clark, propagation analyst for Press Wireless Inc., who said he had checked Nelson's hypothesis and found it worked. But, he added, he had also found that if *all nine* planets were included in the calculations, it worked even better.

Co-operation between rival business firms seems to be closer than it often is between scientists, for Nelson immediately adopted Clark's suggestion, and the two kept in touch from then on. By 1967, Nelson had been able to refine his techniques and achieve an astonishing accuracy rate of 93 percent out of a total of 1,460 specific forecasts.

How does he do it? He describes his work as the study of "simultaneous harmonic relations." The rules of the game for severe storms, he says, are that one of the four inner planets (Mercury, Venus, Earth or Mars) must be at a "hard" angle (0°, 90° or 180°) with another planet further from the Sun than itself. But that is not enough. If it were, Nelson says, "shortwave radio would have a hard time surviving." In addition, at least two other planets must be in an angular harmonic relationship with the first pair. The main angular harmonics used by Nelson were 360° divided by 2, 3, 4, 5 and 6 to give 180, 120, 90, 72 and 60 degrees. These angles in turn have their subharmonics; 90° divided in like manner gives 45, 30, 22.5, 18 and 15 degrees, for example. But such angular subharmonics become significant only when there is a "hard" basic angle with which to combine them.

Pythagoras and Kepler would surely approve of Nelson's methods. Here is both the geometry and the music of the spheres being put to practical use, and not by a medieval astrologer but by a twentieth-century New York communications expert.

Nelson reckons that Mercury is the trigger planet for 90 percent of all magnetic storms. These, he has found, tend to begin after Mercury has moved into a hard angle with another planet. He has also noted that factors other than angles must be taken into account, including planetary perihelia and nodal points (when planets cross the plane of

The Imperfect Sun 53

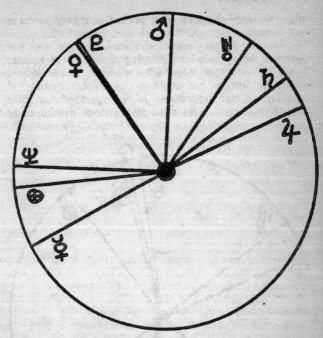

Fig. 3-1
J. H. Nelson's heliocentric chart for 23 March 1940. A severe magnetic storm took place between this date and 27 March, within this period Mercury being in opposition to both Jupiter and Saturn; Venus being square to Mercury and Saturn and in conjunction with Pluto; and Mars being square ot Neptune. Jupiter and Venus, the planets with the strongest vertical tide-raising force on the Sun, were both close to perihelion. Altogether, seven planets were either in a square or an opposition at some point over the five-day period.

Earth's orbit round the Sun), especially the point midway between ascending and descending nodes, which he regards as "a very sensitive area in space."

Is all this beginning to sound like medieval astrology?

Nelson's work is an example of what astrology may once have been and still could be: the study of the celestial motions and the *correct* interpretation of their terrestrial effects. His first book was published by the American Federation of Astrologers, and to many astrologers he is a heroic pioneer—the man who proved astrology to be a valid science.

In fact, he did nothing of the kind, and he does not appreciate being called an astrologer, with some reason, though he has acknowledged his use of certain astrological techniques. But he has made it clear that the study of planetary angular separations is only part of his solar forecasting method. This must also include careful study

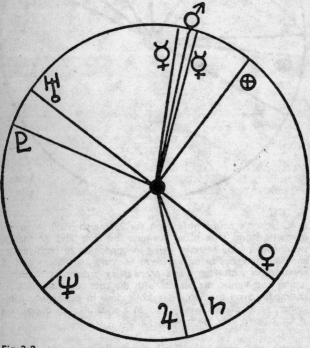

Fig. 3-2
Chart for 12 November 1960, date of one of the severest solar disturbances since March 1940, which Nelson correctly predicted for NASA more than a year in advance. The storm lasted from the 12th to the 16th, radio communications being blacked out on the 13th. Mercury was at perihelion and also in conjunction with Mars; Venus was square to Earth and in opposition to Uranus, which was also square to Earth. There were also trines (120°) between Mercury and Venus and between Earth and Saturn.

of sunspots and geomagnetic activity indices, as well as analysis of signal quality.

Nelson's success as a solar weather forecaster soon came to the attention of NASA, which asked him in September 1959 to predict for the month of November 1960. He did

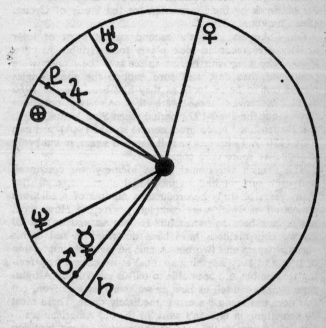

Fig. 3-3
Chart for the greatest ever recorded cosmic shower date of 23 February 1956. Mercury and Mars were in conjunction while passing through their descending nodes. Jupiter and Pluto were in conjunction and square to Saturn. All four of the planets nearest the Sun were on the same plane with respect to the ecliptic, which is extremely rare. "The fact that this rare event was associated in timing with a very rare cosmic ray event causes one to wonder if there might be some physical connection between the two," Nelson says. (Note: also rather rare was the solar activity that preceded this event, the Wolf number rising from 26 to 260 between 8 and 20 February.) Charts adapted from *Cosmic Patterns* by J. H. Nelson, by permission of the author and publishers, the American Federation of Astrologers. For key to symbols, see p. 10.

so, and forecast a severe solar storm for that month, which is exactly what there was—one of the worst in 20 years. The planets' positions at the time showed a host of squares, oppositions and conjunctions. An equally successful prediction for NASA in September 1963 earned him a letter of commendation, and he has also been awarded a gold medal for his work by the Foundation for the Study of Cycles. (See Chapter 6.)

On 4 August 1972 the second largest burst of solar activity ever recorded took place. John Gribbin had the idea of checking with Nelson to see how the planets were positioned that day, and sure enough the same angles turned out to be in force as they had been for the 1960 event. Moreover, these were the only two occasions throughout the 1960-1972 period when Sun, Venus, Earth and Saturn lined up to produce one square (90°) and one trine (120°). Twice lucky? If it happens again, it will begin to look like more than coincidence.[33]

The French astronomer Henri Mémery was convinced that each sunspot had an individual character, and after many years of study he produced a number of predictions almost all of which were completely wrong.[34] How, then, has Nelson been so successful? He has assured us that all his working methods have been described in full in his various papers and two books, and he gives no impression of trying to keep anything secret. Yet we cannot help feeling that he has not been able to tell us everything. A tightrope walker can tell us how *he* walks across a canyon, but this does not mean we can immediately do it. There must be something in Nelson's work of the old American tradition of "flying by the seat of the pants"—using intuition derived from long years on the job rather than in the research laboratory. Nelson has never been a research scientist, and has not published his results in the detail scientists demand. We must also report that two researchers known to us have tried without success to replicate his chief findings. Even so, these still stand unshaken.

In 1975 it was stated in a NASA publication that "individual eruptions from the solar surface have proved impossible to forecast," although "much effort has been expended in the field of solar-event forecasting," and it is admitted that "no major solar-particle events" (such as proton storms) "occurred during an Apollo mission." No mention is made in this publication of *any* of the scientists whose work we have summarized in this chapter.[35]

What does this mean? Was the letter of commendation NASA wrote to Nelson a forgery, a mistake, or what? Were Dr. Head's claims ever disproven? Were Apollo crews just lucky, or were they trained to spot approaching solar storms and steer round them? Nelson's own comment on the above quotations was simply: "This often happens in large organizations wherein the right hand does not always know what the left is doing."[36]

We like to think the remarkable success of the Apollo programme was due to more than luck, especially in view of the fate of Soviet cosmonauts Dobrovolsky, Volkov and Patsayev, whose Soyuz XI spacecraft returned to Earth on 30 June 1971 after 24 days in space. Since what we believe may be the true story of the end of this tragic mission is not generally known, it is worth summarizing here, to illustrate the fact that a precise understanding of the invisible forces of space is a matter of life or death.

Soyuz XI lost radio contact with the ground as it passed through the ionosphere on its way home after a trouble-free mission. This is normal: communications are impossible while a craft is passing through this layer of high electrical charges that produce very strong interference. But Soyuz XI remained silent.

The spacecraft landed automatically without problems, and the recovery team rushed to open the door. Inside, the three crewmen were smiling peacefully, as if relieved to have arrived safely.

But they had not arrived safely. All three were dead.

It was a national tragedy, far more serious than the crash of Soyuz I that had killed V. M. Komarov in April 1967. The Soviet authorities issued a series of different explanations. First they said the men had died from explosive decompression, although autopsies revealed no internal haemorrhage as would be expected. Next, cause of death was given as thrombosis, or blood clots on the neck, though no cause of these were suggested. Other official stories claimed there was a leak in the cabin, while finally it was simply stated that the men had died of exhaustion. (Smiling?)

Dr. Gültekin Caymaz, a specialist at the Ataturk Sanatorium in Ankara, Turkey, thought otherwise as soon as he heard the news on the radio. As it happened, he had been measuring atmospheric electricity daily in 1971, and had only just published a couple of papers on air accidents due to increases in it. These, he noted, had also caused an increase of deaths in his own hospital. The true cause of the cosmonauts' deaths could, he thought, be deduced from the *rigor mortis* grins on their faces. Atmospheric electricity can lead to alkalosis, a pathologically high alkali content in blood and tissues that in turn leads to respiratory arrest. This, due to the increase of CO_2 in the tissues, causes a contraction of the lips into a deathly "grin."

Dr. Caymaz at once alerted the Soviet, American and West German embassies and the Turkish Air Force. What steps the embassies took is not known, but the Turkish military authorities at once arranged a series of tests on the effects of atmospheric electricity inside aircraft. These revealed a sharp rise in DC voltage inside the planes when the grounding wire was removed. (We give more details of these tests in Chapter 4.) Although no charge can accumulate in a closed metal box, a spacecraft passing through the ionosphere will cause an enormous difference in voltage between its interior and the surrounding atmosphere. A spaceman would only have to touch the metal of the ship with a bare hand, and his body would be drained of electricity, causing instant death.[37]

In an article published the previous year, Dr. A. B. Cambel of Wayne State University had suggested fitting spacecraft with large magnets which could modify approaching plasma flow fields and, incidentally, enable radio links to remain open during re-entry. Space capsules would thus be protected by their own magnetic fields, just as Earth is shielded from the full wrath of the Sun by its magnetosphere.[38]

Could the Soviets' deaths have been foreseen and avoided? It seems possible, though we do not have the data on which to base a firm conclusion. One thing is clear: at this early point in the Space Age we can expect plenty more surprises, pleasant or not, so that anybody who even thinks he can predict the weather of interplanetary space deserves a careful hearing.

This brings us to the question of interactions between the Sun and the weather here on Earth.

Sun and Weather

Here we have a mystery as baffling as that of the sunspots. Again, we have plenty of good evidence, many promising clues, and a colossal pile of research to dig through. But we have very few answers. In view of the significance of "body weather," to be mentioned in the next chapter, it is necessary to establish just what is known about Sun-Earth interactions. For as we shall see, there is weather not only above and around us, but also inside us.

Thanks to satellite and space probe data, we now know quite a lot about how the solar input varies. We also know that there can be a time lag of up to two days at least after a solar flare before its effects reach Earth, whether in the form of radiation or energetic particles. And since we know the solar output to be cyclic, we should expect to find cycles of events on Earth to match it. The unmistakable link between Sun and magnetic conditions on Earth has encouraged scientists to look for other such links, which has proved far more difficult than might be imagined.

It has been known for more than a century that magnetic storms interfere with electric telegraph communications, and it has also been found that they can affect telephone lines, underwater cables, high voltage transmission lines, and even gas and oil pipes. All of these can now be well protected from solar attack, yet power failures still occur, and while we are not suggesting that the great New York blackouts of 1965 and 1977 were due to the planets, it may well be that some power failures are. David Williams has had some success at power-failure forecasting, and much interest has been shown by major power companies, at least in the U.S., in what are known in the business as SICs, or solar induced currents. These, incidentally, have been found to increase with increasing northern latitude.[39]

If the Sun finds its way into our buried power cables, surely it must affect our weather? Herschel, in 1801, was probably the first to ask the question whether variations in solar activity were reflected in the seasons of Earth, but the question has yet to be answered conclusively. "Many attempts have been made in the past to correlate solar and terrestrial phenomena," *Nature* wrote in 1939. "There is no direct meteorological effect of any serious importance."[40]

Yet, three decades later, we find an article in the same

magazine declaring that "the apparently very strong connection between the solar cycle and the weather is not generally appreciated." The author, Dr. J. W. King of the Radio & Space Research Station in Ditton Park (England), states that important features of the climate, including droughts, are dependent on the solar cycle "to such an extent that significant progress could be made in forecasting the occurrence of these features if some account were taken of the expected levels of solar activity in the future."[41]

No such account was taken, at least in Dr. King's own country, for three years later (in 1976) Britain was caught with its reservoirs dry, and was forced to ration water in many areas during the worst drought since records were first kept. (Altogether, 1976 was quite a year for strange weather, as we will describe in Chapter 4.) The problem became so serious that *The Times* (17 August) wrote an editorial on it, which contained an interesting remark. When Britain's Water Resources Board produced its future policy recommendations in 1974, it seems the only data it used were average annual rainfall figures for the years 1916 to 1950, a period now generally admitted to have been one of the most anomalous in modern times. "Taking Nature for granted," wrote *The Times*, "the planners contended only with the variables of consumption, costs, and the politics of reservoir-building."

Nature, however, cannot be taken for granted, and in 1976 even the source of the Thames dried up. Crops failed, food prices soared, the stock market and the pound sterling slumped alarmingly, and enormous forest fires kept firemen sleepless for weeks on end. Some Italians came over to make a film, because England looked more like drought-stricken Italy than Italy did. Had the drought continued into September, Britain, already in drastic economic shape, might have been brought to a total standstill by a force even more obdurate than its trade unions—nature. It is unfair to claim, with hindsight, that the 1976 drought should have been forecast, but it is fair to claim that it could have been regarded as quite probable, so that steps to alleviate it could have been taken long before they finally were (under the direction of the Minister of Sport). There is plenty of evidence to support Dr. King's suggestion that in future the Sun's behaviour should be added to the list of forecasters' variables.

Dr. King is evidently a cricket enthusiast. Digging through *Wisden's Cricketers' Almanack*, he has found some curious correlations between high scores and sunspot counts. Cricket is a game that depends more than most on the weather: play is stopped as soon as rain falls, and runs are harder to score on damp wickets. King has discovered that of the 28 occasions when individual cricketers have scored 3,000 runs in a season, 16 have been years of maximum or minimum sunspot counts, while the only five years when more than one batsman passed the 3,000 mark were *all* within a year of minimum or maximum, as were 13 out of 15 seasons in which a batsman scored 13 or more centuries. This indicates that a relation between solar activity and rainfall is quite probable, at least during the summer months. (The year 1976, incidentally, was a solar minimum year.)

Further evidence comes from studies of the crop-growing season, precisely defined as the number of days per annum in which temperatures above ground exceed 5.6°C at Eskdailemuir in Scotland. Charts compiled from 20 years of data show that the season is 25 days longer near sunspot maxima than near minima. The season tends to be longest up to a year *after* the peak of the sunspot cycle, suggesting that a direct causal relationship is unlikely. However, not only do peak sunspot and peak growing-season years coincide (plus or minus a year) far more often than they should by chance, but the cycles show an obvious similarity in both amplitude and phase, that is to say a year of a comparatively low sunspot number peak will see a similarly shaped peak in the growing season curve. That this should also be due to chance is scarcely conceivable, for the correlation has appeared in the last six cycles.

Equally striking correlations turn up between sunspot numbers and some rainfall indices. Data for the period 1697-1970 show that abnormal spring rainfall tends to occur at each extreme of the solar cycle, years of maxima giving dry springs and of minima wet ones.

Dr. King warns us not to get too carried away by the sunspot cycle. It is tempting to assume that sunspots hold the answer to all of the mysteries of life, which is most unlikely, although a study of them can lead to promising clues elsewhere. The solar cycle, King says, is only a "crude index" of solar activity; he feels more attention should be given to the solar wind. After all, it is the wind that reaches Earth, not the sunspots themselves. Space probes have established that during high sunspot periods, the speed of the solar wind increases, whereby more particles from the Sun force their way into our atmosphere, causing the Forbush decreases mentioned in the previous chapter.

And what happens when they get here? An important clue is to be found in the way that radioactive elements move down from the stratosphere to our atmosphere, and at Zugspitze, high in the Bavarian Alps, Dr. Rudolph Reiter spent four years studying and monitoring this process. He found that there is a relation between the increase in the level of two such elements (Phosphorus 32 and Beryllium 7) in the lower atmosphere and . . . solar flares! Reiter feels that the presence of these elements so close to Earth shows to what extent masses of stratospheric air are pushed down and mixed with that of the lower atmosphere. More such masses reach us, he has found, two to three days after a solar flare, and it seems that very slight increases in radiation produced by a flare can have far more influence on the atmosphere than was previously suspected. Solar disturbances, in fact, cause increases in the intensities of low-pressure troughs, and these in turn are directly responsible for the kind of weather that is inflicted on us.[42]

Dr. King has started a rumpus in meteorological circles by suggesting that incoming solar-wind particles may be connected with meteorological pressure changes caused by the Earth's magnetic field at certain specific latitudes in the 75° to 79°N band, which is where the high-pressure ridge is formed that largely dictates what kind of weather is in store for the northern hemisphere.

He seems to have been the first scientist to spot the similarity between the contours of the geomagnetic field and those of atmospheric pressure distribution, and to look at the weather problem from a new angle, coining the word *magneto-meteorology* to describe what he is doing. His ideas have clashed head-on with those of forecasters who use mathematical models that simulate atmospheric conditions, and it remains to be seen whether the link between magnetism and weather can lead to better prediction.

That long range weather forecasting will be possible when we have learned how to forecast solar flares accurately was actually the conclusion of a 1965 report by the National Engineering Science Co. (NESCO).[43] There does seem to be a similarity between geomagnetic field contours and the average way the wind blows in the lower atmosphere, and if it can be shown that there is also a link between Earth's *upper* atmosphere, where the initial blast of solar energy is received, and the lower, it may be that when the rhythms of the solar cycle are fully understood we will be able to know exactly what kind of weather is being mixed up for us well in advance.

Scattered clues are beginning to come together. A study of rainfall in Ethiopia, for example, has shown a strong and consistent correlation between the sunspot cycle and rainfall levels in Addis Ababa.[44] But the peaks and troughs of the rain cycle have *preceded* those of the solar cycle by about 1.3 years (for the peaks). Here is the sunspot mystery in a nutshell: how can the spots cause rainfall in Ethiopia, or anywhere else, if the rain cycle consistently peaks first? The correlation could be chance coincidence, but if that idea is ruled out on statistical grounds as the least probable (except perhaps for the idea that rainfall causes sunspots!), we must look for a third factor, and any search for a factor external to both Sun and Earth must start with the nearest suspects—the planets.

If Sun/weather links were as plain everywhere as they seem to be in Addis Ababa, the mystery would have been solved long ago. But they are far from plain almost everywhere else, due to a great extent to local variations in altitude and also to the latitude of the region in question. An interesting study comparing rainfall patterns in two different parts of the U.S.A. shows there to be an inverse relation between rainfall variations in the northeastern area and in the high plains states, with clear indications that the latter receive an extra dose of rain near times of solar minimum activity. Predictions for droughts and floods up to the year 2000 have been made on a basis of past variations in these states, but we shall have to wait and see how accurate they were.[45]

Our weathermen are on the whole quite good at telling us where today's weather is going to be tomorrow, but they are often helpless when it comes to making accurate predictions even a short time ahead. Could it be that they do not look far enough afield for possible causes of terrestrial weather?

One scientist who believes this is so is J. M. Wilcox of Stanford University, who urged in 1976 that regular daily observations should be made, with the help of orbiting spacecraft, of the outermost reaches of Earth's atmosphere, including the magnetosphere that extends thousands of miles into space. He also urged that detailed studies of the interplanetary field should be made throughout an entire solar cycle. Only with such information, he said, could we fully understand how Sun and Earth weather interact. In a rather pathetic comment on the state of the information explosion of the Space Age, Wilcox pointed out that some such data is probably already available, but nobody has yet reduced it to a form in which it can be of any use to the forecasters.[46]

Thus Herschel's intuition served him well when he speculated on the effects of those "more or less copious emissions" of the solar beams. It is now quite clear that the Sun is not the perfect and constant deity we once thought it must be. The recent discovery that not even its speed of rotation is constant reminds us of how much more remains to be learned about the most important object in our lives. Even the concept of it as a hydrogen bomb fuelled by nuclear fusion has been questioned. It will be some time before the definitive textbook on the Sun can be written.

Meanwhile, there is much to be learned from study and analysis of its visible behaviour, and of the possible influence of the planets on this cyclic and therefore potentially predictable behaviour. The motions of the planets *are* predictable, and if it can be shown conclusively that these motions are involved in the production of sunspots and solar flares, whether directly or indirectly, then our solar system will suddenly become a huge clock with a plainly readable face. This will be a tremendous breakthrough, the most important since it was established that the Earth moves around the Sun.

It has certainly not yet been proved that planetary positions cause the appearance of sunspots, and thus drive the solar cycles, but a mounting pile of evidence points in this direction. British scientist Geoffrey Dean, in an extremely important book that was written simultaneously with ours, has provided some of the most compelling evidence of all. He summarizes it as follows:

"The many attempts to correlate solar and planetary periods are unconvincing because they ignore phase, a consideration of which leads almost immediately to the answer. Very briefly, my findings suggest that the Sun is reacting to resonance peaks in cycles of momentum change synchronized mainly to the motion of Saturn and of Jupiter and Neptune combined, each timed with respect to the solar equator. Over 300 years of sunspots they indicate the phase and timing of each maximum and minimum with an average error of less than one year with no significant drift.

"They appear to indicate the Maunder minimum and all the other minima over at least the last 4,000 years. They are entirely compatible with the findings of Nelson, and they provide an explanation for the hitherto inexplicable phenomenon of latitudinal passage. My model differs from earlier models in that resonance is considered, and cycles are timed with respect to the solar equator."[47]

Now, after this brief glance at the many-body solar system, which we hope was not too brief to give an idea of the complexity of the inter-relationships involved, we come down to earth and look at a few of the ways in which man interacts with the invisible forces around him.

Our universe is an electromagnetic machine, and man might well be renamed *Homo electromagneticus*, for he is indeed an EM system: it is therefore not surprising that man and Universe must interact, as they do to an extent far greater than we can consciously perceive. It is surprising that little inquiry has been made until very recently into the actual mechanisms of such interactions.

In the following chapter, we look at some of these interactions, touching on the subject of mutual feedback between man and environment, as the one modifies the other only to be modified himslef as a consequence of his actions.

The American meteorologist H. H. Clayton, a pioneer sunspot-watcher of the early twentieth century, once declared:

> To the changing position of the sun in the heavens all human life is adapted, and with the surging flames that heave and roar on its surface are entwined both human weal and woe.[48]

We shall see to what extent he was right.

PART TWO

FOREGROUND

Chapter 3

HOMO ELECTROMAGNETICUS

PERHAPS THE most important direction that science is now taking is towards the investigation of "inner space." We went to the Moon before we had finished exploring the molecule; and the microcosm, or "inner cosmos," within us is as fascinating and quite as mysterious as the more familiar macrocosm of planets, stars and galaxies. Astronomy is the oldest of the sciences, yet the scientific study of the process of life itself, by comparison, has barely begun. What distinguishes life from a collection of inanimate chemicals? What are the forces that shape and guide the growth of living beings? What exactly *is* a living being, anyway? What is the purpose of it all, if any? These are questions that have concerned theologians and mystics for centuries, but it is only recently that science has come close to providing some of the answers.

Man mapped the heavens six thousand years ago, but it is less than two centuries since the electromagnetic nature of life was discovered, after Luigi Galvani in Bologna had observed his famous frog's legs twitching on two metal wires. Although Galvani misinterpreted his own discovery, it opened the way for others to lay the foundations of the scientific study and measurement of living organisms. Then, with the discovery in 1820 by the Danish physicist H. C. Ørsted that electricity and magnetism were inexorably linked, a new field of research became open: the study of biological effects of EM fields.

Then an important discovery was made—the living body itself produced electrical impulses. By the mid-nineteenth century Du Bois-Reymond had proved that the nerve force was electrical, and then Helmholtz measured the speed of nervous transmission, finding it to be about 35 metres per second in humans. Thus the modern science of neurophysiology was born.

In 1924 Hans Berger discovered the electromagnetic nature of brain waves and became the first to measure some of them. Today, the electroencephalograph (EEG), which reproduces brain rhythms on a chart, is familiar in both hospital and laboratory. It has now been joined by the less familiar magnetoencephalograph (MEG), which measures the *magnetic* waves from the brain discovered in 1969 by David Cohen, when he showed that the human brain sets up a magnetic field that can be detected a few inches outside the head.

Since man, like the cosmos in which he lives, is an electromagnetic system, everything he does is governed by EM processes. We are made of atoms and molecules, and our bodies contain roughly equal numbers of positive and negative charges. Electricity is used by the nervous system to send messages, by the brain to compute information from such messages, and by the various chemical compounds in the body to react with one another and keep the body functioning. Even in daylight, we are using electricity to read a page of print like this one, as our eyes pick up light waves that have travelled from the Sun to the paper, and convert them to photochemical and biochemical pulses to be sent along to the higher centres of the brain for processing. When we come to the end of the page, the brain must send more electrical pulses to arm and finger muscles to tell them precisely what to do. Such signals are carried via the nervous system, and as with any other EM signal there is no reason why they should not be disrupted or interfered with by other fields outside the body of roughly the same frequency. We go into this in more detail in later chapters.

Man is also a living radio transmitter. Signals run around inside our bodies, and we radiate EM waves into space much as a broadcasting station does, though human radio waves are much weaker and lower in frequency than artificially generated ones. Every time our hearts contract, they send out a wave of one to three cycles per second (cps.), or 3 Hz. (Hertz) in radio language, and these can be measured on another familiar piece of hospital equipment, the electrocardiograph or EKG. And now that we know there is a detectable magnetic field associated with the heart's electrical activity, the MKG or magnetocardiograph has taken its place next to the EKG. Muscles also emit waves (Galvani would be glad to know) both magnetic and electric, of anything from a few Hz. to a few thousand, depending on the size of the muscle and the force exerted. These can now be picked up at a distance, as can *magnetic* waves from the heart, without the use of contact electrodes, on machinery developed in the U.S.S.R. by Professor Pavel Guliayev (who has also invented an "electroauragraph").

The human nervous system is a phenomenal piece of engineering, even for these days of miniature solid-state circuitry. It contains at least three separate systems: sympathetic, parasympathetic and peripheral, and the total number of neurons or nerve cells, is in the region of 20 *billion*. (That is, 2×10^{10}.) Yet all those miles of connections are so carefully routed, intertwined and packaged that we are hardly aware of them except when they go wrong. For all the marvels of modern technology, man has not yet equalled nature in the design of compact, efficient and self-repairing machinery.

From the electrical point of view, the human body is a collection of wet leather bags partially filled with liquids, with some air spaces in between. As a conductor, it is poor compared with the metals normally used to conduct electricity, but very good compared with the substance that surrounds it: air. We can expect therefore that, like any other geometrical object with higher conductivity than its surroundings, the human body will distort the field lines around it, as long as it is in contact with the negatively charged earth.

In recent times, man has begun to modify his environment drastically in terms of electricity, magnetism and acoustics. Houses of wood, stone or brick are giving way to steel-frame structures. Thus we are abandoning buildings made of good insulators in favour of those that are good conductors. Moreover, our homes and offices are surrounded by transmitting stations, antennas, wires and cables, while inside them there are yards and yards of circuits feeding appliances that we often leave running day and night, such as refrigerators and heaters. Are we being affected by all the live wires around us vibrating away at 50 or 60 cycles per second? Are we polluting ourselves electromagnetically? (Relevant here is the vitally important subject of air ionization, which has only recently been given attention by scientists. We mention this in the following chapter.)

We are, according to researchers of West Germany's *Verein für Geobiologie* (Geobiology Union), after their investigations into EM pollution. It is easy to dismiss the "geobiologists" as a bunch of cranks who run around with strange radio sets and divining rods performing rites of electromagnetic exorcism, but evidence in support of their claim is piling up. It is only the last two or three generations in many parts of the world that have had AC wiring in their homes, so it is too early to tell what long-term effects this may be having. But if the geobiologists are right, the staunchest sceptic will soon be having second thoughts before building his house near a radar station, radio transmitter or high tension cable. Moreover, they say, in addition to such artificial EM hazards, the ground under our feet has its own anomalies: places where the natural magnetic field seems to go haywire. The Germans call these "geopathogenic," or "geopathic" zones.

Some buildings, they claim, are veritable *kranken Haüser* (sick houses), in which natural vibrations fall into the uncomfortable range for both animals and humans. To simulate the effects of such acoustic or EM vibrations at extremely low frequencies, two German researchers carried out a simple experiment. They took some volunteers, sat them in chairs and proceeded to shake them up and down from one to twenty times per second. (British TV viewers may have seen something similar done on the BBC *Tomorrow's World* programme.) At certain precisely measured frequencies, the subjects complained of pain or discomfort in practically every part of the body. They suffered breathing difficulties, muscle contractions, headaches, pains in the chest, back and genitals, and a general sick feeling over the whole range. Certain aches and pains correspond to certain frequencies (Fig. 4). In these experiments, only mechanical vibrations were involved, but EM and acoustical vibrations are being set up all the time around us, as we shall see in Chapter 4. Just because we

Homo Electromagneticus 73

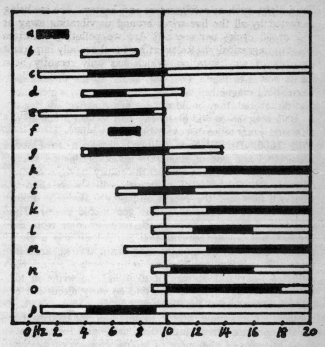

Fig. 4
The results of the chair-shaking test. Subjects felt slightly (hollow bar) or greatly (solid bar) affected as follows:

a. Shortness of breath
b. Difficulty in breathing
c. Hyperventilation
d. Chest pains
e. Arbitrary muscle contraction
f. Resonance of jaw
g. Abdominal pains
h. Muscle tension
i. Back pains
k. Headaches
l. Throat complaints
m. Speech problems
n. Rectal irritation
o. Bladder irritation
p. General uneasiness

(Courtesy Resch Verlag, Innsbruck).

can neither see nor consciously perceive them, we should not assume that they have no effects on us.[1]

But before we come to present-day research in this area, there is a 200-year-old mystery that needs to be reinvestigated.

Magnetism to Mesmerism—And Back Again

Magnetism is one of nature's more baffling phenomena. The Greeks discovered, 2,500 years ago, that a mineral from the Magnesia region of Thessaly had curious properties: it could somehow attract pieces of iron and even pass on its power to them. That was about all the Greeks did learn about magnetism, and, as with electricity, which they also discovered, they could not see any practical use for it. So both magnetism and electricity had to wait a couple of thousand years to be discovered all over again.

In 1600, Dr. William Gilbert, physician to Queen Elizabeth I and one of England's first experimental scientists, rediscovered what thirteenth-century navigators had noticed: that the poles of a compass will always align to what we now call the North and South Magnetic Poles (which are not the same as the geographic poles). The seamen had assumed the magnetic force to come from the heavens, but Gilbert realized that Earth itself was a giant magnet. Even so, he thought magnetism was an animate force. "It reaches out like an arm clasping round the attracted body and drawing it to itself," he wrote in his *De Magnete*, in which we also find an early discussion of "magnetic treatment" for illnesses.

Scientists nowadays prefer to think of magnetism (of which there are three kinds: dia-, para- and ferromagnetism) in terms of alignment of atoms, creation of dipoles, spin and revolution of electrons, and the space or *field* in which its effects can be detected. Yet Gilbert's description still serves with regard to its behaviour, and his belief that it was a living force was shared by many. Paracelsus had been convinced early in the sixteenth century that the human body had magnetic qualities, while a century later the Flemish alchemist J. B. van Helmont and a Jesuit priest named Athanasius Kircher were suggesting that magnetism could actually heal wounds.

Similar claims were made for electricity long before this was shown (in 1820) to be linked to magnetism, and also long before either electricity or magnetism was put to any other practical use. Rudimentary static electricity generators were in existence by the mid-eighteenth century, and electric shock treatment was being given at London's Middlesex Hospital in 1767, using the "Leyden jar" capacitor. A book published in Denmark in 1754 contained illustrations of friction generators, and also gave case histories of patients treated for various ailments from joint pains and menstruation difficulties to epileptic fits.[2] A few years later, John Wesley was asking whether electricity was "a mere plaything, or the noblest medicine yet known." Then Mesmer arrived on the scene.

Franz Anton Mesmer (1734-1815) is often dismissed today as an eccentric charlatan and exploiter of suggestible women, whose career ushered in the Age of Credulity that culminated in Christian Science and Spiritualism. Some even hold him responsible for what they see as a mass relapse into medieval ignorance, witchcraft and black magic.

Yet Mesmer was neither occultist nor charlatan. He was fully qualified in both medicine and philosophy, he was a skilled enough musician to impress Mozart, and he devoted his life to curing the sick. More than anybody else, he tried to "deoccultize" unusual healing methods by bringing them into the open and inviting inspection from any scientist with the time to spare.

For his medical degree at the University of Vienna, Mesmer wrote a Latin dissertation called *The Influence of the Planets on the Human Body*. He suggested that space was filled with psychic fluid or ether, in which tides could be caused by the planets. These tides were responsible for health; sickness being caused when the flow of the natural life-giving force was interrupted. Such ideas were certainly not original; at least seventeen treatises or books on magnetism appeared before Mesmer's debut as a healer, and there is clear evidence that he borrowed liberally from at least one of them for his dissertation.[3]

Mesmer began work as a healer at the suggestion of a Jesuit named Maximilian Hell, and was probably also influenced by another popular priest-healer, J. J. Gassner, whose combination of medicine, exorcism and laying-on of hands won him a huge following. For himself, Mesmer claimed no more than the discovery of what he called "an agency influencing the nervous system, and a method of applying it to heal the most difficult types of malady." He called this agency *animal magnetism*, and he saw it as a fluid comparable with both electricity and gravity that could be stored and transmitted by human operators ("magnetizers") to both living and non-living matter.

Somehow or other, he must have cured at least some people. To treat the hordes who flocked to his house, he built his famous *baquets*, large tubs of water which he magnetized along conventional lines by placing lodestones in them and giving patients amber or glass rods to hold. The confusion that still surrounds Mesmer's name today arises from the fact that he seems to have practised two things at once; electromagnetic therapy and healing by suggestion. The former has had to wait nearly two centuries to be fully rehabilitated, whereas the latter has flourished ever since in various disguises, from hypnotism to spiritual healing with its "magnetic passes" based on Mesmer's own method of passing the hands from head to foot of the patient a few inches from the body. Spiritualists call this cleansing the aura. Mesmer was the common ancestor of Christian Science and psychotherapy: it was the healing ability of one of his American devotees, Phineas Quimby, that inspired Mary Baker Eddy to found the former, while the techniques he popularized influenced a long line of more scientifically-minded doctors from Braid, Esdaile and Elliotson to Charcot, Bernheim, Liébeault, Breuer and Freud.

Like many dowsers and faith healers of today, Mesmer did not fully understand what he was doing himself. Had anybody taken the trouble to investigate him properly, we might know more than we do about his early days, and there is much we would like to know. It may be that he had intuitive glimpses of the value of weak EM fields in treating disease, although there were no instruments in his early years that could have measured them. (Perhaps people were more sensitive in consequence than we are today, with machines to do our thinking for us). It is unfortunate that Mesmer neglected the scientific side of his work, on which it must have been based, when he found he could get results with just his hands and his eyes, and no actual *magnets* at all. Little was known about the true nature of electricity or magnetism when his ideas were formed, just as very little is known today about what "mesmerism" really is and how it works. Dr. E. J. Dingwall notes that our knowledge of the power behind it has advanced little in two centuries; to say that it is "all imagination," he points out, explains nothing. "It merely postpones investigation." Serious investigation of Mesmer's methods and views remains postponed, despite their similarity to much contemporary research into the EM "life energies" of man, to be discussed later in this book.[4]

Homo Electromagneticus 77

Fig. 5
Strange devices from the early days of electrobiology included the Electric Helmet (top, left) and the Body Battery (bottom, left) which apparently came complete with metal crown, a forerunner of the Faraday Cage. The Volta Cross (top, right) already contains a hint of science to come, while the D'Arsonval Cage (bottom, right) was the precursor of a wide range of more elaborate modern appliances that make use of magnetism for therapeutic purposes.

If the Mesmer mystery remains unsolved, at least we now know a good deal about ways in which living beings can be influenced by the EM environment, for good or bad, as man seems to have been instinctively aware long before his technology found ways to measure such effects.

When Galvani's frogs' legs began twitching, many assumed that electricity itself was the mysterious life force, and the quacks lost no time in moving in. Some incredible contraptions came on the market, from Dr. Owen's Body Battery and Elisha Perkins' Patented Metallic Tractor (just two bits of metal that earned him a fortune) to all kinds of copper and zinc belts, bracelets and corsets, and the Volta Cross, which was said to remove pain when hung around the neck.

Some of these contained a hint of science to come, or perhaps of ancient knowledge awaiting rediscovery. The Volta Cross, for instance, became a kind of battery when soaked in water or vinegar, as the cloth separating the two plates (of copper and zinc) absorbed the solution and emitted a weak light. Copper bracelets can be bought today at any London chemist's, and many have claimed relief for rheumatic and other pains after wearing them.

Other devices from the age of electro-biology, metallotherapy, animal magnetism or whatever have evolved into accepted medical equipment. The French scientist D'Arsonval was one of the first to use alternating current for treating patients: he put them inside an "electric cage" and ran a high-frequency current through its wires, thus setting up a magnetic field which, he said, improved the metabolism.[5]

Coming to twentieth-century Denmark, with one of the world's most advanced medical services, what do we find being used in Copenhagen's Rigshospital to treat various forms of body pain, but magnets once more! Dr. K. M. Hansen spent four years in the 1930s studying the effects of magnetic fields on such ailments as lumbago, sciatica and headaches. Using methods even simpler than those of Mesmer, she took electromagnets with up to 14 kilos lifting power and pressed them hard against her patients' bodies. Of course, an element of suggestion may have been at work here, for the magnets were left in position for up to 15 minutes, long enough for suggestible patients to become "mesmerized" by them.

And what do we find at a conference held in Moscow in 1969 on the subject of the effects of magnetic fields on biological organisms, at which more than 100 papers were read, but a report entitled "The effect of *magnetized water* on the condition of certain internal organs of laboratory animals." (Our italics)[6]

Dr. S. W. Tromp of the Biometeorological Research Centre in Leiden (of which more in Chapter 4) has pointed out that "magnetizers" in the style of Mesmer may be influencing their patients not only by suggestion, but by electrostatic stimulation, especially when they practise "contact healing" by placing their hands on the patient's body. Dr. Bernard Grad of McGill University, who has studied contact healing methods in his laboratory, speaks of a "life force" emanating from the hands of certain people, a hypothesis supported by his own work on plant stimulation and healing of wounds in animals. One of us (Playfair) has been given a demonstration of what was called "energy transfer" by a Brazilian healer. He felt a strong vibration in his arms after his fingertips had been held by the healer for a few moments.[7]

We strongly urge readers not to mess around with powerful magnets without qualified advice. Recent research has indicated that magnetic fields can actually *retard* healing processes under certain conditions, by affecting enzyme activity and causing a decrease in the formation of antibodies. Yet there are also indications that by careful control of field strengths it may be possible to control unwanted cell growth, as in certain types of tumour. Several papers read at the Moscow conference mentioned above dealt with this interesting topic.

A possible physical basis for mesmerism and hypnotism is suggested by the fact that people can be put to sleep by pulsating magnetic fields. By circulating a current around a subject's head, using frequencies close to the natural rhythms of the brain, a state of electronarcosis can be induced and the subject put into deep sleep. The time will come when domestic electrosleep machines will replace the sleeping pill: such machines are already on the market in the U.S.S.R., though they may meet some resistance from the chemical multinationals in the West![8] Chinese doctors have recently combined ancient and modern skills by stimulating the traditional points of acupuncture not with needles, but electrically, anaesthetizing patients to the point where they can perform open-heart surgery with the patient wide awake. If humans can have their states of consciousness altered mechanically, is it not possible that Mesmer, his successors and predecessors were also making use of mechanical means, even if they could not understand the process themselves—being forced to postulate etheric forces, animal magnetisms, and so forth?

In an ingenious experiment carried out in the Stuart, Florida, hospital where he works as a surgeon, Dr. E. Stanton Maxey has shown that human brain rhythms can be altered without the target person being aware of what is going on. Soviet research has indicated the possibility of altering brain rhythms by the application of weak waves, and Dr. Maxey eliminated all possibility of suggestion by concealing his equipment and beaming magnetic waves from a toroid generator hidden, without the subject's knowledge, under the pillow. The EEG chart showed that rhythms in the subject's left brain hemisphere altered at once, remaining altered as long as the artificial magnetic field was applied.

Dr. Maxey believes that the intrusion of weak magnetic waves into our brains may help explain the causes of some accidents ascribed to pilot or driver error. Waves that can pass through metallic structures, such as cars or aircraft cabins, may influence mental activity in certain people, prolonging their reaction times and decreasing their sense of awareness: in other words, creating ideal preconditions for an accident. We return to this theme, with more evidence, in a later chapter.[9]

Weak is Strong

Anybody watching seventy-year-old Karl Wallenda as he calmly walked across a London street in 1976—on a tightrope stretched between two office tower blocks—and doing a headstand in the middle for good measure, must have been aware of the need on such occasions for perfect balance.

Life is a tightrope phenomenon, only possible because of delicate balance between permitted extremes. All organic life has evolved in "close co-operation," as Tromp puts it, with the geological development of Earth. If our planet rotated more slowly, days and nights would lengthen and we would alternately freeze and burn. If its axis were to tilt less than its present 23°, we would have no seasons, and vapour drifting to the poles would create vast continents of ice. If our atmosphere were thinner, millions of meteors would crash into us every day. If we were closer to the Moon, tides would cover whole continents, eroding mountains in the process. Any major change in the mean distance of Earth from Sun would knock us off our tightrope for ever, while Soviet geophysicist M. I. Budyko reckons that a drop of less than 2 percent in solar radiation would be enough for our whole world to freeze over. We are as delicately balanced as any of the Flying Wallendas: any factor that can upset our balance, however slightly, is therefore of something more than academic interest to us.[10]

Many living beings have a sensitivity to changes in their natural surroundings that can surpass that of our most elaborate instruments. Until recently, for instance, it was widely believed that no magnetic field of less intensity than that of the Earth (about 0.4 gauss on average) could possibly have any biological effects. However, Professor Frank A. Brown Jr. of Northwestern University has shown that fields so weak as to be virtually undetectable by man are enough to make snails and worms change direction and crabs change colour.[11] Californian researchers have found that rats can detect x-rays, probably through receptors in the backs of their heads.[12] In Holland, work with sharks and rays has revealed that these ferocious beasts are extremely sensitive to AC electric fields, the source of which they can locate without any visual clues.[13] Smaller fish also know where they are going even when their sense organs have been interfered with: in one experiment a blindfolded lobster found its way to within ten metres of capture site from 200 m. away, and similar tests have been done using salmon with their noses bunged up.[14] The celebrated electric fish *Gymnarchus niloticus* responds to magnetic gradients on its body as weak as 0.15 microvolts.[15]

Pigeons are famous for their ability to find their way home even when scientists do their best to make things difficult for them. It is now well established that birds do use the geomagnetic field as well as Sun and stars for navigational purposes, apparently following its contours much as we follow road maps. This must mean that they are sensitive to very thin changes in it, an idea first suggested in 1882 (if not earlier) that has taken a long time to become accepted.

The pigeon was reluctant to give up its secrets, as Dr. W. T. Keeton of Cornell University found after a series of ingenious tests. He stuck tiny magnets on the backs of their necks, setting up a field strength of 0.45 gauss, enough to put them off course if geomagnetism were all that guided them. Two groups of birds, including control birds with nonmagnetic metal on board, were released up to 31 miles from their lofts on clear and overcast days. On the clear day, the pigeons could evidently see for ever and flew home without difficulty, but on the overcast day they were definitely disoriented, suggesting that they prefer to use the Sun for navigation when they can. But then Keeton put magnets on some young birds that had never been away from home, and found that although they were put off course even on a sunny day, a few still managed to get back to port. It seemed clear that birds can manage without either Sun or magnetic field, but not without both.[16]

The matter now seems to have been settled; Frank R. Moore of Clemson University has obtained direct visual evidence that bird flight can be affected by geomagnetic field fluctuations, while psychologist Michael Bookman of MIT has managed to train pigeons to locate food at the end of an artificial tunnel by using only magnetic cues. He also spotted a curious detail. Pigeons only seemed to sense the field when they started fluttering, hovering or jumping about in the tunnel leading to their dinner. Birds that simply walked along the tunnel failed to show evidence that their magnetic sensors were working.[17]

Bookman admits that the mechanism of magnetic sensitivity in birds remains a mystery, but two other scientists think they have solved the problem, and told delegates at a Paris conference in 1975 just how birds sense magnetic stimuli. The answer seems to lie in the most characteristic attribute of birds—their feathers, which apparently act as piezoelectric transducers in the audio frequency range (1-20 kHz.) and also as dielectric receptors in the EM microwave region, or between ten and 16 kHz. Bird feathers, it appears, act not only as selective fibers for EM information, but also as energy transducers and transmission lines. The piezoelectric effect has been traced to the protein keratin, which is also to be found in animal horns and hooves as well as in human skin, hair and nails. Later in this chapter, we shall present independent support for this claim.

The properties of this protein are surely worth examining in more detail. Pigeons apparently use the underside of their wings for magnetic sensing, which might explain why it has also been noticed that they seem to lose their sense of direction after a heavy snowfall, though they regain it within 24 hours even when the snow it still on the ground. The pigeon seems to have a number of emergency navigation methods at its disposal, and it would be wonderful if man had similar abilities. Such a suggestion is not entirely unreasonable.[18]

Much is now known about how plants, as well as birds, fish and mammals, respond to the EM environment, especially natural or artificial magnetic fields, thus adding "magnetotropism" to the many existing tropisms, or ways in which plants grow in response to external stimuli. A botanist, Professor Leslie Audus, has shown that very strong artificial field gradients can evoke a response from a plant, though he makes it clear that he found magnetotropism to have little if anything to do with plant growth under normal conditions. Recent research from the Soviet Union, however, suggests otherwise: Aleksey A. Titayev has found evidence that the Earth's magnetic field can exercise a regulating influence on the growth of mushrooms, while experiments held at Kursk have shown that magnetic anomalies can have a positive influence on the fat content of cows' milk.[19]

Soviet scientist A. S. Pressman has done a lot of work on effects of electric and magnetic currents of all strengths on living systems, and written an important book on the subject. He has found that EM fields can have marked effect on living tissue even when the power applied is so weak that no thermal effect at all can be detected. He believes that some of the phenomena usually ascribed to psychic faculties, or extrasensory perception (ESP), may be explained by presently unknown interactions of very weak EM field effects. Such interactions, he thinks, must be informational rather than energetic, since it is plain that it is not only the amount of energy involved that governs the effects of EM stimuli on life. In some cases, the reaction on the part of the organism actually *decreases* as the stimulus increases. The informational role of EM fields, he says, needs to be examined further.[20]

It is interesting to note in passing that maximum permitted field strengths for radio, radar and power stations in the Soviet Union have been set at levels far lower in many cases than those still permitted in Europe and the U.S. This suggests that the Soviets are more aware than we are of the side-effects of EM fields.

Now we come to the classic case of "how weak is weak?," the story of the Gurvich onion, a fifty-year-old mystery to which the solution has only recently been found.

In 1923 the Russian histologist A. G. Gurvich discovered that when he pointed the tip of one onion root at the side of another, cells would divide faster than usual on the second root. This, he decided, must be due to what he called *mitogenetic* radiation. This was an exciting idea, implying that some kind of radiation, perhaps ultraviolet light, was involved in the basic process of life: cell division. More than 500 papers appeared on the subject over the ensuing decade, mostly in Germany, but attempts to repeat the "Gurvich Effect" failed, and the whole subject was eventually swept under the carpet.

There the story might have ended, for some features of the alleged effect made it difficult to accept as real. For one thing, only *weak* rays would produce the radiation, it was claimed, and if the cell culture was too small, the chain reaction of division would simply stop. This made no sense to biologists at the time, but looked at with hindsight it suggests the true mechanism involved.

The original problem was twofold: was there a Gurvich Effect, and if so, could it be measured? Claims were made in the thirties that UV emission from cells had been detected with gas-discharge counters, but it was not until 20 years later, with the introduction of the photon-counter photomultiplier, that real progress was made, aided by the development of cryogenic techniques in the 1960s, enabling photo-detectors to be cooled to very low temperatures. In 1962 mitogenetic radiation was confirmed to exist, and there followed a new spate of interest, doubtless due in part to the work of Gurvich's daughter, one of the first to publish results.[21]

Proving the radiation to exist, however, was not to prove

that UV emission and absorption is a *controlling* mechanism of the cell. It could be merely an insignificant statistical fluctuation, though Gurvich had insisted that UV light actually induced cells to divide, a claim supported by Professor L. L. Vasiliev of the Leningrad Institute for Brain Research, who claimed he could cause mitogenesis by electrical stimulation of nerve preparations. Then, in 1974, a group of researchers in Novosibirsk headed by Dr. V. P. Kaznachayev caused something of a stir by announcing that two isolated cell cultures could *communicate* by UV radiation! They also claimed that artificial UV radiation by itself could stimulate cell division, and to find out how information got from one cell to another, they set up a test designed to show how cells would radiate energy under stress.

They put two identical tissue cultures in separate sealed glass containers, and placed a quartz barrier in between them. (Quartz, unlike ordinary glass, allows ultraviolet radiation to pass through it.) Then they contaminated one culture with a lethal virus, and noted that the *other* culture also began to suffer, as if it too had been given a fatal dose. Next, they took away the quartz and inserted an ordinary glass shield. Result: the uncontaminated culture went on growing normally while the other one died. This test is said to have been repeated 5,000 times with a high rate of success.

Examinations with an amplified electronic eye showed that while normal cells emit uniform fluxes of light waves (photons), the radiation pattern changes when a virus is introduced, with some diseases having their own characteristic radiation pattern. One of the most surprising claims of the Novosibirsk group is that the photon flux carries information from a diseased cell to a normal one, disrupting the latter's functioning as if it too were infected. Once this "photon code," as they call it, can be cracked, the Soviets believe they may be able to identify disease at cell level. "We feel," they conclude, "that we may then learn to effect healing by altering the photon flux before it contaminates neighbouring systems."[22]

This claim has some fairly wild implications. A Norwegian doctor, V. Schjelderup, has even suggested that when Kaznachayev managed to destroy a culture screened from the virus by quartz, it was not the virus that killed, but an EM *representation* of it! Does this mean that an

EM "virus pattern" could be broadcast over TV channels (by mistake or on purpose) to infect cultures, or people, at a distance? We hope not. Yet it is not altogether unreasonable to suggest that EM plays a greater part in the shaping of living forms than is generally admitted.[23] We return to this subject in Chapter 12.

As for the key to the mystery surrounding Gurvich's original claims, it was found to be absurdly simple: the instruments being used to measure the alleged mitogenetic radiation were themselves absorbing it instead of measuring it!

For a number of reasons, there is still something of a credibility gap in many branches of science between the U.S.S.R. and the west, especially in areas that border on parapsychology or the paranormal. It was encouraging, therefore, when two western biochemists announced in 1974 that they had been able to provide "limited support" for some of Gurvich's claims by detecting a weak chemiluminescence in the UV band from a yeast culture. Though more cautious in their conclusions than the Kaznachayev group, they did however recommend reinvestigation of the mitogenetic phenomenon.[24]

It may seem strange that a weak signal has an effect whereas a stronger signal has none at all, though this is quite a common phenomenon. Nature has her own set of limits, and it seems that interactions involving EM waves can take place only when certain limits are observed. Since our receptor systems are both temporally and frequency timed, stimulation must be within a certain range if it is to be transduced.

Kaznachayev's suggestion that light may be used for healing brings to mind the equally remarkable evidence presented by John Ott of Sarasota, Florida, who runs the Environmental Health and Light Institute there.

During the winter of 1968/69, the U.S. suffered a bad Hong Kong flu epidemic, and Sarasota county health authorities reported that 6,000 people there, five percent of the population, caught it. A hospital had to close down when 61 of the nurses got it, as did a supermarket and a club. However, at the Obrig laboratories, not far from Sarasota town, not one of the 100 employees missed a single day throughout the epidemic period. Nor did they have any special vaccines. How come?

Obrig is a large manufacturer of contact lenses, and was the first U.S. company (we believe) to design a new building to make use of full-spectrum lighting, with special plastic window panes that transmit UV light in all office and factory areas. Obrig workers are in fact enjoying open-air light indoors. They all seem to like it, and production is said to have risen 25 percent since the new system was used.

To prove his claim that natural light—that is, undiluted light as it comes from the Sun—is better for us than what filters through normal window panes, John Ott investigated the influence of various wavelengths of light on spontaneous tumour development in a group of inbred mice. Under pink fluorescent light, he found that mice lived an average 7.5 months. When "daylight" fluorescent light was used the average survival period was 8.2 months, and mice kept under ordinary window glass lived an average 9 months. Those enjoying full-spectrum sunlight kept going for 15.6 months—more than twice as long as the first group, which might be called the "office worker" group.

Full-spectrum lighting does not seem to have caught on as yet, though patrons of a Chicago seafood restaurant can now enjoy their lobsters under it. Now that solar energy is being looked at as a possible source of home heating, perhaps the day is not too far off when we can also have standard domestic sunshine at the touch of a button.[25]

We have seen just a few of the ways in which living beings can be affected by weak forces that in some cases cannot be measured at all. Some of the instances we have mentioned may make the following section seem more plausible, for what we are about to discuss is not only one of man's oldest and most respectable activities, but also one of the least understood.

Dowsing

The art of finding hidden objects or substances, usually water, by walking up and down holding a forked twig probably goes back 7,000 years. Legend has it that it originated after it had been noticed that tree branches seem to dip towards water. In medieval times, "magic wands" were cut from trees and ritually blessed, while by the seventeenth century dowsing was widely practised throughout much of Europe. In the Yakuta region of Siberia it is still a custom today to sit in the shade of a tree and shave a twig when there is an important decision to be made. This induces calm and concentration, and, we suspect, also produces dominant alpha rhythms in the brain, of which more in Chapter 9.

We know what dowsers can do, at least sometimes, but we are not sure how they do it; and as a rule, nor are they. Like hypnotists and faith healers, they go through an established ritual, and continue to do so because they find it works. It is often alleged that dowsing has no scientific basis, though like many other unexplained phenomena it *must* by definition have a scientific basis if it exists at all, even if that basis has yet to be established.

There can be no reasonable doubt as to the validity of the dowsing method. There is too much evidence to be dismissed: much of it from British Army officers of the Royal Engineers (who are not noted for their gullibility), many of whom have been and are very efficient dowsers. Let us mention just one example: in 1952 work began on a £14 million military headquarters near Mönchen Gladbach in Germany, and the engineer in charge, Colonel H. Grattan, happened to be a dowser. In the face of negative advice from local geologists and strong opposition from the building contractors, the Colonel went ahead and dowsed on his own, eventually locating an abundant source of good quality water wholly independent of existing networks in nearby towns. This case has been written up in great detail and published, and it is clear that Col. Grattan's dowsing abilities contributed considerably to the success of a complex engineering project.[26]

What exactly makes the dowsing rod twitch? The immediate cause is movement in muscles of the dowser's hand induced by stimuli from his brain. But are these stimuli voluntary or involuntary? Why does the rod sometimes move *against* the muscular tension applied? How can it react even when the dowser has no knowledge of what lies underground? J. Cecil Maby, a British dowser who was also a qualified scientist, was one of the first to insist that dowsing must have a physiological basis, although he admitted some of his fellow-dowsers often acted as though their faculty were a psychical one. He was convinced that on-site dowsing (as opposed to map-dowsing, which is beyond our scope here) was an electromagnetic process, whereby the human nervous system interacts with weak EM fields radiating from water or minerals, and responds to barely measurable impulses from the target.[27]

Dr. Tromp, also a fully qualified scientist, states categorically that dowsing is a purely physiological phenomenon due to a still unknown supersensory mechanism in the human body. His views deserve serious attention in view of the vast amount of research he has done into practically every branch of human activity. His *Psychical Physics* lists some 700 references to dowsing dating back to the sixteenth century, a list he describes as "far from complete"![28]

More recently, Dr. Z. V. Harvalik, an engineer working for a U.S. Army agency and vice-president of the American Society of Dowsers, claims to have identified the receptors in the body by which dowsers receive and process their sense-impressions. After a series of laboratory tests using various forms of metal shielding, to block off parts of the body in turn, he reckons to have located two important receptor areas: the suprarenal glands (over each kidney) and a spot in the brain near the pituitary and pineal glands. If these discoveries can be confirmed beyond doubt, they will be the first real breakthrough ever, from the point of view of the scientific investigation of dowsing.[29]

We can already list at least four possible mechanisms by which magnetic fields could be transduced into living systems. They could interact directly with the membranes of nerve cells, which could lead to cell modification in strong fields. Electromagnetic fields (that is, AC magnetic fields) could affect the motion of ions, which are involved in nervous conduction. There could be a direct magnetic effect on the water in biological tissues. And finally there could be an equally direct magnetic effect on the iron in red blood cells. The dowsing process may still be a mysterious one, but it can no longer be said to defy any kind of scientific explanation.

Dowsers believe that all matter emits a form of radiation, and that they can train themselves to become sensitive to it, the rod, twig or whatnot being merely a means of exteriorizing their reception of it. Harvalik believes that human beings can respond to magnetic fields as weak as 10^{-12} gauss, a barely imaginable strength impossible to register instrumentally, even on the Squid superconducting magnetometer. Having seen Harvalik in action, as we shall describe, we have reason to give his views the same respect that we give those of Dr. Tromp.

The recent work of men such as Harvalik and Tromp prompts us to re-examine some of the work of the 1930s, especially that of some of the German geobiologists and their forerunners, to see if their claims now seem more reasonable. The term "sleepwalker" (used throughout this book as a *compliment*) is highly apt for the dowser at work: he is obeying his inner voices rather than the logic of his conscious mind, which is how human knowledge has been advanced so often in the past.

Some of the most interesting claims of the geobiologists concern ways in which the ground beneath us can apparently make us sick. Dr. E. Jenny, director of a children's hospital in Aarau, Switzerland, carried out a long series of tests between 1932 and 1945, observing the behaviour of mice in cages six metres long placed partly over a "dowsing zone" (that is, a zone where dowsers can register radiations from below ground, usually due to the presence of water). Over a period of more than 8,000 mouse-nights, Jenny noted that four times as many mice chose to sleep outside the zone than inside it, and that they tended to nest as far away from it as possible, as if aware that it was bad for them. It was bad for them: 13 percent of mice forcibly confined to the zone developed spontaneous carcinomas, while control groups remained healthy and fertile. Repeating these tests under laboratory conditions, German researchers have been able actually to induce tumours in 100 mice out of 120 merely by using high-frequency low-intensity oscillations.[30]

How about people? Are some buildings bad for us? Should doctors give patients' bedrooms and offices a checkup now and then as well as the patients themselves? Perhaps they should, for the geobiologists claim that serious disease can result from living over a "geopathic zone." Many hospitals have "unlucky" beds, in which more patients seem to die than should by chance distribution. It is well known that there are places in modern hospitals where electrocardiograms cannot be made because of disturbance from electrical apparatus, which sets up a "humming effect." Steel beams and metal pipes can concentrate AC fields from electric wiring into narrow zones—the geobiologists' geopathic zones—and back in

1939 Cecil Maby was seriously suggesting that such zones could be a factor in serious disease.

Baron von Pohl spent much time in the early 1930s travelling around Bavaria with his dowsing rod and noting the coincidence of cancer patients and homes built over points of *Erdausstrahlung*, or "earth radiation." G. Rambeau went even further. "We have looked," he said, "for the house which does *not* lie over geologically disturbed areas and is also occupied by somebody suffering from long-term cancer, but have been unable to find such a house."[31]

In 1973, geobiologist J. W. F. Stängle retraced the footsteps of Pohl and Rambeau, not with a dowsing rod but with a scintillation counter to measure Earth radioactivity. He found relative maxima in the town of Vilsbiburg exactly over the spots marked by Pohl forty years earlier. He also found that radiation intensity was up to five times greater in a geopathic zone than it was in a nearby area.[32]

Geobiologists reckon that the surface of the Earth is criss-crossed by a network of "active strips" about two metres wide, unhealthy or geopathic zones developing where such strips cross. Anybody who lives or works over such a spot may feel effects that will pass just by moving bed or desk a few feet to one side. This is an extraordinary claim involving the simplest of solutions to the most serious of problems, and it deserves further study. Recent German research indicates that there may be something to such claims. H. Petschke has shown, after numerous experiments, that blood clotting times can be affected by geopathic zones, when control groups a few feet away are not, and it has been suggested (but not demonstrated, as far as we know) that human red blood cell counts can be similarly affected.

Electrical resistance of the Earth can be measured by driving spikes into the ground and passing a small current through the ground between them. Petschke noticed that when this was done within a geopathic zone, a maximum of conductivity was found corresponding exactly to the position of the "active strips" (perhaps due to water content). Such a difference in conductivity could influence air ionization over the zone as well as the flow of air and ground currents. How such anomalies could cause cancer is far from clear, but the suggestion seems worth following up.[33]

Prospective house purchasers might do well to insist on a survey not only of the house but also of the radiation count in and around it. A good dowser should be able to do this. Anybody who has noticed parts of his house where plants refuse to grow and cats will not sleep might try this form of house-diagnosis.

The time may also come when every garden shed will contain a geiger counter. Tromp has recorded an interesting example of nature following a dowser's predictions. The Dutch like tidy gardens, and in 1952 a privet hedge 100 m. long was planted all round the garden of a new house built over a spot known to local dowsers as potentially geopathic. By the following year, the hedge was growing nicely except for a seven-metre stretch on one side of the garden, right over the previously identified zone. Soil samples were taken, and fertilizer applied, but the stretch died and had to be replaced. In 1954 the replacement strip also died, and yet another replacement had to be planted, again after much ploughing and fertilizing. Third time lucky: the ill-fated strip finally grew, and when dowsers checked, they found that the geopathic zone had shifted—which such zones sometimes do on their own.[34]

In 1975 author and film maker Francis Hitching took a dowser and a pair of scientists to a remote hillside in the Welsh countryside to verify a claim by the dowser, Bill Lewis, that a standing stone contained power originating in the ground and rising up it in a spiral coil. First, Hitching asked Lewis to mark the position of the spiral with chalk, and then the two scientists, Professor John Taylor and Dr. Eduardo Balanovski, went over the surface of the stone with a portable gaussmeter, an instrument that measures magnetic field strengths. There were indeed indications of something anomalous in the stone, and it seemed Lewis may have been right, though Taylor was cautious about jumping to conclusions. If repeatedly confirmed, he thought the results could be "very remarkable indeed."[35]

If the human organism can detect extremely weak signals and interpret them in a useful way, as seems possible, the next step is to get the dowser into the laboratory. Some of the most useful recent work is that carried out by Tromp in the 1940s, when he found that a blindfolded dowser with cotton wool in his ears could detect the presence of an artificial magnetic field, provided sudden variations in field strength were created. He also ran a series of tests using an electrocardiogram as indicator of dowsing reaction to both natural and artificial EM fields, as subjects were driven in cars over rivers and canals. Another of Tromp's discoveries is that a dowser can register the presence of infrared radiation and electrostatic fields from another human body—without the body being present!

We all know the feeling of entering a room and sensing that somebody has just left it. Such impressions are usually due to obvious clues like perfume or heat, but Tromp has found that a measurable "electrostatic aura" also remains for hours or even days. If the bed or chair occupied by the person is grounded, the "shadow" effect disappears. It is tempting to think that some types of "ghost" phenomena may be caused by the presence of such remanent EM aura effects.[36]

Scientists have shown more willingness to demolish Tromp's work, or to ignore it, than to try and repeat it. In 1955 a German doctor attacked it with some violence, refusing to consider that dowsing had any physical basis at all.[37] A few years later, however, French physicist Professor Yves Rocard rose to Tromp's defence with his report on some ingenious experiments, including perhaps the only dowsing test ever carried out at an oasis in the Sahara desert. Rocard tried to find out what part of the body acts as receiver of the dowsing stimulus by standing his dowser in front of a magnetic coil fixed to a pulley, so that it could move up and down without the dowser seeing it. He concluded that the upper part of the body was more sensitive than the lower, and finally zeroed in on the crook of the elbow, finding that dowsers could not dowse when magnets were placed there. Another ingenious idea of Rocard's— some years before talking to plants became popular—was to measure the electrical potential of a tree as its roots reacted to water being pumped into the dry ground. He noted a rapid variation, and also found that voltmeter readings dropped sharply after nails were hammered into the bark. "It is apparently the reaction of the tree to its wound," he says.[38]

One of us (Hill) was given an informal but impressive display of an unusual dowsing ability in 1976 when Dr. Harvalik, on a visit to Copenhagen, showed that he could act as a human direction-finder and locate the position of a radio station. A small transistor was tuned to a local station, the site of which was known neither to Hill nor to Harvalik, who began to move slowly round in a circle, holding his dowsing rod. This gave two maximum deflections at points 180° apart, and when the radio set was rotated until the signal faded, it was found that he was accurate to within 5° of arc.

Harvalik's most remarkable claim is that he can detect brain and heart waves at distances of up to 20 feet. He has found that fear and sex-related "exciting" thoughts produce the strongest magnetic emissions: a claim easier to accept today than before 1965 and 1969, when magnetic heart and brain waves respectively were first measured instrumentally.[39]

"A dowser will be successful," Harvalik says, "the more he is able to detach himself emotionally from the dowsing activity or from any other 'exciting' thoughts." Three hundred years ago another dowser, William Cookworthy, made the same point—long before anything was known about brain rhythms. "The rod must be held with indifference," he wrote, "for if the mind is occupied by doubts, reasoning or other operation that engages the animal spirits, it will divert their powers from being exerted in this process . . ." This, he said, was why best results came from peasants, women and children, who hold the rod "without puzzling their minds with doubts and reasonings."[40]

A contemporary dowser speaks of the need to "shut the brain out and simply rely on the mind," and if we replace Cookworthy's animal spirits with magnetic fields and the process of "shutting out" with a shift to a dominant alpha brain rhythm, the dowsing process begins to look far more plausible, if still not wholly explicable. It may be that the extreme weakness of the EM signals involved has led scientists in the past to underestimate man's ability to register them. Once again, we are reminded that weak can be strong.

What happens when radio waves penetrate the human skin? This (minus the question mark) was the title of a 1974 article by Dr. Andrija Puharich, and it is a good question to ask now that we know man to be far more sensitive to his invisible environment than he is consciously aware.

Puharich, well known for his association with Brazilian healer Arigó, fork-bender Uri Geller, and strange mushrooms, is respected in scientific circles as a prolific inventor with more than fifty U.S. patents to his credit. He knows a good deal about unusual methods of applying electricity to the human body, and one of his most interesting inventions is the TD-100 Transdermal Instrument—a deaf aid for people with nerve deafness. Sufferers from this cannot hear normally however loud they are shouted at; they have no intact nerves to carry acoustic signals to the brain. But Puharich, together with J. L. Lawrence, found that when *radio* waves are connected to a plate attached to a person's forearm and modulated by human speech, the waves can be "heard" inside the deaf person's head, with suitable amplification. They have found that molecules of keratin protein in the skin, when stretched in the presence of AM carrier waves, become radio wave detectors. (Keratin, as mentioned earlier, has been identified as the substance through which birds detect electromagnetic fields. See reference 18.)

Having found that man can be thus provided with an artificial extra sense, Puharich began to speculate on the nature of processes commonly labelled extra-sensory, by which information is exchanged between human beings by mechanisms outside the known senses. Working with Uri Geller, he has found that what is popularly known as ESP (extrasensory perception) can be enhanced by the application of radio energy. After being "energized" by a TD-100 machine, Geller is said to have scored ten hits out of ten in receiving mentally transmitted single-digit numbers. Moreover, Geller seems to have projected the actual *number* on to Polaroid colour film!

Claims such as these are of limited value unless they are repeated, and unfortunately subjects like Geller are hard to find, and equally hard to lure into the laboratory. Moreover, this is a field beset with allegations of fraud. As our instruments improve, however, the opportunities for cheating decline, and Puharich has provided enough information for anybody to try and repeat his work.[41]

Are we all walking organic radio sets? We know that we emit radio waves, but can we receive them without help from instruments such as the TD-100? We all have the same proteins in our skins, including keratin, and evidence is coming in that other parts of the body contain the necessary components of an EM wave receiver.

In 1962, Soviet radio engineer Bernard B. Kazhinsky published a book with the challenging title *Biological Radio Communication*. This was promptly translated by the U.S. Air Force intelligence service (which has a lot of interesting unpublished books lying around), and was rated as classified material for several years. Much of the book deals with paranormal biocommunication, Soviet newspeak for telepathy and clairvoyance, and Kazhinsky insists that EM waves play a part in such phenomena. He points out in some detail that certain nerve groupings and loops look suspiciously like radio valves, triodes and pentodes, antennas, inductors and capacitors.[42]

In the west, several researchers seem to be arriving at a similar conclusion along widely differing paths. One of the most promising areas of investigation is the phenomenon already mentioned briefly of piezoelectricity, which was discovered to exist in bone by Japanese scientists in 1957. It is the process by which mechanical work can be converted directly into electrical energy through crystals, as in early radio receivers that were known as crystal sets. A recent invention using this process is a cigarette lighter that works by mechanically compressing a piezoelectric crystal, generating a voltage of 20,000 V resulting in an electric "flint" that never wears out. The process can also work the other way round: by absorbing EM waves at the right frequency, mechanical motion can be produced by the crystal, which is how some ultrasonic waves are gathered. Scientist Joe Weber has also used piezoelectric crystals in his huge cylindrical gravitational wave detectors.

New York researchers Robert O. Becker and Andrew Marino have shown that the piezoelectric effect in bone and other tissues is not due to mineral content, but to an organic compound: the structural protein collagen. They are especially interested in the possibility that piezoelectricity may be a biological transducer, capable, as they put it, of converting environmental stimuli into biologically recognizable signals capable of controlling growth. In other words, outside forces can perhaps be converted into signals inside the body. This is an idea that needs further examination.

It was Albert Szent-Györgyi, the Hungarian-American who won a Nobel Prize in 1937 for his work on metabolism, who first theorized that cells and other biological components might have electronic solid-state physical properties such as semiconduction, the quantum mechanical process the understanding of which led to most modern radio and computer components. Dr. Becker in his turn has been researching the influences of EM fields on life processes for more than a decade, and has published several dozen scientific papers on his own use of these fields in treating bone fractures and promoting rapid healing of burns and skin ulcers, in addition to studies of electronarcosis and electroacupuncture. Following Szent-Györgyi's lead, he has posed two questions:

Do bioelectronic control systems exist in living organisms, based on electronic conduction mechanisms, that regulate important life processes? And if such systems exist, can they be perturbed in a clinically useful fashion by application of appropriate levels of externally generated EM energy?

His answer to both questions is an unqualified yes. It is already established, he says, that EM forces can be used to change three fundamental life processes in mammals, and possibly also in man; by stimulating bone growth and partial multi-tissue regenerative growth, and by influence on the basic level of nerve conductivity and function. All these effects, he points out, seem to be mediated through perturbations in naturally pre-existing electronic control systems. All are produced by low levels of continuous or pulsed direct current administered directly to the organism, which seems to show a power/density relationship centred at an extremely low power level, according to recent research.

But how can radio waves at levels so low that no heat can be measured from them have any possible effect on our bodies? Dr. Becker thinks this enigma can be explained by assuming that radio energy is converted within the organism into some sort of direct current. We have already seen that the human body performs the same functions as a wide variety of electrical devices. Now it seems we can even act as current rectifiers, converting AC from outside into DC inside our bodies—for good or ill.

Many of us are made aware of the electric and magnetic forces around us only when we get a painful electric shock or are struck by lightning. Electricity is dangerous stuff that can cause serious wounds and easily kill us. But with proper application, it can also heal, and much research into new applications of "electromagnetic medicine" has been undertaken since the dawn of the Space Age.

One creature responsible for the discovery of a new form of EM healing is the salamander, the cold-blooded lizard that has the strange ability to grow a new tail if it loses its old one. Many other animals, not to mention man, must envy the salamander and wonder how it performs this feat and whether man will one day be able to do the same with his arms, legs or other organs. Such an idea is no longer theoretically impossible.

When a salamander grows a new tail, it makes use of a peculiar process that adult man no longer seems to have. A mass of primitive cells gradually forms itself into a complete multi-tissue extremity that has been appropriately *organized*. Electric forces are definitely at work here, for it has been found that when biological tissue is damaged, a "current of injury" flows between the damaged tissue and the healthy area. The former is charged positively in relation to the latter, so the current flows, as we would expect, from positive to negative—that is, away from the damaged area and towards the healthy one.

The question now arises: if damage results from positive current, why could not healing be *induced* by a negative current? That is, by reversing the flow of current from healthy to damaged areas, could normally slow processes of recovery not be speeded up? Research in the 1960s indicated that this was possible in some amphibians, and studies on animals soon followed, of regeneration of fractured bones with implanted electrodes, and of healing skin ulcers and burns by applying direct current of 1 to 3 milli-microamps directly to the area affected. In 1972 Becker reported in *Nature* that partial limb regeneration could be achieved in rats—bone, muscle and all—by applying the right kind of current.

An important feature of electrically-stimulated regenerative growth is that no scar tissue is involved. This raises hopes for heart patients, for when somebody has a heart attack, part of the muscle dies, scar tissue forming as the damaged area heals itself. Often this causes no serious problems, but it can reduce the heart's efficiency as a pump to the point where at present there are only three remedies: heart transplant, mechanical heart, or grafting of extra blood vessels on to the existing organ. None of these is wholly satisfactory for reasons too numerous to mention here.

But, Dr. Becker says, some animals capable of growing new limbs can also regrow portions of heart muscle. Moreover, the control mechanisms for each type of regenerative growth appear to be the same. So, he argues, coming to the crucial point, if we could gain access to these controls in an effective fashion, we would be able to bring about the repair of damaged heart muscle *by growing new heart muscle* instead of scars. Similar applications, he adds, can be envisaged in many other areas of clinical medicine.

This exciting prospect brings to mind the search for the "biological organizing model" behind all living forms postulated by the Brazilian parapsychologist Hernani Guimarães Andrade. His hypothesis has been described in detail in a previous book by Playfair,[43] and will be mentioned again in Chapter 12. Briefly, it concerns the existence of an energetic structure that serves as a mould, or model, for all qualities and properties of living matter.

Dr. Becker's discovery that bones can heal faster by passing current through the area around them is a real breakthrough in the hunt for the organizing models of biosystems. Some German researchers have done away with contact electrodes, using tightly wound coils to produce strong magnetic fields, in the tradition of D'Arsonval and his cage. Patients can be placed inside such coils as they lie in bed, and need have no apparatus in direct contact with their skin. Thus they can be treated while asleep, and even serious cases of broken bones seem to respond to such painless treatment, in which fields of up to 100 gauss (50 Hz.) have been used.[44]

Magnetic therapy devices have already been put on the market by several West German firms. Clearly, there is a future in this kind of thing, but before everybody with a broken bone starts playing around with powerful magnets, we must stress that until more is known about possible links between magnetism and cancer, such treatment is best left to the specialists.

Another promising new use of magnetism is for diagnosis. David Cohen, a pioneer in the study of biomagnetism, has shown that it is possible to measure minute quantities of magnetic particles inside the human body, and has produced magnetic field charts showing the distribution of 500 micrograms of asbestos cement in a man's lung, and only 100 mcg of iron oxide in the stomach of another man who had just eaten some tinned beans. There seem to be many potential uses for this new diagnostic method, especially in industries where workers are exposed to para-, dia- or ferromagnetic metals.[45]

The solid state neural control system that Dr. Becker has developed is of course subject to outside EM influences, and Becker was one of the first to indicate the Earth's magnetic field, or geomagnetic field (GMF), as a possible collaborator in the external control of the human body. In an often quoted paper, he and two New York colleagues reported in 1963 that a significant relationship had been found between fluctuations in psychiatric disturbances, as measured by number of admissions to psychiatric hospital wards, and GMF intensity. There was only one chance in about 10,000 that their findings were due to chance alone. This study, involving more than 28,000 hospital admissions, confirmed an earlier investigation in Denmark and Switzerland that lasted five years and covered 40,000 cases. (In Chapter 4 we give more examples of correlations between GMF variations and various human conditions.)[46,47]

The GMF is not fixed and immutable. Its poles shift, its intensity changes all the time, and it is now known that every so often the poles actually change places, as do those of the sunspots after each 11-year cycle, as a rule. Several such pole-reversals within the last 3 to 4 million years have been confirmed through studies of rock and ice samples, and we may be due for another one around the year 4000 in view of the fact that the GMF has been getting steadily weaker since early in the nineteenth century.

During such a reversal, there must be a period of time when GMF intensity at certain points of the Earth's surface falls to zero, and palaeontologists have noted that many species of fossil seem to have disappeared at such times, new species turning up soon after the reversal period is over.[48] This might help explain some of the crucial evolutionary stages through which our hairy ancestors passed, and provides one more possible theory to account for the disappearance of the mighty dinosaur, the extinction of which has yet to be explained. It is tempting to speculate on possible connections between GMF aberrations and genetic mutations, but for the time being we will point out only that it is quite possible that the history of life on Earth may have been governed throughout its whole span by a geomagnetic influence so subtle that it is only recently that we have been able even to measure it accurately.

There is much more to be said about the interactions between man and his environment both visible and invisible. We have yet to touch on the question of how natural cycles interact with those of our bodies and minds: this will be done in Chapter 6.

In the following chapter, we return to the subject of weather and climate, mentioned at the end of Chapter 2, and look at some of the ways in which fluctuations (some of them cyclic) in our environment leave their mark on the pattern of our daily lives. The most important feature of weather and climate we want to introduce here is that part of them which we do not perceive consciously. We are all aware of such things as temperature, humidity and barometric pressure, but there is more to weather than these; it has an invisible component in the form of electromagnetic properties of which we are usually consciously aware only when we are caught in an "electric storm" with its dramatic lightning effects.

In fact, however, *every* change in the weather is accompanied or preceded by changes in the invisible EM forces. Such meteorological factors require special equipment to measure them, and although our daily weather reports do not as yet include such items as air electricity or GMF strength readings, there is some evidence to suggest that they should. For these, among others, may be the physical and measurable, perhaps even predictable, means by which the mysterious forces of the cosmos could be finding their way into our brains and bodies.

It is time we knew more about them.

Chapter 4

ELECTRIC WEATHER

THE YEAR 1976 was one for the record books. Nature seemed to have gone mad. Sixteen inches of rain fell in a single day in Hong Kong, while Moscow and parts of the U.S. were flooded to a standstill. Meanwhile, Europe sweltered under its worst drought since 1727, major earthquakes seemed to be erupting almost daily,* as a bewildered and slightly alarmed public turned to its weather men for comfort. The experts, however, were as confused as everybody else, the unusually dry September they predicted to follow an almost rainless August turning out to be three times wetter than normal. (At the end of August, members of England's Indian community held special rainmaking prayer sessions, during one of which the first rain of the month actually did fall.)

The following year was not much better. On 19 January 1977, snow fell in the Bahamas for the first time in recorded history. Temperatures in New York were the lowest for 108 years. Orange growers in Florida had to light bonfires in their fruit groves. Tornado winds of 110 mph swept across the Kamchatka peninsula. The Niagara Falls froze stiff. Three million Americans were put out of work in February as temperatures plunged to minus 40°C, accompanied by 70 mph blizzards, and whole towns, such

---

* Earthquakes are far more common than most people realize. There is one somewhere, on average, *every thirty seconds*. However, preliminary data suggest that there were many times more major quakes in 1976 than usual. See also Chapter 5.

Electric Weather 103

as Buffalo, N.Y., almost vanished entirely under snowdrifts. Yet all this time temperatures in Alaska were 30 degrees *above* normal, while in Britain people were still being told to save water as sudden floods caused rivers to overflow.

What did it all mean? Had nature gone mad, or had man simply lost touch with it? Surely, whatever nature does must be natural, and it is up to us to try and understand it. While all this extreme natural violence was going on, those tedious passages in the textbooks suddenly became important. What exactly *were* solar flares doing to us? How were we really being affected by fluctuations in the geomagnetic field? What else, in addition to the weather, was being influenced by forces which nobody understands and of which few have even heard? To what extent are we at the mercy of such forces, and how can we protect ourselves against them? What exactly is going on?

British TV broadcaster Michael Barratt had a golfing date in Bracknell, Berkshire, one afternoon in 1976. Since this is where the British Meteorological Office is situated, he thought he would call a friend there to ask how the weather was shaping up on their local course.

"Rain all day," was the reply, so Barratt dressed for a wet day's play. But the Sun shone all day long. There must be something wrong somewhere, he thought, if they cannot get the weather right on their own doorstep on the day! Later, a man from the Met. Office admitted on BBC's *Nationwide* programme that even in this age of weather satellites, computers, mathematical models and statistics, a good deal of the forecaster's job is still done by guesswork, with one forecast in five quite likely to be wrong.

Meteorology, the study of rapid fluctuations in weather, is a much maligned profession, though not the target of as many jokes as it was a few decades ago. Things have improved a lot: four forecasts right out of five is not bad going, for although daily predictions are based on precise measurements from stations all over the globe and from satellites in the sky, they can never be 100 percent accurate. Our ecosphere—the surface of Earth and the atmosphere above it—is large, complex, and subject to sudden and violent fluctuations. The weather is yet another many-body problem, and when such a problem has to deal with nearly-invisible bodies, it becomes a daunting one to tackle.

If meteorologists have problems predicting tomorrow's (or today's) weather, climatologists—whose job is predicting trends for the future—have an even worse time. As we mentioned in Chapter 2, the period on which most modern climate study has been based until recently was one of the most abnormal ever known in modern times. Moreover, as Professor B. J. Mason, Britain's Met. Office director, has pointed out, statistics can hide more than they reveal.

To make matters worse, climatologists seldom agree with each other. Different tendencies are constantly being identified that suggest totally different effects. In 1976, for instance, two authorities announced simultaneously that the weather in the years to come was going to get hotter and colder. The World Meteorological Office (WMO) in Geneva warned that increasing amounts of carbon dioxide in the atmosphere from oil and coal burning would force temperatures up by several degrees centigrade. Oliver Ashford, WMO planning director, said that a subtropical, humid climate in parts of Europe was conceivable, as rainfall distribution was drastically altered by high carbon dioxide levels. Meanwhile, melting Arctic ice might raise far-northern temperatures and boost agriculture in high latitudes. Greenland as a bread-basket?[1]

On the other hand, American scientist-inventor Iben Browning thinks Earth's climate has entered a cooling cycle, and that we are wrong to believe that the warm and stable climate of the past 50 years is in any way normal. Normal, says Browning, is a climate that is "just terrible!," and the trouble is, we are reverting to that kind of "normal." He foresees a long period of highly erratic weather ahead, and warns: "Unstable times are hard times. You see them throughout history. The rules change. People with food tend to keep it for themselves. The others become very hard to compromise with when their babies are starving." We have, he says, entered one of the unstable periods.[2]

This, if true, will have alarming consequences. The balance of world power will be affected. The Soviet Union's grain belt is in the southern part of the country, and if the average line of good growing conditions were to shift southwards, it would shift out of the U.S.S.R. altogether. The United States, on the other hand, is lucky in that its average good-climate line crosses the centre of the country, and could shift a little either way without bringing ruin in its wake. If one of the superpowers were to become permanently dependent on the other for its basic commodities, all kinds of interesting political situations would develop. Therefore it seems in everybody's interest to try to get climate prediction right.

It may sound far-fetched to suggest that changes in climate or weather could reshape the political map of the world, but the Central Intelligence Agency does not think so. In 1976 it released a report drawn up two years previously and based mainly on the work of a University of Wisconsin team headed by Reid Bryson. We are, said the CIA, entering a mini-Ice Age. Average temperatures in the northern hemisphere, where most international power struggles take place, are expected to drop by one degree (C). And, this, the intelligence community thinks, will be enough to lead to Global Upheavals.[3]

The chief factor in the climate of Britain and NW Europe is the band of wet wind that helps spread a generous amount of rain over fertile farming areas, which at present can produce enough food to support three people per hectare. That single degree drop will lower the ratio to two people per hectare. Further drops in the temperature do not bear thinking about.

Other countries will suffer worse fates. Canada: grain production halved. Soviet Union: Kazakhstan lost for ever to grain growing. India: major drought every four years (monsoons are already starting to behave oddly) and only enough food for 75 percent of the *present* population. China: major famine every five years . . .

Luckier parts of the world include Iran, due for a revival of agriculture on the Iranian Plateau, and the North African coast, which will come to the rescue of Europe after the Common Market grain surplus has been wiped out.

All of this might lead to the collapse of whole nations, both economically and politically. This would be followed by what the CIA ominously calls "militarily large-scale migrations of their peoples."

It may never happen. Or it may. It could be that rival forecasts of a hotter and a colder world will balance each other, and the weather will stay as it is: inconstant and unpredictable. Nobody, however, should ever make the mistake of taking the climate or the weather for granted.

Weather Control

Professor Mason, whom we mentioned earlier, thinks man has a long way to go before his influence becomes great enough to do any serious damage to the natural climate. The general tone of a talk he gave to the Royal Society of Arts in 1976 was: "Don't panic—our computers have climate under control." Nevertheless, there are a number of ways in which man is messing up the atmosphere. Debate still goes on over the "greenhouse effect" caused by carbon dioxide from fossil fuel burning, the effects of cooling the atmosphere with dust particles, and the damage done to the ozone layer by supersonic aircraft. All these effects are real, but opinions differ as to whether they are significant, and with all respect to Professor Mason, we cannot help thinking of the journalist's motto: "Believe nothing until it is officially denied."[4]

With our ecology poised as it is on a tightrope, we cannot afford to ignore anything that might upset the balance. Yet not content with Nature's possible reshaping of our weather patterns, man is doing his best to add to it. In 1976 a group of NASA scientists came up with the (literally) brilliant idea of sending artificial suns and moons into space to reflect extra sunlight on to the dark side of Earth, thereby boosting food production, adding to energy resources, and probably keeping millions of people awake at what used to be night as well. These "solettas" and "lunettas," aluminum-covered plastic satellites, would provide the heat and energy of 160,000 full Moons. Just what havoc such a scheme might wreak in an Earth already far out of step with Nature will be easier to imagine after we have discussed biorhythms and bioentrainment in later chapters. (Once the soletta is in orbit, all it will need is for some enterprising joker to launch a magnifying lens into concentric orbit, focus its rays on the office of the man who thought up this scheme, and melt it and him into slag.)[5]

Meanwhile, there is something worse to worry about. The Ryan Aeronautical Co. has long urged NASA to look at the effects of nuclear weapons tests on the weather, now that more is known about energetic interactions between the ionosphere and Earth's lower atmosphere. Studies have indicated that winds tend to follow geomagnetic field contours all over the world, and since it is fairly well established that this field responds directly to alterations in the ionosphere, it follows that nuclear weapons are affecting our weather to some extent by distorting the field and its gradients after injecting swarms of charged particles into the ionosphere. The Soviet authorities are known to have asked the U.S. not to test nuclear weapons in the upper atmosphere near space-probe launch dates, so as to avoid possibly lethal concentrations of such particles around the Van Allen belts. After the Soyuz XI episode, they have good reason to take no chances.[6]

There may be peaceful uses of weather manipulation. Arthur C. Clarke, in *Rendezvous with Rama*,* describes an artificial planet which has a completely artificial ecology, part of which is an "electric wind" that heralds changes in the weather. However, if there are any evil uses to which an invention can be put, the history of man indicates that they will be found. We can already see military strategists licking their lips at the thought of producing clear sky or cloud cover to order, or wrecking a country's food supply by inflicting an artificially made drought upon it. Yet, in the long run, it may turn out that trying to understand nature will be a better investment than trying to manipulate it.

Weather Cycles

To say that the Sun influences our weather must be an understatement. Although there are some heat sources under Earth's surface, which are important in connection with volcanic eruptions, it is the Sun that provides virtually all our energy. Yet the "solar constant," the amount of energy received by Earth as radiant energy, is not in fact constant at all, as was once thought, any more than is the Sun itself. The variation is small, but it makes prediction difficult, since the amount of solar input varies globally and locally at any given time. To study *cyclic* variations in this input, we need hundreds of years of accurate data, which we have not got. We can, however, already identify a number of cycles that should help make climate prediction on a long-term basis easier in the future.

Paleo-climatological records for the last million years reveal, for instance, peaks in sea-level fluctuations of the global amount of ice at 100,000, 40,000 and 25,000 years,

---

* London: Victor Gollancz Ltd., 1973.

with shorter peaks at 2,500 and between 400 and 100 years. These can be attributed to secular variations in Earth's orbit around the Sun. Nature is a good record keeper; Danish scientist Willi Dansgaard has analysed the isotope content of ice core from Greenland (ice is thought to contain an accurate record of air temperature variations), and found peaks at intervals of 177 and 117 years over the past thousand years. The first figure is of special interest to cycle-watchers, being very close to the interval between grand planetary alignments, and also to the long-term solar cycle of 179 years.[7]

Shorter cycles indicating an extraterrestrial driving of terrestrial phenomena are constantly being uncovered. Two California geophysicists have studied rainfall patterns in Los Angeles since 1900, and have noticed that in each of that city's two rainy seasons, rainfall is concentrated into the first half of 27-day periods, although the phase of the cycle tends to change from year to year. Data from San Diego and Santa Barbara, cities with lower rainfall than Los Angeles, revealed similar results.

These researchers believe that the rainfall cycle may be related to cyclic latitudinal oscillation of the jet stream, the high-speed wind that originates in the upper layers of the atmosphere. This oscillation, they thought, might be indirectly responsible for local oscillations in rainfall. They duly noted some very small effects on rainfall associated with both full and new Moons, and also took possible solar effects into account, but found indications that the phase of the rain cycle was *not* being driven by Sun or Moon alone. "If the periodicity is of solar or lunar origin," they concluded, "then its phase must be determined by another mechanism." Similar oscillations in rainfall have been noted on the other side of the world, in New Zealand, after a study of 33 years of data there.[8]

Yet the Sun-weather mystery is far from solved. For every study claiming a correlation, along comes another to pour cold water on it. A detailed analysis of data from no fewer than 343 weather stations around the world published in 1977 failed to uncover any precise correlations at all on a repeatable world-wide basis, though the authors do not rule out the possiblity that cyclic Sun-weather connections do exist.[9]

We return to the subject of cycles, solar and other, later in this book. Next, we look at ways in which climate and weather have been found to affect human beings, often without their being aware of any such influences at work.

Body Weather

"In 1929, reputable physicians did not discuss weather," recalled the late Dr. William F. Petersen, the highly reputable physician* who pioneered the new science of biometeorology—the study of the effects of weather and climate on living beings on both long and short term bases. Petersen once had a paper on polio rejected by a scientific journal "because it contained too much about polar fronts," and at the start of his career his university even refused to buy him simple apparatus and equipment for his research. Yet Dr. Petersen held fast to his thesis that man is "weather-conditioned." "With the establishment of this interrelation of weather (i.e. the inorganic end) and the organic group (i.e., the human)," he wrote, "the final link has been forged which validates the thesis of 'man as a cosmic resonator,' for weather, in final analysis, reflects but changing tides in energy release from the Sun."[10]

Petersen himself did much to establish the interrelation referred to. From 1927 to 1940, covering an entire solar cycle (peak to peak), he analyzed the weight/length ratio and blood pH of 34,000 babies born in Chicago, and found that the figures went up and down in time with those of the sunspots. He also noticed that babies conceived in October tended to be heavier than those conceived in April or May, and even speculated that there was a periodicity in the production of genius in the population that also kept pace with the solar cycle. His work makes an interesting comparison with that of the Russian scientist A. L. Chizhevsky, to be mentioned later.

Some effects of weather on man are direct and obvious: when the Sun shines and the birds sing, we tend to feel cheerful, and when it rains all weekend we catch cold and feel gloomy. Extreme heat and cold can seriously disrupt our daily routine, while in places like Asunción (Paraguay) the heat and humidity make it difficult to have a daily routine at all other than one of alternate cold showers and cold beers.

Temperature and humidity are perceived immediately and consciously. Less obvious, though just as real, are the

---

* Professor of pathology, University of Illinois, from 1924 to 1942.

effects of weather and climate that we do not perceive consciously. The official definition of biometeorology, given when the International Society for Biometeorology (ISB) was founded in 1956 at a Unesco symposium held in Paris, was ". . . the study of direct or indirect correlations between the geophysical and geochemical environment of the atmosphere and the living—plant, animal and man." Environment was meant in the wider sense, to include the macro-, micro- and cosmic environments that influence the various physical and chemical processes at work in Earth's atmosphere.

Headquarters of the ISB is at the Biometeorological Research Centre in a suburb of Leiden, Holland. Its director is the indefatigable Dr. S. W. Tromp, who has also found time to handle the European side of the Foundation for the Study of Cycles, which we mention in Chapter 6. In 1963, Dr. Tromp published the first comprehensive survey of biometeorology (a work of 991 pages and 4,400 references), and things have moved so fast since then that the first volume alone of a work summarizing subsequent research runs to more than 1,000 pages.[11] The *International Journal of Biometeorology*, of which Tromp was first editor-in-chief, keeps pace with new developments in this important new interdisciplinary field.

Biometeorology is concerned not only with the effects of such traditional factors as temperature, pressure and humidity, but also with such geophysical factors as electromagnetic and gravitational fields, waves and configurations, including atmospheric potentials, spatial electric charges, aid conductivity and radioactivity, vertical currents, propagating and standing EM waves, disturbances in natural geoelectric and geomagnetic fields, and variations in such distant factors as planetary tides. These have been termed *biotropic* factors.

Some people are extremely sensitive to changes in the weather, reacting strongly to the approach of a cyclonic or low pressure zone. Such persons are known as *meteoropaths*, and they can feel effects even when indoors at a constant temperature. A meteoropath can climb a hill, causing a pressure drop of about 30 mm. of mercury— about the same as for a common cyclone—without feeling any ill effects if there is no meteorological disturbance around him. Both temperature and pressure must therefore be ruled out as crucial factors that disturb the meteoropath, and the same goes for humidity. What, then, can the influence be to which he responds? It must be an elusive property outside the traditional factors listed above.

The search for indicators that might explain meteoropathic phenomena has revealed some possible culprits, such as atmospheric waves of very or extremely low frequency (VLF and ELF), and other non-propagating atmospheric electrical fields. Direct biological response to EM waves is implicit in the explanation of biometeorological influences, which is hardly surprising in view of some of the material we discussed in the preceding chapter, although such a response has only recently become accepted by more than a handful of scientists. We now look at some of the evidence for such responses, and then introduce the subject of low frequency fields, which are a lot more important and exciting than they may sound.

The idea that man is affected in subtle ways by his environment is not a new one. The Danish philosopher Johan Tetens (1738-1807), a forerunner of modern psychology, noted in the eighteenth century that climatic factors could influence the way people thought and acted; while a century later Henry Thomas Buckle, in his two-volume *History of Civilization in England*,* saw history as a constant process of mutual interaction between man and nature. Yet only recently has twentieth-century man begun to identify the mechanisms of such interactions, or even to accept that there are any.

"There is hardly an organ in the human body which escapes the effects of changes in the meteorological environment," Tromp says, stressing the fact that such effects are reflected both directly and indirectly in man's *mental* processes, which respond to the subtlest stimuli from the physical environment. He lists the main receptors for such "biotropic" stimuli as the skin, respiratory tract (throat and lungs), eyes, nose, nervous system as a whole, and the brain, especially the hypothalamus.

The skin reacts to thermal and ultraviolet radiation (hence suntan) and also to changes in its acidity. Thermal radiation can also lead to alterations in the hormonal function of the pituitary, thyroid and adrenal glands, blood sugar, albumin and globulin levels, cell composition and

---

* London: J. W. Parker & Son, 1857 and 1861.

functions of liver and pancreas; while UV radiation affects at least eight body functions, including protein metabolism and gastric secretion.

The human throat and lungs are affected not only by temperature and humidity, but also by air ionization (of which more later in this chapter), acidity of the air, increased or decreased oxygen pressure, trace elements such as ozone, and pollution by gases, particles or aerosols. Many of these factors also affect the eyes and nose: direct stimulation (through the nose) of the rhinencephalon, or olfactory region of the brain, can alter a number of important body functions. We shall mention more such man-weather interaction mechanisms later.

Hippocrates insisted that medicines had different effects according to the time and season they were taken, and we now know that biotropic effects on people also vary widely according to the time of day or year, and also to the person involved. Such effects can last long after the conditions that caused them have ceased to prevail. Tromp has observed that a period of only one hour of contrasting atmospheric conditions is enough to upset physiological conditions for some time after the former have stabilized themselves.

One of his first large-scale research projects in biometeorology was a four-year study of the effects of weather on 200 mental patients. These, some of whom were potentially violent, were studied on a 24-hour basis, their state of restlessness being rated in relation to their "normal" state for each period of day and night. Patients from seven different situations were involved, and when all the results were gathered in and studied, some curious facts came to light.

For each of the four years of the study, overall restlessness peaked between November and January, with secondary peaks in July and August. Correlations were also noticed between day-to-day weather changes and patients' behaviour: unrest increased with the passage of warm air masses and decreased with that of cold ones. Oddly enough, when the weather was bad in the conventional sense, with heavy rain or snow, the nursing staff were more affected than the patients.

Dr. Tromp believes that some of these behaviour fluctuations can be related to the fact that schizophrenics have poorly functioning thermoregulatory mechanisms in the hypothalamus, or thermal brain. Tests using hot-water foot baths have shown that they take far longer than normal people to respond to temperature changes, and since the thermal brain is directly responsible for several physiological processes, the effects of a malfunctioning hypothalamus will be felt in many ways.

A normal person can stand violent changes of air temperature and adapt quickly to them if he is prepared for them. Flying from London to Rio de Janeiro in February can mean a sudden air temperature rise of up to 40°C. Yet human *body* temperature cannot rise or fall more than a few degrees without our lives being in danger. Ideally, our bodies like to stay at normal temperature, and part of the job of the hypothalamus is to keep them there. It also has an important influence on the rhinencephalon, part of the emotional brain which, in turn, plays a key role in man's unconscious moods, instincts and emotions, though it is also involved in consciously perceived feelings, especially the sense of smell. The slightest upset to the hypothalamus, then, can lead to all kinds of distress.

The Suicide Season

Dr. Tromp and his Leiden colleagues have also uncovered some interesting facts concerning suicides, for which rates in Holland are among the lowest in Europe, at around six per 100,000 of population per annum. Even when known attempted suicides are included, as they should be on this type of survey, the rate is less than 1.5 per day even for the largest town in the country. So it would be remarkable if up to five Dutchmen killed themselves, or tried to, on the same day in the same town: even more so if such "cluster days" turned up again and again in several different towns, on days when nothing particular happened that might be thought partly to blame. Yet this is exactly what has happened. Going through records for 1954 to 1970, Tromp found no correlation at all between suicide cluster days and temperature, wind speed, hours of sun and rain, or days with long-wave magnetic impulses. What he did find was that cluster days coincided again and again with periods in which there were sudden sharp changes in weather and temperature. It seems the change is more upsetting than the actual weather, and this provides a clue to the link between suicides and restless schizophrenics.

In the case of suicides, Tromp believes that a sudden change of thermal input, up or down, can disturb the regulation mechanism of the potential suicide, and provide the trigger that tips the balance between tendency and action. He is not suggesting that any kind of weather conditions can *cause* suicide, since he has noted similar weather conditions on cluster days and on suicide-free days. A person must already be a potential suicide, for reasons unconnected with the weather, in a state where very little incentive is required to make an actual suicide attempt.

Breakdown of our heat regulation system in the hypothalamus has been suggested as a cause of restlessness among schoolchildren, also of road and industrial accidents. The hot and dry *foehn* wind of northern Switzerland is known to cause all kinds of distress, including migraine, mental depression, irritability, dizziness, decreased self-control and reaction speeds, leading to a rise in accidents and general crime, including suicide. Here another factor enters: that of air ionization, which we discuss in more detail at the end of this chapter, though the suddenness of the onset of the wind is also relevant here. Dr. Tromp believes that *foehn* conditions very probably increase both incidence and degree of any pathological condition. Less obvious conditions than these, it seems, may do the same.[12]

Seasonal variations in suicide rates continue to baffle sociologists. It is now generally agreed that rates for both suicide and attempted suicide peak in April and May, with a lesser peak around November. There is no generally accepted explanation for this. Professor Erwin Stengel, author of a recent survey,[13] admits that "seasonal increase of the suicide incidence is still mysterious." At first sight it might seem that suicide rates rise with rising temperatures, but no correlation has been found between hotter springs and early summers and increased suicide statistics. Nor is there any evidence that people are more suicide-prone in countries with extremes of hot and cold.

The French sociologist Emile Durkheim, author of one of the first modern studies of suicide, was convinced that temperature had nothing to do with it. The physical environment, he said, has no effect on the progression of suicide. "The latter depends on social conditions." It was all a matter of greater intensity of social life. This conclusion is particularly surprising, because Durkheim devoted a whole chapter of his book to "Suicide and Cosmic

Factors," amassing a good deal of statistics before deciding that there were no cosmic factors involved! He noted, for instance, that suicide rates vary enormously with latitude, being more than eight times higher between 50° and 55°N than between 36° and 43°, and exactly double the rate for latitudes beyond 55° Even more interesting was the discovery that suicide rates closely follow seasonal variations in the length of the day, and that far more suicides were committed during daylight hours than at night.[14]

Tromp's hypothesis that sudden changes in the weather can tip the balance from potential suicide to an actual attempt, successful or not, is entirely consistent with some of the facts Durkheim seemed to have ignored. (Though Durkheim deserves full credit for his thorough examination of the many factors that determine potential suicides.) Recent evidence, based on more accurate statistics than were available in Durkheim's day, suggest that weather can trigger potentially pathological conditions as well as such drastic acts as suicide.

Obviously, there are a great many factors besides the weather that can lead to somebody wanting to kill himself, and some of these factors are quite surprising. A Californian sociologist has discovered, for instance, that deaths on his state's excellent roads shoot up by an average 9 percent in the week after a suicide story has appeared on the front page of a local newspaper. Following the most celebrated suicide of the period under study, that of Japanese novelist Yukio Mishima, road deaths rose by 18 percent. Suicide, it appears, is infectious, and the suggestion that there is a "suicidal component" in motor vehicle accidents is worth taking seriously. To most drivers, it is already fairly obvious.[15]*

There are indications, however, based on apparently sound evidence, that sudden changes in the weather can have a number of distressing effects. A study carried out in three Dallas hospitals in 1946-1951 showed an increased frequency of acute myocardial (heart muscle tissue) infarcts, a common cause of heart failure, during periods of sudden weather change caused by inflows of polar or

---

* In Brazil, where road deaths are among the world's highest, a popular sport in the 1960s was "Paulista Roulette," in which young São Paulo drivers would tear through red lights at night. The consequences were frequently fatal.

tropical air masses. Later, a correlation was also found between weather changes and coronary artery disease, and again it was not the weather that did the damage but the sudden change in it.[16]

Sharp fluctuations in solar activity can apparently have the same effect. In 1959 Marcel Poumailloux, a physician, and R. Viart, a meteorologist, announced a very striking (*"vraiment impressionante"*) correlation between increase in heart failure, solar activity peaks, and points of geomagnetic agitation. They criticized the authors of the Dallas report mentioned above for confining their attention to the immediate weather and overlooking the "broader geophysical context." In their own report, the French researchers mention six specific periods in 1957 and 1958 (the year of greatest solar activity yet recorded) when infarct incidence rose exactly at times of sharp fluctuation in solar behaviour. They also recorded a drop in heart failures during a four-month period of relatively calm Sun. As to the possible mechanisms involved, they thought the sudden increase in blood clot formation must be the point of contact between Sun and man, noting that spontaneous and "apparently inexplicable" variations in clotting times were often reported even in normal patients. They do not mention the work of the Japanese scientist Maki Takata, which had then been going on for 20 years, revealing an unmistakable link between solar activity and certain properties of human blood, which we describe in more detail in Chapter 6.

The work of Poumailloux and Viart has made little general impact in the West, but it is fully supported by numerous reports from the Soviet Union and Eastern Europe. Surveys made in Irkutsk between 1956 and 1964, and a study of more than 1,500 cases in Kiev in 1966, revealed similarly strong correlations between solar activity via the geomagnetic field and certain heart and blood disorders. Nor are such effects confined to the sick: after 24,000 tests on a group of healthy people in the 20 to 40 age group, M. S. Kaibyshev found a link between the frequency of their heartbeats and fluctuations in the Earth's magnetic field: specifically, alterations in direction of its vertical component. In Poland, S. Panek studied a group of schoolchildren between 1955 and 1958 and found that even rate of growth could be correlated with geomagnetic

Electric Weather 117

activity, and there are also indications that women's menstrual cycles are influenced by it.[17]

There is no limit in sight to the number of ways in which weather and climate might be affecting us, consciously or otherwise, and therefore there may be any number of ways in which we are guided to some extent by external forces by having internal bodily functions triggered by them. It now appears extremely probable, at least to Soviet scientists, that the Sun affects our invisible environment and that this in turn affects all of us almost all of the time.

For centuries, country folk (and grandmothers) have been able to foresee changes in the weather by "feeling it in their bones." Recent research in Germany by Dr. Hartwig Schuldt (M.D.) lends considerable support to the idea that man is a living barometer. Moreover, he is a precognitive electromagnetic barometer. In a simple experiment, Dr. Schuldt has shown that the human body reacts to changes in the weather on an unconscious level. He asked his subjects to hold two metal electrodes, one in each hand, while a current from a small battery was passed through the circuit, so that conductance (or inverse resistance) could be read on a meter and noted down.

After making more than 10,000 such measurements, Dr. Schuldt noticed a remarkable fact: although fluctuations in individual subjects were widely variable, on days of rapidly changing weather, such as that caused by the approach of a cold or warm front, *all* subjects showed a sudden drop in conductance. He used some of his patients who had had limbs amputated for his experiments, and noticed that on such "weather-sensitive" days, they would complain of pains in their "phantom limbs." Discussing this experiment with Hill in 1977, Dr. Schuldt said that this was one of the most striking correlations he had ever seen. We already know that electromagnetic precursors of weather changes can be detected in advance: now it seems, by extending Schuldt's methods, it should be possible to record approaching weather changes and compensate for them, which might bring much relief to patients whose ailments react unfavourably to such changes.[18]

As for the actual mechanisms involved in these subtle interactions between man and his invisible environment, these have been a mystery until quite recently, when, largely as a result of the U.S. Navy's Project Sanguine (designed to find ways of communicating with submarines), much attention has been focused on a part of the EM spectrum previously generally ignored, or at least regarded as without any possible biological significance.

ELF Fields

If all EM and acoustic waves were visible, we would find ourselves in a jungle as apparently dense as any tropical rain forest. For the Earth's ionosphere is crammed with energy states of countless frequencies and intensities, and because the effects of such EM states are not normally perceived by our conscious minds, with the exception of that tiny band around the middle of the EM spectrum we call "visible light," we tend to forget about them or to assume they are harmless. As for acoustic or sound waves, to which we are relatively far more sensitive, there is plenty of evidence to indicate that they, like many forms of EM radiation, can be far from harmless.

The energies in the extremely low frequency range (ELF), both EM and acoustic, are of special interest to us because they come within the range that includes most of the known emission frequencies of parts of the human body, from just above zero to about 100 Hz. Because they have such long wavelengths, ELF waves are extremely penetrating and difficult to shield: they can travel round the circumference of the Earth without losing much energy. In other words, we cannot get away from them.

In the previous chapter we saw some of the ways in which ELF fields influence the behaviour of living beings. They have been linked to a number of effects, from emotional disturbance and psychiatric disorders, acute attacks of glaucoma, blood clotting, leucocyte counts, heart failures and changes in oxygen metabolism, to some extraordinary behaviour by bees, to be mentioned later. In Chapter 6, we shall see what can happen when impinging waves originating from external ELF fields are organized into regular cyclic behaviour; but first we must find out how they originate in nature as well as in our laboratories.

The question of how ELF fields are set up in the atmosphere has intrigued researchers ever since it was first raised in the 1950s. The principal natural source now

Electric Weather 119

seems almost certainly to be the weather, especially thunderstorms. (Waves from distant quasars also have ELF components, but it is not yet known how important these may be in relation to those that originate nearer home.) When lightning strikes, forms of EM waves or pulses called atmospherics (*sferics*) are produced, and while their frequencies extend well into the VLF band (up to 10 kHz.), they also show dominant ELF peaks at 30 to 100 Hz.

In 1952, W. O. Schumann of Munich University published an important paper on the mechanism by which ELF and VLF waves are set up in the space between Earth's surface and the ionosphere. This interspace constitutes a concave spherical cavity resonator—a conductive sphere surrounded by a dielectric—and Schumann found that when ELF wavelengths come close in length to that of the circumference of the Earth, a resonant system, which produces the "Schumann resonance," is set up. Power spectra of these resonances reveal amplitude peaks at 7.8, 14.1, 20.3, 26.4 and 32.5 Hz.: all within the ELF range.[19]

Herbert L. König, also of Munich, has found after making radioelectric measurements of natural ELF signals that they tend to vary from summer to winter, also within the 24-hour period, usually being strongest in early afternoon and weakest in the small hours of the morning. He also believes that inversion layers (cause of the horrendous smogs of Los Angeles, São Paulo and other cities) sometimes play a part in ELF field production, though some low-lying and electrically conductive air masses can act as a shield to ELF signals. Some ELF field fluctuation may also be caused by rain, and if we add to all these natural sources those produced artificially by power and telephone cables, or by defective household appliances, it is clear that we are seldom far from some kind of ELF EM field.[20]

As for what can happen to living beings in such fields, some of the evidence is quite alarming. In one German experiment, honey bees were confined to an area containing a field of 50 Hz. at 6,000 V/m. (volts per metre). The bees became not only very restless, but so aggressive that they began to attack each other. Broods inside the hive were destroyed, newly populated hives were abandoned in three days, while one whole hive inadvertently committed mass suicide by blocking the entrance to their hive—their normal defence against attackers from outside—and suffocated.

Nobody is suggesting that people are going to start behaving like this, but it is worth noting that the conditions described above are very similar to those existing underneath high-tension power lines. We recently learned of a group of people in southern England who became seriously upset after a power cable had been installed, running right over their rooftops. They complained of headaches, general irritation and even skin rashes, and were especially worried about possible harm to their children.

Another recent German experiment has shown that rats kept in an artificial 50 Hz. field at 15 kV/m. for fifty days had their heart beats slowed down under the ELF field influence, the effect building up with repeated exposure. The purpose of this test was to find out whether the stringent regulations in force in the U.S.S.R. regarding work near high voltage cables were justified. It seems they are.[21]

An enormous amount of research has been done into possible effects of ELF fields on human organs, and while most of this is too technical even to summarize here, it seems clear that our hearts and circulatory systems can be influenced in a number of ways, especially in disorders related to abnormal blood-clotting times, flocculation and sedimentation rates.[22]

When we turn to ELF acoustic or sound waves, however, there is no doubt at all as to the impressive physical effects these can have.

Infrasound

Professor Vladimir Gavreau, of France's Centre National de la Recherche Scientifique, became seriously interested in infrasound after being exposed to intense low-frequency acoustic waves issuing from a defective ventilator. "As a result," he recalls, "our ear drums were periodically compressed—a process that was both exceedingly painful and potentially dangerous." Everything in his laboratory would start vibrating like mad, and yet no actual sound could be heard from outside.

The human ear can only register sound, or acoustic frequencies, of about 20 to 20,000 Hz., any frequency below or above these limits being known as infrasonic or ultrasonic respectively. A curious feature of infrasound is that it can be highly directional, thereby apparently contradicting traditional laws of acoustics, which state that low frequency sound should propagate in all directions. Looking into this mystery, Gavreau found that infrasound could make a room of the right dimensions resonate in the ELF range, though not a smaller or larger room next door to it. So the potential danger lay not in the infrasounds themselves, but in the resonances they set up.

In the interest of science, Gavreau offered to serve as human guinea pig. Placing himself in the path of the offending ventilator, which was emitting infrasound at 7 Hz., he found that his head soon began to throb, making the most simple intellectual activity impossible. This frequency lies close to the border between alpha and theta brain rhythms, which are characterized by absence of conscious effort, and by trance or sleep. He was apparently undergoing a form of "bioentrainment," a subject to which we devote a whole chapter (8): that is, his brain was being temporarily driven by the external source that he could not consciously detect.

Gavreau found he could render the infrasound harmless by playing loud taped music, thus reducing his ears' sensitivity. He was worried, though, about the effects infrasound must be having on people living near large industrial plants: not the workers inside them, but people living near enough to receive infrasonic waves in their "quiet" homes, where ears would be more sensitive to them. He believes that these "hidden" infrasounds can produce fatigue, dizziness, nausea and irritation, and can help cause allergies and even nervous breakdowns. "Paradoxically," he comments, "workers in noisy factories are less exposed to these effects than their counterparts in neighbouring comfortable 'acoustically treated' houses: the apparent silence in such houses increases the sensitivity of the ear and infrasounds penetrate more easily to the semicircular canals and to the brain." The term *deafening silence* may be literally correct.

Investigating infrasound is a tricky business, as one of Gavreau's colleagues, R. Levavasseur, knows to his cost. He built a huge whistle with 1,000 times the acoustic output of a police whistle, and was bold enough to test it himself. "This," Gavreau notes, without going into details, "proved sufficient to make him a life-long invalid." Yet oddly enough, infrasound can also be beneficial. Gavreau found that a man who had lost his sense of smell recovered it after a blast at a whistle emitting sound at 155 decibels. (The limit of tolerable noise is about 80 dB, except for rock music addicts.) The intense vibration of his nasal cavities had somehow restored his lost sense. This method seems worth examining—but very carefully.

Other sinister inventions developed in Professor Gavreau's laboratory include a form of "acoustic gun" that emits a coherent beam of sound, the acoustic equivalent of a laser. Gavreau himself had a go with it, and afterwards became aware of a painful vibration inside his body. "Presumably, if the test had lasted longer than five minutes, internal haemorrhage would have occurred," he notes calmly. However, he goes on, "unanimous and vociferous protests from other members of nearby laboratories have since put an end to further tests of this nature." We hope there are peaceful uses to which Gavreau's diabolical inventions can be put, such as fog dispersal, and we also hope he is still alive. He does not seem to have published anything lately.[23]

Like ELF electromagnetic waves, infrasound can travel great distances. When Krakatoa erupted in 1883, killing 36,000 people and destroying two-thirds of the island, infrasonic waves travelled an estimated three times round the world. The Lisbon earthquake of 1755, probably one of the strongest ever, caused ripples on lakes in Norway. Tornadoes and magnetic storms are also thought to give rise to infrasound waves, and it has been suggested that the Earth is being continuously bombarded by them: a curious feature being that such waves seem to come from different directions according to the time of day.[24] It is also thought that infrasound can bring on attacks of general depression, fear and foreboding. Animals are more sensitive than man to such "sounds," hence the numerous reports of "fight or flight" reaction prior to earthquakes that could be caused by advance waves inaudible to man. (We give some examples in the following chapter.)

Michael Persinger, a Canadian scientist who is an authority on ELF field effects and is also keenly interested in phenomena commonly named "paranormal," reckons that many so-called precognitive experiences can be attributed to unconscious perception of ELF sound waves. In a survey of 128 cases he classified as precognitive, he found that 23 percent involved a situation in which the person concerned apparently saved his life by "following a hunch," and in more than 60 percent of these, the precognized event took place within seconds or a few minutes after avoiding action had been taken.[25] This might help explain numerous "psychic" stories of miraculous escapes from falling bombs during World War II: even the British prime minister, Winston Churchill, is said to have ordered a room cleared just before a bomb fell on it. The real mystery is why animals and humans interpret such advance signals as they apparently do, and yet again we are reminded how little we know about the workings of the human "mind," if behaviourist psychologists will pardon the word.

Infrasound is still largely a mystery. Experiments have shown that rats will learn to select levers that put a stop to low-frequency stimulation, and Persinger points out that inherent properties of such stimuli are enough to cause an "avoidance response," though why this should be so is far from clear.[26] It has been suggested, by University of Illinois researchers, that infrasonic waves generated by storms can even affect the behaviour of people far from the storm area. Examining records of 1,500 primary schoolchildren and 1,000 accident insurance claims, they found that both accidents and school absentee rates rose when there was a severe storm some distance from Chicago, while the city itself enjoyed normal weather.[27]

The most commonly experienced adverse effect of a low-frequency oscillation is seasickness, which is usually caused by the roll of a ship at extremely low frequencies. One can also be "seasick" on land. During the construction of the new Forth Bridge in Scotland, strong winds caused one of the towers to oscillate at an estimated wavelength of just over one metre at a frequency of 0.22 Hz. Workers on the tower felt so ill that they had to stop work, an essential safety precaution since one of the most immediate effects of "whole body vibration" is blurring of the vision. Human tolerance to such vibration, incidentally, is reckoned to be lowest in the 4 to 8 Hz. band.[28]*

Anybody who lives or works in a building high enough to sway appreciably and suffers from an inexplicable ailment might find it worthwhile to examine his surroundings

---

* See Fig. 4 (p. 73).

from the ELF acoustic (or EM) aspect. It might help to check all schoolrooms and university lecture halls for the presence of ELF waves, since a susceptible student might find it physically impossible to concentrate or even stay awake on the receiving end of such vibrations.

Each of us has suffered from what we assume to be acoustic pollution. Playfair well remembers struggling to keep awake during lectures on Russian history given in a physics laboratory in the days before Cambridge University gave the Slavonic faculty its own building. It was not the fault of the lecturer, who addressed the same group in another room on a different day. It was not the fault of Playfair, who kept awake during all his other lectures. One wonders what this kind of thing may be doing to overall educational standards.

Hill has also suffered repeatedly and painfully from the effects of what appears to be an unwanted resonance. His apartment is on the fifth floor of a six-storey building, directly over a filling station. When a large truck drives in, leaving its engine idling while waiting in line, he often feels a sharp pain in the back of the head, presumably caused by the resonating steel frame of the building and subsequent acoustic waves inside it. Something like this may be happening to many urban dwellers, whose headaches are all the more distressing for having no identifiable cause.

A California woman, whose case was investigated at length by an electronics engineer, was made physically ill and almost driven "mad" because of her extraordinary hypersensitivity to EM waves, which she could actually *hear*. It is interesting to note that so-called lunatics often complain of noise inaudible to others. Some of these unfortunate outcasts may be just as sane as anybody else, only far more sensitive to their polluted sound or EM environment.[29]

Audible sound can be just as harmful as infrasound, though we all develop resistance to it to some extent. The Mabaan tribe of Sudan have no firearms or drums. Nor do they have deafness, hypertension or cardiovascular diseases, provided they stay in their homelands. When they move to Khartoum they tend to develop all these ailments, and a study of them has indicated that the noise factor is as important as those of general stress, diet or change in social environment.[30]

Noise is more than just annoying. Prolonged exposure

to it can significantly alter blood cholesterol and plasma cortisol levels. Soviet researchers have linked excessive industrial noise with high incidence among workers of circulatory, digestive, metabolic, neurological and psychiatric problems. It can also bring about changes in the skin's electrical potential, the skeletal tension of muscles, and gastro-intestinal motility.

It can also, of course, make you deaf; by gradually raising the hearing threshold, which is the lowest level of noise the ear can register. Cinema managers in northern English industrial cities turn up their amplifiers higher than those in other parts of the country, presumably because most of their patrons have lost their ability to hear soundtracks reproduced at levels acceptable to a London audience. This is reliably reported as a demonstrated fact.[31]

Yet noise can be put to good use, the most offbeat of which in our experience comes from Tibet, where apparently it is, or has been, used to raise blocks of stone! We would hesitate to mention this but for the remarkably detailed evidence provided by the late Henry Kjellson, one of the pioneers of Sweden's aircraft industry, who has left a very precise description of how Tibetan monks build walls on high rocky ledges. It is based on first-hand evidence, and we have also been able to obtain his original drawing of the event, which is reproduced here, we believe, for the first time outside Scandinavia. According to Mr. Kjellson, here is what happened:

Blocks of stone measuring 1.5 metres square were hauled up to a plateau by yaks, and placed over a specially dug bowl-shaped hole one metre in diameter and 15 cm. deep. The hole was 100 metres from the sheer rock wall on top of which the building, presumably a hermitage of some sort, was to be built. Sixty-three metres back from the stone there stood nineteen musicians, spaced at five-degree intervals to form a quarter-circle, in groups, as clearly shown on Kjellson's drawing. Measurements were taken extremely carefully, using a knotted leather thong.

Behind the musicians, about 200 priests arranged themselves so that about ten stood behind each musician. The instruments involved were drums and trumpets of various sizes. (Kjellson gives the exact dimensions of the 13 drums and six trumpets that made up this unusual orchestra.)

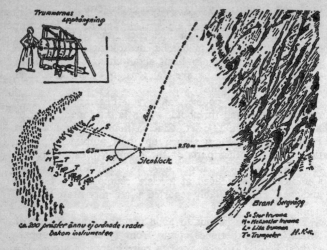

Fig. 6
Monastery construction, Tibetan style, according to Swedish aircraft designer Henry Kjellson. The steep mountain side is on the right. In the centre is the stone block, and on the left are the priests and musicians. S = big drum, M = medium drum, L = small drum, T = trumpeter. Inset shows method of suspending drum, and gives an idea of its size. As shown here, Kjellson says, the 200 priests are waiting to take up their positions in straight lines of 8 or 10 behind the instruments, "like spokes in a wheel." Unlikely as it may seem, this operation has an intriguing precision, made slightly more credible by Kjellson's meticulously detailed description. (Courtesy of L. Kjellson).

Then, at the command of the chief priest, the music began. The beat was set by a gigantic drum weighing 150 kilos and slung from a specially built frame so that it was off the ground. (See inset in illustration.) Two monks took turns at each trumpet, blowing a total of two blasts per minute. All six trumpets were pointed towards the stone on its launching pad, and after about four minutes of what must have been indescribable racket (since the meticulous Kjellson fails to describe it), the stone rose into the air, wobbled slightly, and then, as the noise from trumpets, drums and chanting priests increased, followed a precise parabolic course of some 400 metres up to the top of the cliff. In this way, we are told, five or six blocks were lifted in an hour, although some of them apparently broke upon landing. Accidents will happen even in the Tibetan construction industry.

Kjellson makes a bold attempt at an explanation for all this, based on the example of the German "whirlwind cannon" developed during World War II. This, like Gavreau's invention referred to above, was the acoustic equivalent of a laser, concentrating a narrow beam of intense sound energy strong enough to knock down a brick wall from a distance of 500 yards. By focusing such a device over a heavy stone, Kjellson thought that a low pressure zone could be created, so that the atmospheric pressure under the stone would do the lifting, which would explain the hollow cavity dug by the Tibetans, whose musicians and monks apparently formed a sort of human "acoustic laser."[32]

Kjellson not only provides far more technical information than one is usually given for paranormal events, but also claims that the whole operation was filmed by a fellow Swede who managed to shoot two different levitations from different angles. Should this film ever turn up (Kjellson's family knows nothing about it), it may help explain the mystery still surrounding the building of places such as Machu Picchu in Peru (which Playfair visited in 1963). We are still asked to believe that this astonishing city was built by no more than unlimited manpower and fibre ropes, though how one could haul 200-ton blocks of stone up a sheer 1,000-foot mountainside even today, and then carve them as if they were made of butter, is far from clear.

Implausible as it may seem, Kjellson's account of Tibetan building methods is no less plausible than any other theory of how cities such as Machu Picchu were built. It could be that the high altitude and consequent high ion content of the air had something to do with it, and to close this chapter we take a look at one of the invisible components of the substance in which we live: air.

Vitamins of the Air

There is nothing like a good breath of fresh air. It tastes clean and invigorating, gives us a healthy appetite and helps us sleep soundly. On the other hand, there is nothing worse than the kind of aerial soup many people live, work and choke in. Considering the importance of air, as vital to man as water to fishes, it is surprising how little is known about it.

From the days of the Phoenicians it has been known that some climates are better for us than others. The Romans used certain areas for rest and recuperation because of the quality of the air, and nobody can climb a mountain, stroll along a beach or sit by a waterfall without noticing this quality. Many factors may be involved, such as temperature, humidity, atmospheric pressure and wind—the most easily perceptible components of what we call climate; but there are other factors, of which one of the most important is the presence in the atmosphere of electrically charged condensation nuclei known as *ions*. (Or *air ions*, to distinguish them from ions of other substances.)

Air consists principally of oxygen, nitrogen and carbon dioxide molecules. These in turn consist of atoms, and like any other atom they have a positively charged nucleus orbited by negatively charged electrons. Most air molecules are electrically balanced, and thus neutral, but when one of them gains or loses an electron it becomes charged, negatively or positively (respectively). This process, called ionization, can be caused in many ways: by ultraviolet light from the Sun, radioactivity from the ground, rapidly moving water such as breaking surf or waterfalls, or lightning discharges. Air ions can now also be generated artificially in a number of ways.

Soon after Benjamin Franklin discovered lightning to be an electrical phenomenon (1752), people began to talk about the possibility of an influence of natural electricity on plants, animals and man. In 1777, the Lyons Academy offered a prize for the best paper on "illnesses caused by an excess or deficiency of electrical fluid in the body" and the means for curing such excesses. About this time, Coulomb showed that the atmosphere conducts electricity, however poorly (though he himself misunderstood the nature of this process), while Volta was one of the first to believe in the therapeutic value of atmospheric electricity.

It was not until the eve of the twentieth century that the conductivity of the atmosphere was explained, when Elster and Geitel and (independently) Wilson showed that both positive and negative carriers of electricity exist in air, and the name *ion* was given to such carriers. Yet at least a century before this, air ion therapy was being practised and recommended for a number of ailments, notably eye diseases. The Swedish chemist Jacob Berzelius (who also introduced mesmerism to Scandinavia), described electrical treatments in his 1802 doctoral dissertation, and the German naturalist-explorer Alexander von Humboldt investigated the effects of "galvanic electricity" on many illnesses. Such methods must have had some success, probably due to the attraction of ions from the surrounding air by those early static electricity generators. Once again, the twentieth-century researcher is led to wonder whether anything is really new.

One of the first scientists to carry out research into the newly-identified air ions was a remarkable Russian named Aleksandr L. Chizhevsky,* a versatile modern sleepwalker whose fertile but tragic career will be described in more detail in Chapters 6 and 10. Chizhevsky constructed probably the world's first mechanical ion generator in his Moscow laboratory in 1922, and for the next seventeen years he carried out an enormous amount of research into the effects of air ions on living systems.

Right from the start, it was evident to him that negative ions were good for his animals. Rats would respond after only a couple of hours in a negatively ionized area, showing considerably more sexual activity than a control group outside it. Newly-hatched chicks kept in the negative ion cage would start putting on weight faster than control groups, after only a few days.

Chizhevsky wondered what effects negative ions might have on disease, and initial tests with guinea pigs infected with tuberculosis gave encouraging results: after eight weeks the entire control group had died, while every one of the ionized animals was alive, well, and moreover free from TB. They were in fact healthier than they had been before the start of the test.

Convinced by now that negatively ionized air led to a lasting mobilization of the body's defence mechanisms, Chizhevsky decided to try it on human patients, after

---

* His name is often mistransliterated as Tchijevsky, Tchiyewski, etc., which may explain why it is sometimes hard to find in library catalogues.

claiming remarkable success with a circus chimpanzee, curing it in four months of TB with apparently nothing but daily doses of ions. In 1928 he set up a special ionization chamber, where he gave treatment to 130 TB patients, finding that it improved their overall health and helped them put on weight. (It must be borne in mind that in experiments of this kind involving human patients, there are often inadequate control groups, for a humane scientist will be reluctant to deny anybody a treatment he sincerely believes to be effective. Paradoxically, this sometimes leads to the treatment being overlooked, even if in fact it is beneficial.)

Chizhevsky was never sure exactly why ionization was good for people. He speculated that when we breathe negative ions, we absorb them at once into the bloodstream and set up a current which, in turn, produces other forms of energy, chiefly thermal, mechanical and chemical, thereby giving the immune system a general boost.[33]

Much work on ions was done by many scientists before World War II, and many conflicting claims were put forward. Not all early researchers used statistical or control methods that would be acceptable today: prewar research is of more interest to us when it has been proved to be repeatable, which has not always been the case.

A great deal of modern ion research has been prompted by the weather, especially certain types of hot and dry wind that plague people in many parts of the world. Earlier, we mentioned the Swiss *foehn*. The French have the *mistral*, Californians the *Santa Ana*, Canadians the *Chinook*, northern Italians the *tramontana*, and Israelis the *sharav* or *khamsin*. All mean much the same thing: trouble. These winds cause insomnia, depression, hypertension, breathing problems, allergic reactions and notable increases in accident and suicide rates. Chizhevsky once suggested that courts should take such winds into consideration when sentencing law-breakers under their influence, and we understand that Israeli judges are doing so: the *sharav* being a major problem in Israel, where it causes illness of some kind in 30 percent of those exposed to it.

At the Hebrew University in Jerusalem, Professor F. G. Sulman studied a group of 800 *khamsin* sufferers over a four-year period. He found an unusual amount of a stress hormone called serotonin in their blood, possibly due to the high positive ion content in the air before the onset of the wind. He found he could reproduce most of the wind's effects artificially, and that negative ion treatment brought patients' conditions back to normal. Also in Jerusalem, doctors at Bikur Holim hospital found that negative ions, without any supporting treatment, even antibiotics, terminated spastic conditions in infants suffering from asthmatic bronchitis after much less time than required for conventional treatment. This is perhaps because the ions cause the offending substance (allergen) to precipitate.[34]

Ions, both positive and negative, affect different people in different ways, as do many other things such as tea and coffee, but in general it can be said that positive ions tend to produce ill effects while negative ions are good for us, most noticeably in relieving hay fever, asthma and various allergies. They also seem to stabilize blood pressure, improve work capacity, learning ability and reaction time, and help combat even serious ailments.

Many offices, schools and factories now use ionizers, and they have even been used to calm restless racehorses. In Budapest, a survey of positive ion levels in the streets led to the city police fitting mini-ionizers in their patrol cars. Negative ions have also been found useful in treating burns, and even in stopping people snoring. Dr. E. R. Holiday has called them "vitamins of the air," and he reckons that biochemistry is now discovering the value of air in the way it discovered food 70 years ago, when Sir Frederick Gowland Hopkins discovered the first vitamin and called it an "accessory food factor." Air ions, Dr. Holiday thinks, may be accessory metabolic factors.[35]

Some scientists are reluctant to believe that ions do anything more than just keep the air clean by causing dust or pollen particles to precipitate. Dr. Albert P. Krueger of the University of California, however, believes there is a direct link between air ions and the metabolism of the body, at least of animals. Tests with mice have revealed that high or low concentrations of positive ions raise or lower blood levels of serotonin (5HT), a versatile neurohormone which can induce profound neurovascular, endocrinal and metabolic effects throughout the body. A high positive ion count, Krueger says, increases both production and renal secretion of serotonin in rats, and he points out that a charge-related metabolic response also occurs in humans. We can never say that an effect observed in animals must also occur in humans, but there is always a possibility that it does, and the indications are that the ions we breathe have considerable influence on our general state. Dr. Krueger is disturbed by the fact that many people spend much of their lives deprived of natural air ions, especially city dwellers and factory workers. "The ultimate dimensions of biological changes produced by air ion loss," he warns, "may prove to be as disenchanting as some revealed by Rachel Carson in *Silent Spring*."[36]

Chizhevsky believed that the influence of ions could be transmitted directly through the blood stream. In one of his experiments, he connected the blood systems of two animals, and found that when one only was exposed to atmospheric ions, the other showed the same reaction although it had not breathed the same air.[37]

Dr. Igho H. Kornblueh, the pioneer of ion therapy who died in 1973, suggested in 1957 that the effects of ionization on brain rhythms should be investigated, and it is now established that through it, human alpha waves can be reduced in frequency and increased up to 20 percent in amplitude.[38] Artificial ions have also been found to be good for sportsmen, increasing body capacity and normalizing the metabolic balance of certain vitamins.[39]

A simple but dramatic way of demonstrating the potential importance of ion control is by a so-called reaction test. A common way of finding out how well the nervous system is working is to measure the time-lag between application of a stimulus and the motor response. This time-lag is called the reaction time, and its importance is obvious in any situation where rapid action is called for, such as driving cars, piloting aircraft, working with dangerous machinery, and many forms of sport.

At the 1953 automobile trade fair in Munich, an enterprising local scientist built a machine that enabled visitors to see how they would rate in an emergency traffic situation. All they had to do was jam on the brakes when a warning light flashed, and their reaction times could then be measured, indicating how far they would have travelled before stopping a real car. The machine was a great success, and similar devices are now widely used in driving schools and road safety institutions. In Munich, it was found after thousands of tests that drivers' reaction times could be favourably influenced when negative ions were beamed at them, whereas positive ions made them feel sleepy, inevitably lengthening reaction times. Many motorists now use mini-ionizers, and there are grounds for suggesting they should be made as compulsory as seat belts.[40]

Another accessory that might interest motorists is the portable field intensity meter invented in 1956 by Bulgarian-born Dr. Cristjo Cristofv, who discovered that certain types of plastic seat cover, especially polyethylene, can develop extremely high local negative electric fields of up to 10,000 volts per meter (V/m.), as a result of friction against passengers' clothing. For reasons not entirely understood, the human body reacts to the absence of the natural positive electric field of the Earth, which measures an average 500 to 800 V/m. This may be because a positive field attracts negative ions, which have been shown to have a number of beneficial effects, whereas a negative field will not. Since interiors of cars, aircraft and submarines are spaces largely shielded from the natural environmental electric field because of their metallic structures, they serve as Faraday cages to some extent, drastically altering the EM environment. Numerous tests, some dating back to the nineteenth century, have shown that negative fields sap the vitality of animals, induce fear, cause loss of appetite and reduce fertility, whereas positive fields improve respiration, digestion and general metabolism.

The U.S. Air Force has carried out tests with strong positive fields and found that they help considerably to improve pilots' general alertness and combat fatigue. The addition of a 1,000 V/m. positive field to a small enclosed and shielded space can not only reverse effects of fatigue but raise a human subject's overall performance well above its normal level. Dr. Cristofv has developed a fist-sized "anti-fatigue device" which works off a car battery and generates 3,000 V/m. of positive electric field. Here is yet another gadget that might help save lives on the road and in the air.[41]

This is no exaggeration. On 28 May 1971 three aeroplanes of the South African Air Force crashed into Table Mountain near Cape Town. On 25 August 1975 four F-104 aircraft of the Italian Air Force flew straight into a mountain in Luxembourg. There was no explanation for either disaster. Nor has there been a satisfactory explanation for the large number of Starfighters lost by the West Germans, not to mention all those Bermuda Triangle disappearances.

Dr. Gültekin Caymaz, the Turkish doctor whom we mentioned in Chapter 2 in connection with the mysterious death of the Soyuz XI crew, is convinced that sudden increases in atmospheric electricity can and do cause accidents. Immediately after the Soyuz XI disaster, he undertook his own investigation with the co-operation of the Turkish Air Force. First, he talked to pilots, and learned at once that three of them had had the experience of seeing a fellow pilot flying in formation suddenly drop his head downwards and crash to his death. Three more unexplained accidents.

As it happened, Dr. Caymaz had been very interested in air electricity even before the Soyuz affair. In 1970, he began a series of daily measurements of atmospheric potentials that went on for 15 months. He noticed that whenever there was a sudden high positive voltage, there would be an air crash or a road accident, and often also an increase in haemorrhages among TB patients in the 1,000-bed Ataturk Sanatorium in Ankara where he worked as a specialist in physical medicine. He set about looking for ways in which air electricity could be related to all this fatal crashing and dangerous internal bleeding. A logical start was to observe the effects of aircraft interiors on blood electrolytes (substances that become electrically conducting in aqueous solutions or in blood, such as sodium and potassium). Taking blood samples from six soldiers before and after a one-hour flight, he found an increase in sodium of up to 20 percent, also increases in cholesterol and blood alkalinity (pH), and a decrease in potassium content. The results, though far from uniform, showed beyond doubt that some human beings are significantly physically affected by flying in aircraft. Further tests revealed, as Dr. Caymaz had expected, that atmospheric electricity readings were considerably higher inside aircraft cabins than outside. (Astronaut James Irwin has said that the heart attack he suffered after his trip to the Moon in Apollo 15 may have been due partly to potassium deficiency.)

A simple experiment with a plane on the ground showed that the airfield's atmospheric voltage reading of 6 mV (positive) rose to 12 mV inside the aircraft, reaching 16 mV when the door was closed. Dr. Caymaz was alarmed to find readings close to those that he had previously associated with road accidents and lung bleedings, and at a NATO international conference on aircrew fitness he made the following suggestions:

1. Atmospheric electricity should be measured every day, and the readings broadcast on public radio. People would then be warned that their reflexes would be slower and should drive or fly with extra care.

2. Whenever he flew, Dr. Caymaz said, he took potassium chloride, calcium, magnesium and vitamin C pills to offset possible ill effects of variations in electricity. He recommended pilots to take potassium salts whenever electricity readings were high.

Drivers of heavy vehicles usually attach ground wires to them to drain off excess electric charge into the ground. Car owners might try doing the same for long journeys, though obviously pilots cannot do the same.

Turning from the sky to the ground, Dr. Caymaz noticed that after each of the 90 peaks of positive electricity in the 15-month period of his observations, there would be some kind of disaster. It could involve humans, or it could be a natural catastrophe, such as an earthquake. (We examine these in the following chapter.) On the whole, it seems that atmospheric electricity readings might well be included in our daily radio and TV weather forecasts. One reason why this has not been done is that there are many different ways of measuring natural electricity. (Caymaz's method was entirely different from that of Cristofv, and probably very accurate, since he used a Philips DC millivoltmeter with special filters to remove artificial electrical signals.)[42]

Here is a new approach to accident prevention that needs further study. We cannot assume that any stimulus from the electromagnetic environment passes unnoticed just because we are not consciously aware of it. Another medical researcher who has been looking into aircraft accidents from this angle is Dr. E. Stanton Maxey, the surgeon mentioned in Chapter 3, who has logged several thousand hours piloting his own plane. Maxey points out that the "microclimate" of the aircraft cockpit has received little attention even though factors known to influence it, such as air ionization and modulations in natural EM fields, have been widely researched.

Maxey himself has shown, as we have described, that it is possible to alter the rhythm of a brain by beaming extremely low frequency magnetic waves at it. This, he thinks, may be what is happening at the time of some air crashes later ascribed to "pilot error." The electrical component of the aerial EM environment is shielded from the pilot by the metal of the aircraft, but the magnetic component is not. Since certain types of weather are known to produce ELF EM waves, it seems unquestionable that all pilots are running the risk of having their brains—which are more valuable than any of the instruments in front of them—interfered with by natural forces. Maxey, who holds a number of patents for his aircraft blind landing system, has designed apparatus for recording magnetic conditions inside the cabin and automatically generating magnetic waves to compensate for fluctuations in them and guarantee a stable climate for the pilot. Here is a good example of an interdisciplinarily-minded biometeorologist putting his wide experience of medicine, electronics, aircraft construction and human states of consciousness into practice and helping save lives.[43]

Just how sensitive some people are to ultra-weak environmental stimuli has been shown in a series of tests conducted by Dr. Andrija Puharich with the late Eileen Garrett, who was one of the few "mediums" or "psychics" to welcome scientific study of her strange telepathic and clairvoyant talents, once even commissioning such an investigation herself. In one test Mrs. Garrett was asked to call out when she thought that a random generator, located a third of a mile away and activated by bursts of cosmic rays, was in operation. Although sitting inside a double Faraday cage, she managed to do this to the extent that there was only one chance in a million that her calls were fortuitous. This suggests that she was responding directly to the cosmic rays themselves, for there was no known normal way she could have perceived the field of the generator at such a distance, even without two thicknesses of screening in the way. If sensitivity of this level is possible in humans, may there not be a million other ways in which we are constantly responding to stimuli of infinitely remote origin? Until we have better instruments, more subjects as co-operative as Mrs. Garrett, and a great deal more money for research, we may never know.[44]

In this chapter, we looked at just a few of the ways in which man responds to some of the invisible components of his environment. In the following chapter, we turn to earthquakes. The connection between these and our brains may not be an obvious one, yet earthquakes concern us here as examples of the invisible forces of nature at their most destructively violent and, until recently, unpredictable. By learning more about such events, we may well learn more about ourselves.

We are not suggesting that people are likely to explode, unless epilepsy or cerebral haemorrhages can be regarded as "brainquakes," but it could well be that the forces at work in the timing and triggering of earthquakes are the same forces that influence living systems throughout their lifespans, in ways we may not realize. Earthquakes also interest us because while much of this book is speculative, there is no doubt that earthquakes happen.

By way of a footnote to this chapter, we might point out that although people do not erupt like volcanoes, they have been known to combust spontaneously and be reduced to a pile of ashes. This happens more often than many people realize, and there is an impressive array of evidence for this most bizarre of all manifestations of invisible force, including testimony from firemen, policemen, pathologists, coroners, and even the FBI. There is also plenty of horrific photographic evidence on police files. Official verdicts in such cases, in which people simply start burning for no obvious reason, are often more inherently implausible than the events themselves, which remain unexplained.[45]

From one example alone, we can suspect the role of external or cosmic forces. On 7 April 1938 three people were charred to death at almost exactly the same time, no normal cause of death yet having been suggested for any of them. One was a truck driver near Chester (England), another was a motorist near Nijmegen in Holland, and the third was the helmsman of a ship off the coast of southern Eire. An added curiosity is the fact that when plotted on a map and connected by straight lines, the three death sites form an isosceles triangle with an apex angle of 120°. Given the rarity of this phenomenon, the fact that three took place at the same time, two equidistant from the third, raises the odds against chance to galactic proportions.

The forces of nature, as we have said before, should not be taken for granted.

Chapter 5

EARTHQUAKE

SORRY, BUT quite a lot of California is going to slide into the sea in 1982, plus or minus a year or two.

This is not our prediction, but that of two astrophysicists, John Gribbin and Stephen Plagemann, who have laid their reputations on the line in a way few professionals have ever dared in the past. They are not merely stating that earthquakes *can* be triggered by planetary positions, but have even named the place and date where they *will* be. Moreover, the original prediction was made and published more than ten years before the presumed event, and no statements could possibly be less equivocal than these:

> By disturbing the equilibrium of the Sun, which in turn disturbs the whole Earth, the planets can trigger earthquakes . . . (In 1982) the Los Angeles region of the San Andreas fault will be subjected to the most massive earthquake known in the populated regions of the Earth in this century . . . Los Angeles will be destroyed.[1]

Credit for the original prediction is due to Gribbin, who stated in 1971 that the fault might be triggered in the late 1970s or early 1980s, shortly after the next peak of solar activity, now expected between 1982 and 1984. If, he argued, there really is a connection between solar activity and the Earth's rotation rate (as we mentioned in Chapter 1), then it would be plausible to expect that "the strains in the Earth's interior that would arise could trigger regions of (geologic) instability into earthquake activity."[2]

Such a region of instability is the San Andreas fault, which runs down the coastline of California marking the boundary line of the North American and the Pacific Plates, two of the largest of the 15 or so "lithosphere plates" that make up the Earth's outer crust. This, as the exciting new science of plate tectonics has shown, is not a continuous solid chunk of rock, but a number of separate areas, rather like bits of shell that have been peeled from a hardboiled egg and then stuck back on again.

Why 1982? By this year, the celestial bodies will have moved into what has been rather loosely termed a "Grand Planetary Alignment," which means that all the planets will be in the same segment of sky for a time, each of them moving through a series of conjunctions with one or more of the others. (See Fig. 7.) This happens only every 180 years or so because of the length of the outer planets' periods, and Gribbin predicts that this will stir up the Sun in such a way as to shift its centre of gravity in relation to the solar system centre of mass. This, combined with the fact that 1982 will probably be a year of maximum solar activity, is expected to lead to a period of enhanced flare output that will cause very powerful strains to build up in the Earth's interior. Something will have to give, and that something is one of the very few major fault areas long overdue for a major displacement. So, goodbye California.

Californians are getting rather tired of such prophecies of cataclysm, too many of which have been proved wrong in the past. Only four percent of their property in the Los Angeles area is insured against quake damage, and threats have even been made to sue makers of predictions in future. Nothing hits real estate values harder than a well-publicized earthquake forecast, and the National Science Foundation has been obliged to issue a policy statement on the question, in which it recommends preparing for the next disaster rather than predicting it. "When an earthquake is predicted for a region, its tax base starts to collapse," one of its members has said. The fact that its physical base may also be about to collapse seems to bother him less.[3]

Even so, the Gribbin-Plagemann prediction is good news, except of course for Californians. It has shown that major and specific predictions of natural events can now be made based on plentiful and consistent evidence. It may be remembered as the occasion upon which astronomers, instead of being "so poor" as John Eddy put it (see Chapter

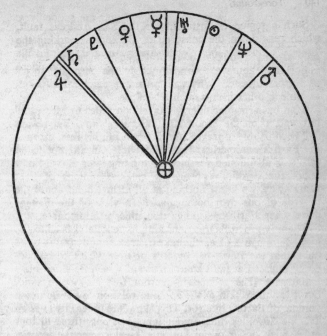

Fig. 7
Much misunderstanding has arisen concerning the so-called Grand Planetary Alignment due in the 1980s. At no time will every planet be in a straight line, but on more than one occasion they will all be bunched together in one segment of sky. Here is how they will be arranged at the beginning of December 1980. Jupiter and Saturn will be in conjunction, with Jupiter exactly square to Mars on 1 December. In November 1982 the planets will be even closer together, in a 64° segment of sky. This alignment may coincide with maximum solar activity.

2) about answering questions the public most need to ask, finally stood up to be counted.

No natural event is more frightening than an earthquake. One of us (Playfair) was sound asleep at 3:03 a.m. on 31 May 1935, when a 30-second upheaval shook much of the city of Quetta (now in Pakistan) to pieces, killing at least 30,000 people—a thousand deaths per second. Had he been living a couple of miles to the south-west, he would not be writing this book. Every aspect of an earthquake is terrifying. Your whole world is destroyed, you are suddenly at the mercy of an unimaginably hostile environment, and you can never live in peace in your own house again. Not only are major quakes often followed by aftershocks, but major fault zones are condemned to the perpetual threat of further disaster. Why can we not predict earthquakes, as we can other events such as eclipses?

Man has sent machinery 200 million miles to Mars, and has walked on the Moon 200,000 miles away. Yet he has still been unable to travel 20 miles beneath his feet, through the Earth's outer crust; Project Mohole, an attempt to do this, was abandoned for lack of funding. For all we know about the workings of distant galaxies billions of light years away, we know nothing at all at first hand about the innards of our own home planet. In view of the damage done when Earth has indigestion, this is unfortunate.

There is general agreement among geoscientists as to the immediate cause of earthquakes: stresses and strains build up within the Earth for months or even years, to be suddenly released at fault zones, most of which lie along plate boundaries. The ground under our feet may seem solid, but it is a mere skin of the apple in relation to the immense volume of Earth's interior. This skin (the outer crust) is an average 20 miles thick under land, but only three to four miles thick under ocean bottoms. Next comes the mantle, some 2,000 miles of solid crystalline mass, and finally the core, the outer part of which is thought to be liquid. Yet, solid as most of it may seem, the interior of the Earth is far from quiet. The "soft" upper layer of the mantle is almost liquid: it consists of rock, metals and magma under tremendous pressures and at very high temperatures. Moreover, there are even tides in the "solid" crust, diurnal displacements of up to 9 inches, which like any other tides are governed by the celestial bodies. These may prove an important clue in the search for the timing mechanism of earthquakes, an understanding of which is highly desirable.

Nature often warns us in advance when an earthquake is about to take place, but until very recently, we have not known what to listen for. The 1935 Quetta quake was very well documented (it is the subject of an excellent full-length book), and when it was all over, survivors exchanged "premonition" stories, many of which were similar to others reported elsewhere. Birds had been seen hopping about on the ground the previous evening, refusing to perch in trees. A British Army officer's dog whined so insistently an hour before the quake that it kept its master awake and probably saved his life. A six-month-old baby, who had never cried at night, yelled and screamed until his parents took him into their bed just half an hour before a falling roof beam squashed his cot flat. His parents' bed was untouched.

Four hours before the disaster, Betty Montgomery (wife of the future Field Marshal Viscount Montgomery of Alamein) was doing some embroidery after dinner when, at about 11 p.m., she suddenly noticed that every strand in the skein of silk she was holding had separated from the rest and gone stiff. "Monty" had no explanation for this, though a dinner guest suggested the electricity of the air must be responsible. (He perhaps remembered that Benjamin Franklin had noted a similar effect on his kite string on the day in 1752 he discovered the electrical nature of lightning.) There were also some well attested cases of precognitive dreaming of the earthquake, while one man kept bringing up the subject of earthquakes during the week before it, repeatedly telling friends he was sure a big one was imminent.[4]

A number of interesting biological and geological phenomena have been observed prior to earthquakes. In Japan, freshwater catfish have a habit of rising to the surface and leaping into the air at such times, and in 1932 a Japanese scientist discovered that this fish has a hard bony outer covering instead of scales, which seems to be able to pick up acoustic and mechanical vibrations and perhaps also EM waves. A piezoelectric effect, mentioned in Chapter 3 in connection with keratin in bird feathers, may be at work here. There are certainly many ways in which Japanese fish appear to feel earthquakes coming. Before shocks, deep-sea fish have been caught near the surface, and sardines have been found in only 20 cm. of water with five times their average food intake inside them. Octopuses have been seen coming ashore "as if drunk," as one scientist puts it, and odd behaviour has also been noted in lobsters, cuttlefish, eels, loaches, pike and mackerel. A correlation has been claimed between daily fish catch totals and subsequent seismic activity, suggesting that fish do their best to flee from what they interpret as a signal of impending danger.

Nature, it seems, does her best to warn us when an earthquake is on the way. In Japan, miners can feel the presence of *chiki* (literally: "air from the Earth") as it rises from below just before a shock. Rats leave not only sinking ships, but also areas about to be seismically disturbed. Well water levels drop, rainbows appear, and magnets have been said to lose their power. Lunatics seem particularly sensitive to oncoming shocks, while shortly before one quake in the town of Edo, an epileptic woman was arrested and expelled —for predicting an earthquake![5]

People, on the whole, do not have a good record as predictors. California, which must have the most self-proclaimed psychics per capita of any location in the world, has yet to produce a reliable earthquake forecaster.* For the time being, it seems, we can rely more on animals. In India, for instance, cows have been observed running around before a quake with their tails in the air, in a typical "fight or flight" reaction state.[6] An interesting eye-witness account of strange animal behaviour comes from the nineteenth-century British naval meteorologist Robert Fitzroy, who found himself in the town of Concepción, Chile, one day in 1835. At 10 a.m., swarms of screaming seabirds darkened the sky overhead. At 11:30 dogs ran out of the houses, and ten minutes later an earthquake totally destroyed the town. More recently, in 1954, inhabitants of the Greek village of Sofiades were saved from possible death when the sudden panic of a flight of storks was interpreted (correctly) as a sign of an impending tremor. According to author Frank W. Lane, whose *The Elements Rage* includes fascinating examples of extremes of natural violence of all kinds, pheasants seem particularly sensitive to advance signals of earthquakes, while there is a general

---

* We do not rule out possible future successful prediction by hypersensitive persons, once the mechanisms of precognition are understood and ways found to develop this strange faculty, which undoubtedly exists. In fairness to California's psychics, we must add that some have better records than others. Writer David St. Clair, an authority on that state's paranormal scene, was told by psychic Clarissa Bernhardt in mid-1976 that she saw prolonged seismic activity in southern California between April and December of 1978, with much of the land between Los Angeles and San Diego underwater by the end of February 1979. This information was given to Playfair in July 1977. (Personal communication from David St. Clair.)

tendency for birds and insects to cease their normal activities.*

If birds, fish, animals, babies and lunatics can predict earthquakes, cannot a well-equipped scientist do the same? Up to a point, he can. Known precursors include changes in ground water levels and in the resistivity of the Earth, gravimetric changes, enhanced aurorae, and changes in piezoelectric surface voltages of up to 10,000 V in rock zones containing large amounts of quartz. The resistivity of the Earth can be measured continuously in seismically sensitive areas—abrupt decreases of up to 15 percent prior to quakes have been noted, as well as increases after them.

Geoelectric and geomagnetic field readings have also shown sharp changes just before tremors. Magnetic changes are usually of less than 15 gamma—a tiny but detectable amount—though they can be higher. When the Alaska earthquake of March 1964 erupted, killing 114 people and wrecking the town of Kodiak, it so happened that a magnetometer was running in a laboratory not far from the epicentre. It stopped running when the quake knocked out the power supply, but the undamaged chart showed a very clear and sudden increase in geomagnetic field intensity of 100 gamma just over an hour before the event. This may have been a signal from rocks undergoing a change in stress, but whatever the cause it showed that "magnetic monitoring" near known fault areas is feasible, as evidence from an earlier quake in Nevada also indicates.[7]

Geoelectric phenomena are primarily electrolytic in nature: that is, they depend on water and the ions and other charge-carriers in it, since most rocks are essentially electrical insulators. Carried through porous rocks, or forced through small channels and cracks between rock masses, such telluric currents flow all round the Earth's crust, and may be linked to ionospheric currents on a world-wide basis.

Another early warning sign of seismic activity is the presence of a surplus of the radioactive gas radon in water above a known stress area. Radon is very unstable, with a half-life of only 3.5 days, and detection of such surpluses is difficult, but possible. Taking all known physical earth-

---

* Newton Abbot: David & Charles, 1966. Page 181.

quake precursors into consideration, it has now been established that the stronger the shock, the longer the interval between precursor and shock will be. Precursors of an event measuring 6 on the Richter scale may be sent out nearly three *years* (1,000 days) beforehand; for a magnitude 5 event the interval is about 100 days, and for a magnitude 4 event around ten days, showing an almost linear relationship.

Other forms of earthquake prediction verge on the bizarre. For instance, a Russian who discovers a snake frozen to death will fear the worst, for snakes will apparently leave their warm holes even in winter, preferring to freeze stiff rather than go underground again prior to a quake. Sudden migrations of rats, cats, monkeys and even donkeys have been reported. The world's first seismograph —Chinese, of course—made use of animals, not real ones but models. Designed and built in the second century A.D., by which time the Chinese had a thousand years of earthquake data on record, this remarkable gadget consisted of a jar containing a pendulum connected to articulated dragons' jaws clamped on to a small ball. When a shock wave hit the pendulum, one of the jaws would open, dropping its ball into the open mouth of a toad below. Whether it fell to left or right indicated which direction the shock had come from! European science was unable to invent anything as efficient as this until the nineteenth century.

Even better than predicting earthquakes would be to prevent them altogether, and this could be done in theory by causing them. This may sound slightly mad, but oil company geophysicists are taking the idea seriously, though we know of no instance of its having been put into practice as yet. Dr. Constantin Roman, a scientist with Amoco Europe Inc., reports that petroleum engineers trying to inject fluid into the reservoir of an oil field have detected minor tremors caused by the sealing faults of the field.

"This effect," he says, "led the scientists to think of the possibility of releasing the strain accumulated along active faults by lubricating the fault plane with injected fluid, and thus triggering gradually a series of minor earthquakes, which might in this way prevent the occurrence of a destructive earthquake." It has been suggested that the San Andreas fault could be "defused" in this way, and though Dr. Roman's idea makes some sense in theory, we would rather he tried it first. He may have difficulty persuading anybody to watch.[8]

More immediate prospects for earthquake prevention are to be found in careful study of animal behaviour, by farmers as well as scientists. Dr. Helena Kraemer of Stanford University has suggested that ordinary farm animals grazing near fault areas could have a valuable part to play in helping short-term forecasts, and she and her colleagues are now keeping a watchful eye on the behaviour of the chimpanzees on their own animal reserve. Acting on a suggestion by geophysicist Bruce Smith, they analysed the chimps' activities over a period in 1976 in which 25 minor tremors were recorded near the notorious San Andreas fault, which runs right down the coastline of California through several highly populated areas. On days preceding quakes, they found, the animals were more restless and less interested in climbing. There was no "experimenter bias," since the chimp data was already in the computer before the quake information was fed in. Dr. Kraemer is optimistic about the possibility of basing predictions on chimp behaviour. "But," she admits, "to prove it conclusively, we will have to predict the next earthquake."[9]

There is at least one recent case on record in which a living creature unquestionably saved many lives; not before an earthquake, but before a bombing raid on the German city of Freiburg. It was the night of 27 November 1944, and a large number of allied bombers was on its way to its target. Suddenly, a duck in Freiburg's Stadtgarten (city park) began to quack so loudly and insistently that it woke the neighbours who, for reasons not entirely clear, interpreted the noise as an air raid warning and went down into their shelter. Half an hour later the raid came and the bombs fell. Everybody within earshot of the precognitive duck was saved, though the poor bird itself was killed.

The survivors were so grateful to the duck that they put up a handsome statue of it in the Stadtgarten, and wreaths are still laid there today. Though this case is often cited as one of "para-normal" sensing, it is more likely that the duck simply *heard* the bombers approaching, picking up infrasonic components of the roar of their engines from perhaps 100 miles away, interpreting them through reflexes conditioned by previous similar experience as a danger signal, and responding with a typical fright reaction.

(Keeton, whom we mentioned in Chapter 3, has found that pigeons can "hear" sounds with frequencies of less than one cycle per second.)

One of us (Playfair) was given a vivid demonstration of invisible force at work while on a visit to the Channel Islands shortly after the supersonic Concorde aeroplane had gone into service. The islanders had been complaining of odd noises in their homes for some time, though they knew Concorde pilots were forbidden to exceed the speed of sound over inhabited areas. It turned out that the pilots were obeying the rules, but that while flying above the speed of sound over the English Channel, they were setting up sonic booms that bounced off the sea all the way up to the ionosphere and down again (as many as four times), still strong enough to make windows rattle and even split the stone wall of a cottage, although no normal sound of the plane itself could be heard. (A four-year-old girl explained the mysterious noise as "window talking.")

If sound waves can travel 600 miles and remain audible to humans, it is quite possible that birds and animals can pick up acoustic or electromagnetic waves from underground movements preceding earthquakes. We seriously recommend residents of fault areas to keep a pet duck.

On 4 February 1975 an earthquake registering 7.5 on the Richter scale* caused much damage in the Chinese province of Liaoning, but allegedly no loss of life. The official explanation was that the event had been correctly predicted, and the whole area evacuated just in time. It had been expected, in fact, since 1970, when the region was declared a potential risk area, and the order to clear out was finally given as seismographs and tiltmeters confirmed evidence from well water levels, geoelectric currents, and strange behaviour in animals. "If reports received accurately

---

* The *intensity* of an earthquake shock is measured on the Modified Mercalli Scale from 1 to 12; the *magnitude* of energy released being assessed according to the Richter Scale, introduced in 1932 by Professors B. Gutenberg and Charles Richter. It does not end at 10, as is frequently assumed, but has no upper limit, the highest magnitude yet recorded being 8.9. Richter himself has rated 1976 as an "exceptional" year. According to him, a quake of magnitude 7 to 8 is "major," while anything over 8 is "a great event."[10] The Richter Scale was modified in 1977. In this chapter, we cite Richter readings. For details of the revision, see *Journal of Geophysical Research* 82, 2981-2987, 1977.

describe the events that transpired," *Earth-Sciences Review* wrote in March 1976, "then the Chinese achievement represents a watershed in the quest for earthquake prediction."

But you cannot win them all. On 28 June 1976, the first of a series of devastating quakes took place, one of which (magnitude 8.2) may have killed half a million people, though the Chinese, like the Russians, do their best to suppress details of domestic disasters. The Chinese forecasters certainly missed all of these, although they did keep the people of Peking living in the open air for several days waiting for tremors that were predicted, but failed to occur. Meanwhile, the *People's Daily* recommended overcoming earthquake effects by criticizing Teng Hsiao-ping, a former deputy prime minister who was out of favour at the time.

The year 1976 was a rare one for major earthquakes, with three or more times the expected number of events of magnitude 8 or more. (Subject to confirmation.) A curious feature of that year's seismic activity was that although major quakes took place all over the world, one of the most notoriously quake-prone regions, Western California, escaped intact. The San Andreas fault stayed put, perhaps enjoying the calm before the cataclysm promised by Gribbin and Plagemann.

Sun, Moonquake and Earth Tide

There is evidence, as we mentioned in Chapter 1, that changes in the Sun's activity are related to Earth's spin and consequently to the length of the day. If a strong solar storm can, in effect, put a brake on a body the size of Earth, the question naturally arises whether such a force could be involved in triggering of earthquakes. Studies of events of magnitude 7.5 or more throughout a complete solar cycle suggest that there is indeed a correlation between Sun, length of day, and earthquake incidence.[11]

Seasonal changes in the rate of Earth's spin seem to be a combination of effects caused by the Chandler wobble, movements of great air masses over the oceans and continents, and the more irregular jerks brought about by tidal forces of Sun and Moon. Such forces can raise or lower ocean levels, while tides in the atmosphere can cause movements of air masses; in each case energy is lost by the Earth through friction so that it slows down. (We are only talking about variations of milliseconds, so we trust there is no cause for alarm.)

However, if some regions of Earth are being compressed by tidal forces while others are being stretched or placed under tension, considerable energy could be trapped in a fault area, and the spatial distribution of such energy must be related to the type of earthquake that could be triggered by tidal forces. For a fault zone under lateral tension, we would expect tidal forces to be strongest when Sun and Moon are directly overhead, whereas for a zone under compression, the shear forces will be at their maximum when either Sun or planets are rising or setting on the horizon.

There is much to be learned about tidal links with earthquakes by studying the Moon. Now that seismographic equipment has been landed on the lunar surface, we can compare possible roles of Earth tides on moonquakes, and of lunar tides on earthquakes. Soviet astrophysicist Dr. Nikolai Kozyrev, of Pulkovo Observatory in Leningrad, has even suggested that *all* major earthquakes can be forecast by observing the surface of the Moon. He bases this surprising claim on the assumption that tectonic processes on Earth and Moon are linked, and that there is a precise relation between the timing of seismic events on each. It is, he says, "as if the Moon were in direct contact with the Earth; as if it were its seventh continent."

Kozyrev has compared data from 630 major earthquakes from 1904 to 1967 with the 370 items listed in NASA's catalogue of transient lunar events for the same period. He has found evidence to support his claim that Earth and Moon events are mutually registered, though with a time lag of up to three days. On 1 April 1969, for instance, he personally observed a transient red spot in the Moon's Aristarchus crater the day after two deep-focus earthquakes.* Then, in September, earthquakes on the 20th and 21st were followed by a glow in the same crater. Going over the records, he found that Moon events—not actual quakes, but luminous spots caused by gas emission—tended to cluster not only just after Earth events but also just before them. Peak "moonspot" activity shows a symmetry

---

* Moonspots are neither new nor rare. Herschel saw 150 of them on 22 October 1790. One wonders if he speculated that, like sunspots, they might be related to terrestrial events.

from two to three days before and after, and when such a peak coincides with high tides, Kozyrev reckons that an earthquake is most likely to happen.[12]

He believes that other celestial bodies will show interrelationships similar to those between Earth and Moon, and he calls for more attention to setting up a permanent seismic observation network on the Moon. Another Soviet astronomer, Gurgen P. Tamrazyan, believes lunar phase and orbit eccentricities to be linked to earthquake incidence, while a NASA report has found a connection between the Moon's position in relation to Earth and the releasing of strains in the lunar interior. Periodicities of 27 days, 206 days and 6 years have now been positively identified in moonquakes, which appear to be strongly connected to some component of tidal stress. All in all, there are many unsuspected links emerging between Moon and Earth.[13]

The same can be said of the Sun. Soviet astronomers and geophysicists have even claimed that overall seismic activity on Earth obeys the same rhythm as that of solar activity. The leading spokesman for this theory is A. D. Sytinsky of the Institute for Arctic and Antarctic Research in Leningrad, who has gone as far as to state that seismic phenomena on Earth *depend* on solar activity. He studied data on 594 quakes of magnitude 6.5 or more, including all those during the 1957-1967 sunspot cycle, and although he could not find any apparent cycle in earthquakes, he did find that the overall amount of energy released in earthquakes reached a peak about a year after solar maximum and again at minimum. He also found that quakes were more likely to take place two or three days after the passage of an active solar region across the central meridian of the Sun. By following the progress of these regions, or solar disturbance centres, he claims to have predicted certain quakes in advance.

Moreover, Sytinsky thinks he knows how all this happens. It is due, he says, to solar corpuscular radiation, and he points out that there is a known correlation between the brightness of comets and the cycle of solar energetic output, which is also due to corpuscular radiation. His claims may sound too good to be true, but there is some rationale behind them, which can be summarized thus:

Particles from solar flares disturb Earth's atmosphere at points where their relatively weak energy triggers the much stronger latent energy of cyclones and anticyclones. This

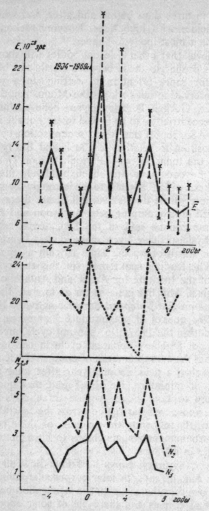

Fig. 8A
A: Sytinsky's chart of Earth's overall seismicity (top) as expressed in units of 10^{26} ergs (vertical column) shows a marked peak shortly after solar maximum. Horizontal line denotes years before and after solar maximum. Charts N_1, N_2 and N_3, show, respectively, annual numbers of earthquakes equal to or greater than 7.0, 7.5 and 7.75 (Richter).

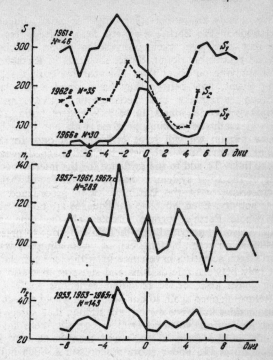

Fig. 8B
B: Distribution of the number of cases (N) in which active sunspots pass through the central solar meridian, and average area (S) of spots in this region in relation to quakes of 6.5 or greater. Horizontal column denotes days before and after passage of spots through the CSM. Charts n_1 and n_2 shows distribution at solar maximum and minimum respectively. (Courtesy of Soviet Academy of Sciences).

leads to a redistribution of air around the Earth's atmosphere, creating a loading effect on its centre of mass. The energetic effect of such a process can be enough to cause polar displacements and fluctuations in the rotational velocity of Earth. Thus, increased solar activity increases the potential energy of the atmosphere, leading to a reduction in rotational velocity and a release of kinetic energy—in other words, an earthquake. This release in turn causes disturbances in the troposphere and helps build up cyclones.

We already know that solar output is closely linked to variations in the Earth's magnetic field. Indeed, this was the first Sun-Earth cyclic correlation to be established beyond reasonable doubt. Soviet scientists in Kazakhstan, after studying no fewer than 174 aftershocks of the 1970 Tyup event, now believe they have found a correlation between earthquake and geomagnetic activity, and if Sytinsky's theories stand up to independent testing, then we will really be making progress with our many-body jigsaw puzzle; we will have evidence not only for Sun-magnctism and Sun-earthquake but also magnetism-earthquake links. To add to the evidence for the interconnectedness of the solar system, it has now been found that the microseismicity of the Earth varies in accordance with solar activity. Even when no earthquakes are taking place, the whole Earth's surface vibrates all the time, such "microseismic" activity being thought to be due to pressure variations on ocean bottoms caused by standing sea waves. Two Czech scientists have come up with close correlations not only between microseisms and sunspots, but also with solar flares and cosmic radiation variations.[14]

Returning once again to our Moon, we would expect that if lunar tides influence earthquake timing, then surely they must also affect volcanic eruptions, since these involve liquids, in the form of rock and magma at very high temperatures and under enormous pressures. Such liquids, like any other, must be subject to tidal influence, and observations made in 1968 by a very lucky scientist give strong support to this theory.

In that year, Dr. Tom Simkin of the Museum of Natural History in Washington had the good fortune to be on the spot (well, within sight of it) when a volcanic caldera collapsed in the Galapagos Islands, triggering a series of seismic events. At the precise moment the collapse began, on 11 June, the tidal force of the Sun was at its maximum, while the Moon was at its closest approach to Earth. Simkin found an exact correlation between seismic activity and positions of Sun and Moon: indeed, for the 42 hours following the initial collapse, aftershocks took place at local tide maxima *and at no other times*.[15]

Other volcano-tide links were spotted in 1973 by two scientists, after analysis of reported times of major land eruptions since 1900, and a comparison of the location of the events with the phase of the various Earth tide components. They found a link between eruption times and the fortnightly component of the tide for the total data set of 680 eruptions, and also found that eruptions were more likely when tide amplitudes were greatest. This confirmed a study carried out the previous year for a single volcano, Stromboli.[16]

Scientists analysing the aftershocks of a September 1966 quake in Truckee found a strong 25-hour periodic component, with more events taking place at minimum tide, while data from the Apollo programme have shown that moonquakes in a certain region occur only at tidal maxima when the Moon is at perigee. In 1972, a statistically significant relationship was found between tide and quake in the western U.S., though it was only marginally significant elsewhere. This underlines one of the problems concerning prediction of seismic events: no two earthquakes need be alike, and there is not likely to be one single cause for all of them. Taking samples for computer analysis from all over the place may only confuse the issue. So many factors are involved that each fault zone must be regarded in the way doctors should regard their patients—as individuals. There may be no point in studying earth tides, for instance, unless one also studies the way they relate to the fault zone in question. Lumping together too much data from too many zones widely different from each other may lead to true correlations being hidden.[17]

John Simpson of Akron State University can certainly be accused of data-lumping after his study of no fewer than 22,561 earthquakes, which probably included every single event between 1950 and mid-1963 of magnitude 5.5 or more. This was a very different kind of sample from that of Tomaschek (to be mentioned below), who took only events of 7.75 and more, or our own modest pilot survey, also to be mentioned below, which took 39 events from a short period of time.

Simpson found two features that are consistent with the tidal hypothesis, though he himself regarded them as insignificant. First, when he plotted his earthquakes according to phase of Moon, he found a peak of 730 events, instead of an expected 625, occurring just after quarter-Moon, with a secondary peak of around 710 shortly before new Moon. The minimum, 560, occurred just after full Moon. We feel it would be worthwhile to see if these clusterings can be repeated when, for example, only earthquakes from one fault area are studied.

The second effect to emerge from Simpson's marathon survey was an apparently significant peak just before lunar apogee, which was just what he was not expecting. After all, when the Moon is closest to Earth, at perigee, maximum tidal forces occurring at such times should coincide with maximum tidal triggering influences. But they did not, although to confuse matters still further, there was a sharp drop just after apogee.[18]

Earlier, we mentioned the Chandler wobble (the slight wobbling motion of the Earth's poles) and its possible links with earthquakes. It has been suggested that seismic activity is related to polar tides by "pumping" of the wobble, which has a period of about 437 days. These polar tides are perhaps stronger than those of Moon and Sun in effect, which could give rise to periods of seismic activity of seven and 40 years. According to this theory, the release of elastic or seismic energy from Earth's interior is the common source of both earthquakes and the wobble, the release process being triggered by the physical processes that produce the observed polar variations.

There must be as many ways to predict earthquakes as there are to skin cats. Much progress has been made: we know more or less *where* nearly all earthquakes are going to happen (about 80 percent are located along the edges of the Pacific), but we do not know *when*. If we knew, thousands of lives could be saved at once just by blowing a whistle or sounding a siren. People are killed in earthquakes because buildings fall on top of them. All they have to do is be in the open air. In the 1935 Quetta shock, old hands who had lived through previous shocks there dashed out of their houses in a few seconds, in some cases moments before heavy rafters fell onto their still warm beds. Heavier sleepers died horrible deaths, buried under brick walls, while others slowly asphyxiated from the choking dust that was all they could breathe while trapped and waiting for help. All it needed at Quetta was five minutes' warning, and 30,000 men, women, children and babies might not have died.

Planets and Earthquakes?

Can astrology help us predict earthquakes? If extraterrestrial forces from the Moon play a part in seismic events in the Earth, as seems possible, cannot forces from the planets also be involved?

The overall record of astrologers in predicting earthquakes, or indeed anything else, is terrible. Roger Hunter, a geophysicist with the U.S. National Ocean Survey, made what seems to have been a sincere and thorough attempt to correlate seismic events with planetary situations traditionally supposed to favour them, and concluded that "throwing darts blindfold at a calendar would give similar results." He found that although 17 major earthquakes took place between January and August 1970, the forecasters in *American Astrology* magazine missed every single one. Of the 16 quakes they did predict, naming only the month and the country, they were right for only three minor events. They also missed the 31 May disaster in Peru, one of the worst of this century, in which 30,000 people died.[19]

Nevertheless, Hunter keeps an open mind. While not believing that planetary forces can cause or even trigger earthquakes, he sees no reason why correlations cannot be found as there are between two clocks each showing the correct time. There is no cause-and-effect link between clock A and clock B: somebody winds them up and adjusts them according to the radio time signal. As we see in the following chapter, it is common for two entirely different things to share similar rhythmic cycles, and it may well be that planets and earthquakes are both "regulated" by a common source. What that source might be is beyond the scope of our imagination at present.

The case for astrology grows stronger when specific predictions are published well in advance, and prove correct. A thorough survey of the almanacs published in the nineteenth century by A. J. Pearce might reveal some impressive direct hits, but these must be evaluated in proportion to the misses. If you predict anything long enough, it will probably happen. Even General Franco eventually died, as predicted by astrologers regularly for at least twenty years.

An article published early in 1976 in *Gnostica* magazine contained specific predictions that have coincided to some extent with observed events. The writer, Arthur Prieditis, said that an "exceedingly malignant" Saturn/Uranus square would cause general upheavals in world politics and economics between August 1975 and June 1977, and when combined with Mercury/Jupiter dissonances, which indeed have been present at some of the strongest quakes of this century, would make three particular days "doubly interesting to watch." These were 2 August and 15 November 1976, and 27 February 1977. Mercury and Jupiter were in opposition on the first two dates, and square on the third together with a Uranus/Saturn square.[20]

We watched accordingly, and noted that there were indeed earthquakes on the first two dates, in the New Hebrides and China respectively, each of magnitude 6.9 (subject to revision). On the third date there was a minor tremor in Hollywood which caused no damage and probably registered about 3 on the Richter scale. A few days later, however, the first serious quake of the year wrecked much of Bucharest, so Prieditis can claim two direct hits and one very near miss. We cannot draw any definite conclusions from his forecast, however, due to the number of variables (date, place, magnitude) involved, although we can say that if astrologers keep up this standard of accuracy in future, the odds against their predictions being due to chance, or throwing darts at a calendar, will lessen considerably.

To be of real value, earthquake predictions must name the place as well as the date, and give some idea of the magnitude. This may one day be possible, but only by using a great many different methods together. Earthquake forecasting is one branch of science that must be interdisciplinary if it is to succeed, and there is no reason why the astrologer should not join forces with the geophysicist, the astronomer, the zoologist and the duck.

Obviously, the planets have to be somewhere at the time of every earthquake, and there are far more of these than many people realize (about two per minute). We usually hear about them only when cities are wrecked and people killed. Even so, some planetary configurations do seem to turn up more often during *major* seismic events than chance should predict. The Mercury/Jupiter dissonance is beginning to look suspicious; it was present at three major events in 1960 (Chile, Iran, Agadir) and also at the 1902 eruption of Mount Pelée. It is interesting to recall that Nelson insists on the significance of a hard angle, such as a square (90°) between Mercury and an outer planet, in connection with his forecasting of bad radio weather conditions. (It must be remembered that Mercury, with its 88-day orbital period, will be at a "hard" angle with another planet every few days. Nobody is suggesting that Mercury/Jupiter aspects *alone* are of any importance at all.)

Incidentally, Prieditis disagrees with Gribbin/Plagemann with respect to the 1982 Great California Disaster, arguing that nothing exciting happened on 5 February 1962, when several celestial bodies were in conjunction, or during the "grand alignments" of 1624 or 1803. (Accurate records of earthquakes were not kept until 1896, but Gribbin has discovered that there were in fact a number of quakes in the then sparsely populated California region in the first decade of the nineteenth century.)

The idea that planetary positions could trigger earthquakes was given some respectability by Dr. Rudolf Tomaschek, a president of the International Geophysical Union who, instead of denying *a priori* that planets could possibly trigger anything, did what Nelson was doing at about the same time—he dug out the old records and looked at them from a new angle.

In 1959 Tomaschek announced that planetary positions *did* seem to be linked to earthquake incidence. His startling claim was based on a study of 134 events of magnitude 7.75 or more, and was published not in a backyard astrological broadsheet, but in *Nature*, which has a fine record for allowing unconventional opinions to be aired. Tomaschek singled out Uranus as the planet to watch. Its position in relation to earthquake epicentres was the same far more often than it should have been by chance, the odds in question being 10,000 to one, which are highly significant.

His findings were promptly denounced as "pseudosignificant" by a mathematician, to which he replied that he was well aware that stresses and strains were the primary cause of earthquakes yet he insisted that the position of Uranus might have something to do with their timing. There was in fact a devastating quake at Agadir in Morocco not long after the publication of Tomaschek's paper. He suggested that if the people of Agadir had read *Nature*, and kept out of doors while Uranus approached meridian (10 p.m. to midnight) after the minor precursor shocks, they might have been safer during the main quake, which struck at 11 p.m. and killed about 20,000 people. "An unbiased approach to these problems, of which the correlations of Uranus are only a part and a first step, may help humanity," he concluded. He also pointed out that there was something special about Uranus: it is the only planet the direction of whose axis coincides with the plane of its orbital revolution, so that any magnetic field issuing from it must have an influence quite unlike that of any other planet.

There is also, it seems, something special about Jupiter, to which Gribbin and Plagemann drew attention with their book mentioned at the beginning of this chapter. In 1977, James Neely announced that after looking at data for nearly 800 earthquakes of magnitude 5 or more, he had found a correlation between the position of Jupiter and the time of eruption for which there was only one chance in 2×10^{39} that it was due to chance alone! This is a figure of such staggering magnitude—two followed by thirty-nine noughts—that, to say the very least, it calls for confirmation. We have not been able to examine Neely's data in detail, but we trust that others will have done this by the time this book is in print.[21]

There was certainly no rush to follow up the claims of Tomaschek's *Nature* paper. Indeed, as far as we are aware, there were only two attempts apart from Neely's mentioned above: our own very modest pilot project and a very interesting study by Theodor Landscheidt.

Dr. Landscheidt has theorized, after a study of the period 1904-1964, that the position of the galactic centre and the apex of the Sun's galactic motion can be related to severe earthquakes. The conception, he suggests, of the galactic mass being attracted by the centre of the Milky Way as the static pole, is a reasonable one. He also believes gravity waves must be involved, and that the solar system as a whole functions as gravitational resonance aggregate. Resonance amplification of gravity waves could be possible when the distances involved in the resonator are close to the lengths of the waves, which must be extremely long. But, Landscheidt says, the Sun *always* functions as an attuned gravitational resonance aggregate when the planets are clustered round it in configurations corresponding to the even oscillations, or harmonic vibrations of the quadrupolar gravitational waves. Then the whole system vibrates like an antenna for such waves from the galactic centre.[22]

This is the theorizing of a sleepwalker in the grand Keplerian tradition, and we must point out that Dr. Landscheidt is not a professional scientist, but a retired

High Court judge. His views deserve a hearing, however: astronomy and cosmology are fields in which amateurs in the past have made significant contributions, and continue to do so today.

While this book was being written, we decided to see for ourselves whether there were any indications that planetary positions coincided with exact times of earthquakes more often than they should by chance. Hill, to whom credit is due for the original idea, took charge of the programming, and using event cards from the Center for Short-Lived Phenomena (to be described in Chapter 6), he took a pilot sample of 39 events magnitude 6 or more, feeding the computer with local time, latitude and longitude of each event. In a few seconds, the computer emitted a printout the size of Sunday's *New York Times*, giving ecliptic longitudes, declinations, latitude and declination parallels, angular distances of every planet from every other: a total of 396 numbers for each earthquake. The two of us looked at a bewildering mass of information and wondered if we, Tomaschek, Landscheidt and all other planet-spotters were just crazy.

We started by checking to see if the position of Uranus during the 39 events looked interesting. No luck: results were close to what we would expect by chance distribution. Then we recalled Nelson's insistence on the significance of subharmonics of 90° (multiples of 15, including 7½) in radio weather prediction. Perhaps they were significant in earthquake prediction as well?

According to chance, two or three planetary angles for 39 events should fall on one of the multiples of 15 ($\pm$ ½°). For the first set of data we tried, we found five. Allowing a margin of error of one degree each way, we got 14 out of 39, and with a margin of 2° we found 20, or more than 50 percent! Next, we took a single event, the Kuril quake of 10 June 1965 (7.2 Richter) and found the 15° multiples piling up one after the other. In our geocentric horoscope, the Moon was 75° from Mars, 120° from Uranus, 119° from its ascendant, and 151° from its node. Mercury was 45° to Venus, 61° to Jupiter and 29° to Saturn. Venus was 15° to Saturn, and Saturn was 29° to the Sun. Mars was 135° from the lunar nodal point. Taking three of our strongest quakes, we found exactly three times the chance distribution of angles among the planets in multiples of 15.

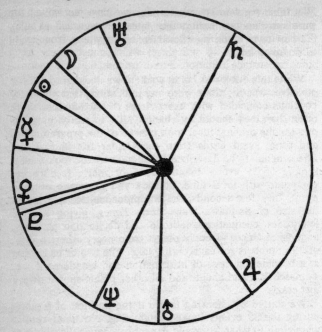

Fig. 9/A
Geocentric horoscope of the 31 May 1935 Quetta earthquake (magnitude 7.5). Sun and Moon are about 26 hours from conjunction, and moreover only one degree apart in declination (i.e. with respect to Earth's equator). Jupiter and Mercury are only one minute of arc from sesquiquadrate (135°), an angle found interesting by Nelson. (Jupiter-Mercury "dissonances" are considered significant by astrologers. See page 158). Note also the near-opposition between Neptune and Saturn, with the Sun almost square to both. Charts like this are of limited value, for they do not show distances of celestial bodies from perigee and apogee or declinations, which may eventually prove just as important as planetary angles. Only a great deal of computer time will tell.

Our only purpose in carrying out this pilot study, as we have said, was to see if there was any indication at all of *any* correlation between planetary angles and earthquake times, enough to make a larger study worth undertaking.

Earthquake 163

We think we were successful. But we have not solved the problem. No two earthquakes must be regarded as alike. The focus of a seismic event can be anywhere from ten to 450 miles below its epicentre, which must surely affect time of surface eruption. Some earthquakes are tectonic, while others are volcanic. Some take place on land, others under the sea. Some take place in areas where such events are commonplace; others erupt in places like Stoke-on-Trent, England, where they are very rare. For the purposes of prediction, we cannot talk about earthquakes as a group

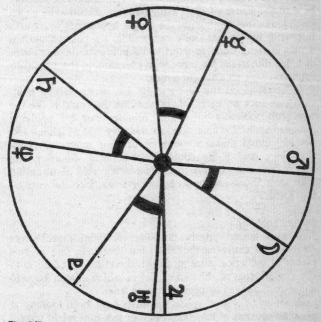

Fig. 9/B
Chart for 18 February 1955, when another earthquake (magnitude 6.2) struck Quetta. Here we find an entirely different but quite remarkable planetary configuration in force, with the Moon and all four inner planets within 10° of opposition to an outer planet. We also find Mars one degree from trine with Pluto and one degree from square to Uranus. It is also interesting to find four pairs of celestial bodies arranged within 3 degrees of 30° apart, as indicated by double lines.

any more than we can talk about people without taking into consideration the characteristics of the individual; and no two people have ever been identical. We have certainly not solved any problems, but we have reason to believe we have helped define part of one correctly.

The idea of external triggering of earthquakes is not a new one. In 1935 Harlan T. Stetson of Harvard University uncovered a correlation he considered "surprisingly striking" between the phase of the Moon and the timing of more than 2,000 deep-focus earthquakes (60 miles or more). He insisted that lunar tidal forces could not possibly be a major cause of such a high-energy disturbance as an earthquake, but he thought the trigger hypothesis was worth following up. It was quite reasonable, he pointed out, to expect a weak body to exert some influence on a massive one. He illustrated his point with an example from his own experience as an amateur sailor.

If he stood on the fore-and-aft line of his sailing boat, with one foot on each side of the line, he could make the boat roll violently in a few minutes just by applying pressure with alternate feet. In this way he, weighing 160 pounds, could make fifteen tons of boat move very considerably. And if he could do that, why should Moon and planets not be able to cause considerable perturbation with their incessant input, however weak, into the core of the Earth?[23]

Happy Ending?

It may sound utterly fantastic to claim that human activity can cause earthquakes, but there are one or two pieces of evidence pointing in this direction, in addition to the experiments mentioned earlier in which oilmen hope to cause miniquakes to prevent major ones.

A group of Scandinavian scientists have been looking at possible sources of low-frequency signals that might trigger earthquakes. In addition to earth tides and tectonic movements, they have been considering a third factor: man-made machinery. A strange infrasonic component has been isolated from spectral analysis of the NORSAR seismic network in Norway; a frequency of 2.78 Hz. also being found to issue from the Hunderfossen hydroelectric plant near the network. This figure happens to be 1/18th of the electronic net frequency (50 Hz.) used in Norway.

Could the acoustical and mechanical vibrations of the huge generators at Hunderfossen somehow be coupled into ground motions? It seems possible. And the more such plants we have, the more generators will hum and thump and generally agitate Earth's crust.[24]

In addition to such obvious threats to the underground environment as deep nuclear tests and heavy quarry blasts, there are others more apparently innocuous yet equally effective. In 1976 Chinese seismologists announced that shortly after a new reservoir had been filled in 1959, seismic shocks began to occur in the region "almost incessantly" until there was a 6.1 magnitude earthquake in March 1962, since when there have been a further 200 tremors. The scientists are convinced that the penetration of water through fissures has had a good deal to do with this series of events, providing us with one more potential man-made source of environmental pollution to worry about.[25]

Something similar, but with far more alarming implications, took place near Denver after the U.S. Army drilled a colossal cesspool two miles deep and filled it with several million litres of liquid waste from the Rocky Mountain Arsenal. This led to a minor epidemic of earthquakes in the area, as the liquid (presumably potentially dangerous) lubricated shifting plates in local faults.

To end this chapter of death and destruction on a more cheerful note, let us think of how we can make the interior of the Earth work for us rather than against us. Heat is constantly being generated in it by radioactive decay, and parts of it reach temperatures of up to 3,000°C. An interesting development in the battle to produce pollution-free energy is the tapping of these vast natural heat resources that lie under our feet. Geothermal energy, as it is called, has been used in Italy since 1903; about a third of the population of Iceland live in geothermally heated houses. Natural heat is also in use in France, Japan, Mexico, New Zealand, the U.S.S.R. and California.

British geologist Ron Oxburgh describes the Earth as "a natural power plant with an output thousands of times greater than all man-made power stations put together." The snag is, of course, that Earth's heat flow is spread so thinly that, in his estimate, it is sufficient to light only two domestic bulbs per acre in Britain, though if it could be channelled and concentrated, it might supply up to 2 percent of Britain's present demand. The more holes we drill in the Earth, the less internal heat will be forced to escape elsewhere. It is surely worth further exploiting the idea of drilling safety valves near major fault zones and getting cheap heat into the bargain.[26]

It is now clear that there are many ways in which influences from outer space affect our inner spaces, both of our Earth and of ourselves. As man modifies his surroundings, for better or worse, so these surroundings modify him in return. This process is not always as obvious as the case of the Chinese reservoir that caused a major earthquake; who knows how many other "natural" events great or small may not be due in part to our own actions?

Now, in the following chapter, we turn to the subject of cycles, of which there are a great many to be found in Nature. We have already mentioned the sunspot cycle of about 11 years, and pointed out that although this seems to coincide with certain terrestrial events, we cannot yet be sure that the one causes the other.

The solar system is a well oiled machine that runs smoothly on its own without requiring maintenance. And since its component parts move in regular and largely predictable orbits, we on Earth inevitably find ourselves caught up in a system of cyclic, or rather multicyclic, activity. We have already found evidence that Sun, Moon and planets influence us in a number of ways; therefore the "cycles of Heaven" deserve closer study. This is no mere academic pastime, nor is it an exercise in numerology. Many aspects of our lives must also be cyclic, and they may very well be linked to celestical cyclic activity. This raises a disturbing question.

Man does not welcome the suggestion that he is not master of his own destiny. This idea of any form of preordained future is abhorrent to most of us. Yet whether we like it or not, we must face the fact that since we live in a multicyclic environment, our lives must be subject to the periodic influences of that environment. If, as seems quite possible, predictable cyclic patterns are involved in events as dramatic as earthquakes, we must try to find out what other effects such cycles may have on our Earth and ourselves.

PART THREE

RHYTHMS OF LIFE

Chapter 6

CYCLES

WHAT IS a cycle? The word comes from the Greek *kuklos*, a circle, and has several meanings today. In our context here it means an event or sequence of events that repeats itself apparently indefinitely. This repetition may be regular and rhythmic, like the 24-hour cycle of night and day, or it may not. Some cycles are extremely slow, like that of the precession of the equinoxes, which takes place only once in 25,827 years as the entire celestial background appears to make one revolution with respect to the Earth's pole. Other cycles are slower still: it has been suggested that the creation of the planets follows a huge cycle of some 80 billion years. On the other end of the scale, we have events that repeat so fast that we have to measure them in terms of cycles per second: wave oscillations at the top end of the electromagnetic spectrum—secondary cosmic rays— run into the frequency region of 10^{24} cps., a figure too enormous to mean much to anybody except a nuclear physicist.

Some cycles are familiar and apparently changeless, such as those of the day, the various lunar months, the four seasons, and the year. Others vary considerably in length, such as the solar or sunspot cycle, which has averaged just over 11 years since it was first measured accurately, although individual cycles have been anywhere from 7 to 17 years long, and there have been times when the sunspots have disappeared altogether. Such irregularity makes its cause far harder to determine than those of daily, monthly or yearly cycles which clearly depend on the relative motions of Earth, Moon and Sun. This tendency for some cycles to seem to wander off course for no apparent reason, and come back into step, is one of the most baffling aspects of the subject.

Another mystery is why one cycle persistently coincides with another when there is no obvious reason why it should. What, for instance, has the abundance of U.S. grasshoppers and Hertfordshire partridge to do with sales of General Motors vehicles or the level of Lake Huron? Yet all these, along with 24 other phenomena, share a cycle of just over nine years, and moreover all their cycles tend to rise and fall together, locked in inexplicable synchrony. The study of cycles is full of such enigmas.

It is only a short time ago that we began to realize that other things than days, seasons and sunspots come round at regular intervals. The modern age of cycle study began in 1875, when a book called *Benner's Prophecies of Future Ups and Downs in Prices* was published, and a lecture was given by an economist named W. S. Jevons to the British Association on the subject of *The Solar Period and the Price of Corn*.

Neither Jevons nor Benner was the first to look for cycles in natural and artificial events. Sir William Herschel had suspected there might be a relation between sunspots and weather, and hence grain prices, but he had too little data to work with. R. C. Carrington, the brewer-astronomer who studied sunspots from 1853 to 1861, published a chart comparing their cycle with that of corn prices, and although he did not pursue the matter, he touched on a question that still mystifies us today: which cycle causes which other cycle? For instance, it is well established that the width of certain tree rings corresponds closely to sunspot counts, but it is not certain which is cause and which is effect. Do sunspots cause tree rings to thicken? Or do thickening tree rings cause sunspots? Or does a third unknown factor cause both?

Jevons, in his 1875 address, foresaw a new age of order in the affairs of man. The sunspot cycle was already suspected by then to have some relation to the occurrence of geomagnetic storms, and probably also to that of rainfall, so it was reasonable to suppose that the periodicity of the Sun's activity might be reflected in harvests, and hence in prices of commodities such as corn. Jevons claimed to have found an 11.1-year cycle in English agricultural prices going back to the thirteenth century, and although he later admitted he had got his sums wrong, he thought he had spotted another cycle—of commercial crises.

Samuel T. Benner, an Ohio farmer, was more successful than Jevons, perhaps because he was more personally involved in the effects of cycles. After his business went bankrupt in 1873, he set about thinking why prices should fluctuate the way they did, and why there seemed to be panics in the market at regular intervals. Why, for instance, should the price of pig-iron go up and down every nine years or so? Benner had no idea, but he forecast that it would go on doing so, as indeed it did. According to Dewey (see below), anybody buying and selling the stuff according to his predictions would have made 44 times more than he had lost, right up to the outbreak of World War II.

Other individual cycles began to turn up as researchers followed leads in a variety of areas. "The thing to hunt down is a cycle," the scientist Norman Lockyer said in 1872. He was referring to meteorology, but cycle-hunting soon began in earnest on all fronts. The astronomer N. R. Pogson thought he had found a link between sunspots, rainfall and grain prices in India; a meteorologist on the island of Mauritius noticed that there were more cyclones in the Indian Ocean at solar maxima, so that more damaged ships would put in for repair, while a study of records kept in Paris showed apparent similarity between sunspot numbers and amounts of rainfall.[1] Such links seemed promising indicators of some grand design in which every component of nature performed like a member of an orchestra obeying some unseen conductor, but the evidence never became quite strong enough to enable predictions based on it to be made, and a scientific hypothesis without predictive value tends to be swept under the carpet.

Fortunately, however, the cycle-hunters kept hunting, and recent evidence suggests that some of the early hunters were on the right track without knowing it. Before we go on with the story, however, a word is necessary about general features of cycles.

The most important of these are frequency, period, phase, amplitude and base level. Frequency is the number of times an event takes place per unit of time, e.g., 100 cycles per second. The period of the wave, or time between peak occurrences, is in inverse proportion to frequency: that is, a wave with a frequency of 100 cycles per second has a period of only a hundredth of a second. Phase refers to a difference in starting point for signals with the same period; for instance, if one 24-hour cycle begins and ends at midday, while another begins and ends at 1 a.m., then the two are out of phase with each other by one hour. If they both begin and end simultaneously at 12:30, they are in phase, or phase-locked. Amplitude has to do with the height or depth of each curve above or below the base or zero line: a periodic curve of voltage going from zero to four volts is of higher amplitude than one moving from zero to three. Finally, base level indicates where the cycle starts from as measured vertically: if the price of corn fluctuates regularly between one and two dollars while the price of wheat varies from two to three dollars in parallel swing, we have cycles of different base level and amplitude, though phase, period and frequency may be the same.

Cycles become particularly interesting when they are found to coincide in phase, especially when they fall back into phase after some period of artificial distortion, such as a major war. It is quite remarkable, for instance, that the index of U.S. industrial production (excluding agriculture) between 1875 and 1931 and the rate of change in sunspot areas over the same period should fit each other like a glove on a hand, as two Harvard researchers found to be the case after they had gone through some of Jevons' early work.[2]

This could be coincidence, or it could mean that the Sun's activity has a direct effect on U.S. industry. (It could also be that U.S. industry has a direct effect on the Sun, but it is hard to see how.) It is more likely that a third factor needs considering, in this case, people. People are involved in industry, both creating demand and providing supply, so if there were a link between solar output and human behaviour, we would expect this to be reflected in industrial performance.

Of course, all cycle synchronies may be coincidence. There is no reason why Canadian lynx and muskrat abundance, Arizona tree-ring widths, English financial crises and barometric pressure in Paris should not fluctuate together in cycles of between 9.6 and 9.7 years, as they do —and in phase with one another as well. But it does seem that coincidence must be ruled out when cycles return to a phase-locked position after a period of desynchronization.

Then again, if all cycles were due to chance, we would expect their frequencies to be randomly distributed, which they are not.

In this century, the man most responsible for putting the study of cycles in all branches of natural and human activity on a solid scientific basis is a remarkable American named Edward R. Dewey. Like Samuel Benner, Dewey became interested in the subject following a disaster. In September 1929 he went to work for the U.S. Department of Commerce, a few weeks before the worst economic catastrophe in history, and one of his first jobs was to find out "why a prosperous and growing nation had been reduced to a frightened mass of humanity" by the great Wall Street crash. None of the economists he consulted seemed able to agree with any other, so Dewey soon lost faith in economists altogether.

At about the same time, Chapin Hoskins, managing editor of *Forbes* magazine, also lost faith in economists, and decided to find out for himself how the economy actually behaved, rather than listen to theories. He soon came up with an interesting discovery: for some reason, people in certain cities would withdraw large sums of money from their banks every three months. Then, he noticed, every group of three three-month periods formed a larger cycle of nine months, and there were signs of an even longer 41-month "withdrawal cycle." Hoskins immediately put his discoveries to practical use, and began to make predictions based on them. He must have done well, for he was soon hired by Westinghouse Electric as probably the world's first professional cycle-hunter.

In 1937 Dewey and Hoskins joined forces as industrial company analysts, and one day, after Hoskins had brought off a strikingly accurate forecast concerning prices of shares Dewey had just bought, Dewey felt the time had come to study the subject of cycles in earnest. There were cycles lurking everywhere, in nature as well as in economics. "The problem had to be attacked on a broad front," he recalls.

Accordingly, in October 1940 he set up the Foundation for the Study of Cycles,[3] bringing together a distinguished group of men from business, government and the universities, including one of the outstanding pioneers of cycle study, Yale geology professor Ellsworth Huntington, and also the zoologist (Sir) Julian Huxley, later to become director general of UNESCO. The Foundation, now affiliated with the University of Pittsburgh, has amassed a list of more than 1,300 cycles, covering everything imaginable from war, climate, employment, stock prices and industrial production to such oddities as tent caterpillar breeding habits and even the output of medieval religious writers.

"Learning how the universe functions," Dewey writes, "is, to my mind, the noblest activity of the human race. It is, literally, reading the word of God." A proper study of cycles, he thinks, will enable man to eliminate war, disease and economic depressions. These are worthy aims, but achieving them is not going to be easy, even with high-speed computers and teams of skilled programmers. Dewey admits that the complexity of the subject has led many to abandon it in favour of something easier, "like the fountain of youth or the lost continent of Atlantis."[4]

Dewey and his foundation are engaged on a gigantic piece of detection. The problem to be solved is the working of the cosmos, no less, and how we are affected by it. The methods used are those of detectives everywhere: collection of every available clue, study of the overall picture that emerges, and pointing to the guilty party. Dewey's great contribution to science is the mass of evidence he has collected and filed in the foundation's library, which is divided into three sections. The first consists of facts on cycles, their subjects and durations; the second records what others have written on the subject; while the third deals with interrelationships between one cycle and another, and with clues to possible causes. The foundation is now well established in Europe through its affiliate, the International Institute for Interdisciplinary Cycle Research, which was set up in 1969 in Leiden. Since 1970 it has published the quarterly *Journal of Interdisciplinary Cycle Research*, and held international conferences in addition to carrying out some interesting original research of its own.

The key word in cycle research is *interdisciplinary*. It implies that the subject is of interest to scientists in all fields, not only to statisticians. Specialization in the sciences has now reached the stage where some scientists are scarcely able to communicate with anybody. It can be argued that the information explosion has made it clearly impossible for, say, a physicist, to keep up with both the

Photo: Scott Hill.

Above Left: A. L. Chizhevsky, Soviet 'father of heliobiology.'
Above Right: Solco W. Tromp, of the Biometeorological Research Centre, Leiden. Below Left: John H. Nelson on the roof of the RCA Communications building in New York. Below Right: Michel Gauquelin, discoverer of important links between sky and man.

Photo: courtesy J.H. Nelson and American Federation of Astrologers.

Photo: courtesy M. Gauquelin.

Franz Anton Mesmer, long dismissed as a charlatan. His methods have yet to be satisfactorily explained.

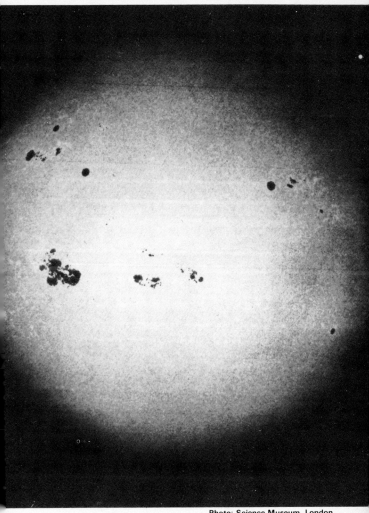

Photo: Science Museum, London.

The imperfect Sun and its Spots. Small white dot in upper right-hand corner indicates relative size of Earth, which would easily fit into even a medium-sized sunspot.

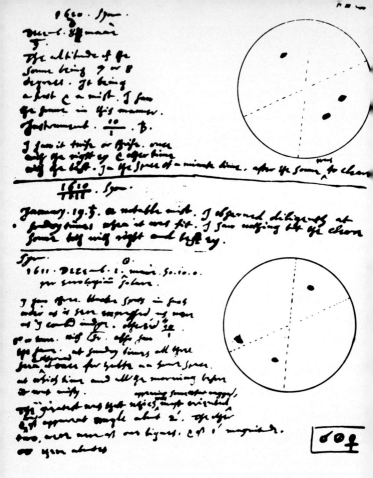

Photo: Science Museum, London.

Thomas Harriot (1560-1621) was one of four men who observed sunspots through their newly invented telescopes around the year 1610. Here is a page from his notes.

Opposite Page: Hitherto unpublished photos from the Playfair family album taken shortly after the 1935 Quetta earthquake in which 30,000 people were killed in 30 seconds. Top: The Beleli bridge, 8 miles from the centre of town. Centre: the remains of the Kandahari bazaar, formerly the main commercial district. Bottom: Indian and British soldiers search for bodies in the wreckage.

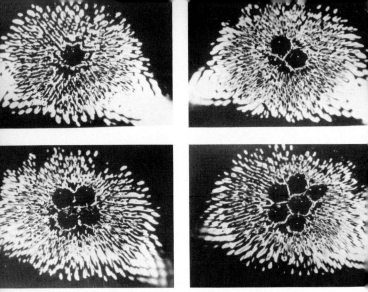

Professor Laithwaite's 'electromagnetic plants.' Top Left: Start of a series of 'flower' patterns made by dish of ferrofluid over a single iron-cored AC coil. Right: bringing the dish closer to the magnet doubles the central 'carpels.' Centre Left: four central 'heaps,' and (right) an intermediate, unstable state of five 'heaps' about to become six around a central core. Below: a multi-seeded 'sunflower' created by very high flow density. See p. 306.

Photos: courtesy E. R. Laithwaite.

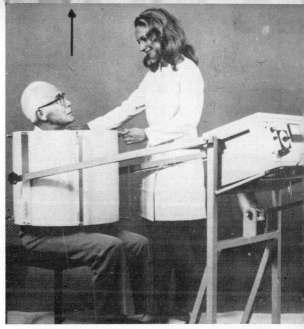

Magnetism used for healing in the twentieth century. Apparatus manufactured by ELEC of Wiesbaden (top) and Pamatron GmbH of Frankfurt (below). Arrow indicates direction of magnetic field.

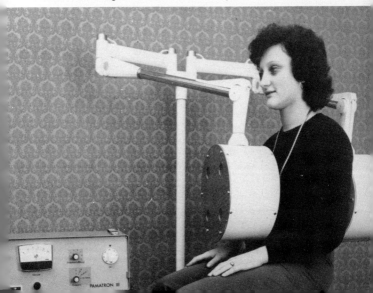

Swedish geo-biologist Bertil Nilsson dowsing a house electromagnetically for geopathic zones.

Photo: Scott Hill

The Psychic Duck of Freiburg. Statue erected by grateful survivors of the air raid of which it warned them in advance.

Photo: Scott Hill

many-body problem and the behaviour of psions or red, white and blue quarks. How, then, has he time for reading about cycles?

But the editor of *Nature*, David Davies, recently had this to say:

> *Time was when biologists, seismologists, volcanologists, oceanographers, meteorologists, climatologists, palaeontologists and astronomers could live almost separate lives. And, needless to say, none of them talked much to biologists. But now we need to understand galactic clouds to speculate on Ice Ages; we need to know about volcanoes when analysing aerosols in the atmosphere; we need to understand continental drift when looking for new resources, and maybe it helps to know about animal behaviour when predicting earthquakes. In the midst of this we have learnt that among the apparently random occurrences of geophysical phenomena there are some tenuous threads suggesting a grander design than simply isolated games of dice . . .*
>
> *We are often told that one of the things that makes science unattractive is the endless fragmentation and specialization of disciplines. Sometimes, however, the boundaries between disciplines melt away when exposed to first-rate thinking. Perhaps a good general example is the way that the concept of the ecosphere has caused many biologists to expand their horizons. The same sort of mind-broadening is going on apace in our study of the physical environment of the Earth.*[5]

Dr. Davies was referring specifically to earthquakes, but his words can apply to many other fields. As has often been said, nature is not divided into departments, like universities! More boundaries between disciplines began to melt with the appearance in 1976 of *Interdisciplinary Science Reviews*, a quarterly devoted to "the integrated work of two or more specialists with different backgrounds and training." By printing papers on such subjects as the interactions of art and engineering, or economics and chemistry, ISR feels it will achieve the effect of a chemical compound of which the properties are greater than the sum of its internal parts. (The appearance of ISR was greeted with a most unfraternal blast from a slightly older brother, *New Scientist*, which while admitting its material to be "scholarly and authoritative," accused its publishers of having no motive beyond profit-seeking and producing yet another unnecessary journal to clutter the shelves of libraries. So much for co-operation in the search for that "grander design".)[6]

Readers of these new interdisciplinary science magazines are now regularly exposed to much first-rate thinking on a wide range of topics. In the first five volumes of the *Journal of Interdisciplinary Cycle Research*, of which Dr. Tromp is editor-in-chief, there were studies of cycles in plants and animals at cellular to population levels, and of cycles in physiology, medicine, entomology, meteorology, geophysics and astronomy, as well as in economic phenomena and in the art of cycle detection itself.

Cycles have been found in almost every conceivable biological phenomenon, from the activities of crabs, pigeons and turtles and the cellular processes of rats and fish, to the growth rhythms of fir trees and mushrooms, the flight of honeybees and the locomotor activity of 40 different species of fruit fly. In human life, cycles occur in so many forms that we are devoting the next two chapters to them. In the physical sciences, new cycles have been unearthed in countless phenomena from varves at the bottom of lakes to inversions at the top of the atmosphere. Isolated facts that would have been of little general interest 50 years ago are now being fitted into a larger picture, thanks in part to another recently-formed institution, the Center for Short-Lived Phenomena (CSLP).

Unconsidered Trifles

Autolycus, the rogue in Shakespeare's *The Winter's Tale*, recalls his father as "being, as I am, litter'd under Mercury ... likewise a snapper-up of unconsidered trifles."

Their most famous successor was an eccentric American, Charles Hoy Fort, who spent much of his life in the New York Municipal Library snapping up and copying down a huge collection of trifles from every source he could lay his hands on. The first of his four extraordinary works, *The Book of the Damned*, came out in 1919 to be hailed as "one of the monstrosities of literature," although another critic claimed it contained the germs of at least six new sciences. Trifles snapped up and published by Fort include accounts of black snow, blue moons, green suns, red rain, unusual things seen in the sky, and especially things seen to fall from it, like showers of sulphur, seeds, frogs, bits of icebergs, and even pieces of flesh. Fort recorded history as nobody ever did, and even allowing a generous margin of error, some of the facts he preserved for posterity still need explaining.

Strange things still happen today, but instead of Fort we now have the more scientific CSLP to record them. The Center, founded in 1968 under the wing of the Smithsonian Institution, is now a private non-profit organization that collects and disseminates information on what it calls "transient events," which may be anything from earthquakes and falls of meteorites to mass beachings of whales and sudden plagues of snails. Whereas Fort merely *collected* odd facts, often from rather dubious sources, the CSLP actually investigates them as well.

Events are recorded by the CSLP's "environmental alert network" of several thousand scientists and students (including Hill), and are then summarized on report cards and mailed to about 2,000 subscribers. Some of these events would have delighted Charles Fort, such as the inexplicable draining of an entire lake in Alaska, a sudden proliferation of Brazil's deadly piranha fish, mass migrations of Ontario dragon-flies, an insect plague in southern Turkey, an invasion of mice in Hawaii, and a host of other once-only happenings such as the discovery of a Stone Age tribe in the Philippines that thought it was the only one on Earth. Whereas Fort relied on the press for his trifles, it is the CSLP that keeps the press informed with its regular supply of neatly typed and well summarized event cards.[7]

The CSLP is not concerned with cycles as such, but much of the data on its cards will be of immense value to future cycle-hunters when a transient event is seen to happen a second time. For instance, if Lake Linda in Alaska suddenly drains itself again, we can suspect that the first draining was no freak event, but part of a cycle. If hordes of whales flop on to Australian beaches every x years instead of just once, another possible cycle may have been spotted. And if lakes drain and whales collapse at specific points on the solar cycle, we will have another "coincidence" to add to the hundreds already on file.

The longer the period of a cycle is, the longer it takes to identify it, especially when there is insufficient good data from the past to work with. We have a hundred years of reliable sunspot figures, but no similar statistics on whales or lakes. These two particular events may have no relevance to anything else. They may never happen again. There was a time when astronomers wondered if the sunspots would ever come back, and it was not until 250 years after their first observation by telescope that they were associated with anything at all on Earth. Had Dewey's foundation and the CSLP existed in the seventeenth century, we would know a great deal more than we do today about practically everything in which cycles play a part.

We come next to the work of two men who showed that cycle study is no mere academic pastime, but a subject that affects every living being on Earth.

The Sunspot Mystery

"To be subjected to cosmic effects, man does not have to be shot into space; he does not even have to leave his home." This was the discovery of the late Professor Giorgio Piccardi of the University of Florence, after many years of research into a subject that does not sound particularly exciting in itself: the chemical reaction of an inorganic colloid, bismuth oxychloride.

The mysteries of the cosmos reveal themselves in unexpected ways, however; to John Nelson while trying to improve radio communication, and to Piccardi while looking for an easy way to descale industrial boilers. Piccardi's initial experiments resembled medieval alchemy rather than modern chemistry. He knew how to get the calcareous crust off the boilers' linings: all that was needed was "activated water" produced by an electrical reaction between a drop of mercury and some low-pressure neon placed in ordinary water. How did this affect the water's chemical composition? Not at all, but it cleaned the boilers, and it has been said that any method of solving a problem is the right one if it works every time, whether the method is understood or not.

This was strange enough in itself, but there were more mysteries to come. The method turned out not to work every time after all. Piccardi noticed that there were occasions when his activated water was less effective although there was no obvious explanation for this. So he planned a lengthy series of tests, beginning in 1951, designed to reveal something of the secret life of water.

Water is strange stuff. It is more dense in liquid than in solid form, unlike most substances; it responds to the slightest variation in gravity; it has been shown to have absorption spectra in the VLF band, corresponding to wavelengths of 30 km.; and above all it is apparently temperamental. What Piccardi found, and proved after hundreds of thousands of tests spread over more than a decade (and subsequently repeated independently by many other researchers), was that chemical reactions involving water take place at a different rate according to when and where they are performed.

He designed three tests to run simultaneously, each to serve as a control for the others, and named them F, D and P tests. The F test was to measure the difference in precipitation rate between the inorganic colloid in ordinary water and in activated water. The same set-up was used for the D test except that a metal screen was placed over the apparatus, while in the P test normal water was used in both tubes but only one was shielded by the screen.

During International Geophysical Year (1958), Piccardi's tests were repeated all over the world, and in 1959 he announced that precipitation rates had been found to relate to electromagnetic shielding, altitude above sea level, latitude, and—solar activity! A test carried out in one town gave a different result from a test run at the same time in another town, and yet another result in the same town but at a different time. Piccardi deliberately avoided using biological substances, because he thought they might tend to compensate for any unusual external stimuli, which makes his discoveries even more intriguing. Previously, a chemical reaction involving inorganic compounds was thought to be as standard and unchangeable as anything in science. Now it was found that virtually no two were alike.

The reactions to the Sun were particularly dramatic. An extra burst of solar radiation would be reflected at once in the precipitation rates in the unshielded test tubes. (Those with screens were designed to keep extraterrestrial radiation away.) Meanwhile, in the shielded tubes, reactions went on steadily regardless of solar activity. Piccardi's shielded P tests were kept up daily throughout an entire solar cycle (1951-1959), and in 1962 he published graphs showing an almost exact agreement between the daily sunspot number and test-tube reaction rates. As with many other correlations with solar activity, sunspot counts actually peaked just *after* reaction times, suggesting once again that both cycles were being controlled by a third factor as yet unidentified. There were also variations in average reaction rates according to season of the year, which led Piccardi to formulate his "solar hypothesis," according to which chemical events on Earth correspond not only to short and long term solar events, but also to the position of the Earth on its helicoidal path across the galaxy.

Not content with establishing beyond reasonable doubt that solar events are reflected in some terrestrial phenomena, Piccardi forces us to take radiation from the galaxy into account as well, for not only did his reaction rates correlate with the solar cycle as a whole and with such transient events as magnetic storms, sudden cosmic ray bombardments and solar flares, but another annual cycle with a peak at vernal equinox (around 21 March) was noted, which could only be due to effects from outside the solar system.

If the structure and behaviour of water are affected by extraterrestrial forces, then all living matter must be similarly affected. Piccardi coined the term "non-living substratum" to describe precisely what is affected: the water and colloids dispersed throughout all living matter including, naturally, that of our bodies which are at least two-thirds water. (Water, by the way, is particularly unstable around body temperature.) Since chemical reactions are taking place all the time inside us, keeping us alive, it is no fantasy to suggest that to some extent we are regulated by cosmic forces in our most fundamental processes.

In 1955 Piccardi and a colleague carried out a series of tests at the Camerata Hospital in Florence to see whether human blood behaved in a similar way to the water used in his P test. After 115 experiments, they found that coagulation time was *always* delayed when blood samples were placed under thin copper screens. The delay averaged at 50 percent, a difference of considerable biological significance. (One hates to think of what would befall a haemophiliac who cut himself inside a Faraday cage.) These findings were subsequently confirmed in Japan, and the correlation of the biological (blood) and inorganic P tests is taken as proven.[8]

The tests have now become standard in many countries as a means of observing extraterrestrial effects on local events. A particularly interesting variation was introduced in a German experiment when it was shown that different types of radiation had marked effects on precipitation rates: infrared increased them, whereas they were decelerated under gamma, x or ultraviolet irradiation.[9]

At about the time Piccardi began his search for the clue to the mysterious behaviour of water, Dr. Maki Takata of Toho University in Tokyo was undertaking similar research into equally strange properties of human blood. He was already well known for the test he had designed to measure the flocculation rate of blood by calculating the amount of the organic colloid albumin in the serum. This test is known as the Takata reaction, and the index of flocculation was always thought to be constant in men but variable in women over the period of the menstrual cycle, which made it useful to gynaecologists.

In January 1938, just after a peak sunspot year, reports began coming in from all over the world of an unexplained increase in Takata reaction times: not only for women, but also for men! Takata looked into the matter at once, studying two men over a four-month period: one in Tokyo and the other 250 miles away in Kobe. Their daily flocculation indices matched each other exactly, which seemed to rule out local factors, and while studying their indices Takata noticed an odd detail: they tended to perk up sharply about fifteen minutes *before* sunrise.

Immediately suspecting that the Sun was involved somehow, Takata carried out tests in an aeroplane flying at over 30,000 feet, to see if this effect would show an increase due to the relatively thinner shielding from solar rays at such heights. It did, and further evidence was provided by measuring flocculation rates during a solar eclipse, which Takata did on three separate occasions, finding that the Moon's path across the Sun effectively blocked its influence and caused the blood indices to decline. This in turn confirmed earlier tests done 600 feet underground at the bottom of the Meiken mine shaft.

Takata's early research was of great interest in showing how the Sun's radiation can be shielded, and without the aid of a mine shaft or eclipsing Moon this is still difficult. We terrestrians, as Gauquelin has said, can regard ourselves as "living in the interior of the Sun." Takata also found, after some 20 years of non-stop observation, that the flocculation index of our blood is virtually locked onto the solar cycle, responding directly to whatever the Sun happens to be doing, whether rising, setting, emitting a flare or hiding behind the Moon. The index also increases with diminishing latitude. "There is no doubt," Takata concluded in 1951, "that solar radiation must contain a hitherto unknown biologically effective component, characterized by its strong penetration power and intensive vital-ionizing ability."

Subsequent research has indicated that the real cause for the sunrise-flocculation index correlation may be the presence of ELF electrical waves set up in the atmosphere shortly before the Sun rises. Experiments have also shown that artificial electric charges produce changes in the albumin similar to those that occur naturally in the Takata reaction, while it has been found independently that the ability of the blood to coagulate quickly is also affected by magnetic fields in the ELF range at 50 Hz.

We have greatly oversimplified the discoveries of Piccardi and Takata, and interested readers are referred to their original material. What is quite clear, however, is that, as Piccardi put it: "all living beings are bound more intimately to the external world than one would think." As for Takata, he summed up his work by declaring "Man is a living sundial!"[10]

The question again arises: *what else does the Sun do to us?* Before we look at some more terrestrial cyclic events in which the solar cycle seems to be implicated, we must emphasize again that we cannot regard the Sun with certainty as the first or sole cause of any of them. Correlations with Earth events would be far more exact than they are if this were the case. It is safer to regard the rotating band of sunspots as a junction, whence energies are beamed at us in a reasonably regular pattern that may one day be wholly predictable. The fact that terrestrial "effects" often precede solar "causes" reminds us that more than the energy produced by the Sun alone is involved; its spot cycle may be a convenient bulletin of its activity, but no more.

It is barely a century since the sunspot cycle was established as our first clue in the search for less obvious cycles. Then came the discovery that the spots were electromagnetic. Next it was learned that man is also an EM entity, related to the cosmos far more directly than our ancestors could have imagined, although many of them were instinctively aware of such a relationship. Something from the Sun actually comes all the way to Earth; rays and particles find their way into our atmosphere and perhaps even into the cells in our bodies, and since, as we described in earlier chapters, both cosmic inputs and the solar output are cyclical in nature, it would be logical to expect many events on Earth to obey similar fluctuations.

We have already mentioned the close correlations between the solar cycle and geomagnetic conditions, chemical precipitation and blood flocculation rates. These are now so well established that they are no longer open to reasonable doubt. The link between the Sun's output and the thickness of some (but not all) tree rings, and varve deposits, also seems virtually certain. There is also good evidence for cyclic phase-locking between Sun and overall incidence of earthquakes in some areas, mental disorder incidence, traffic accidents, overall incidence of heart attacks, telecommunication reception, admission to psychiatric hospitals, in addition to a number of social phenomena to be mentioned in a later chapter.

Professor Chizhevsky, whose work with air ions we mentioned in Chapter 4, did an enormous amount of work on cycle study throughout his career. In one of his books, written in 1936 but published only in 1973, nine years after his death, he provides compelling evidence linking variations in the solar output to a wide variety of diseases, notably epidemics.[11] Such studies were foreseen in the eighteenth century, when W. Hillary published his *Account of the Variations of the Weather and Epidemical Diseases from 1726 to 1734* and I. A. Rutty brought out a *Chronological History of the Weather, Seasons and Diseases in Dublin from the Year 1725 to 1765*, although neither of these authors can have been very knowledgeable about sunspots and their cycles.

Chizhevsky began studying sunspots while still a student, publishing his first paper, on *The periodical influence of the Sun on the biosphere of the Earth* in 1915, when he was eighteen. He not only dug through a huge amount of material, mostly French and German, on relations between climate and disease, but did a great deal of original research of his own. Going over Russian archives for the nineteenth century, he found an impressive correlation between solar activity and epidemics of cholera, influenza, typhus, diphtheria and meningitis. Over and over again, peaks in the incidence of epidemics coincided with peaks in solar activity, and by rounding up all available evidence on periods of solar maximum since the sixth century, he even claimed to have found "incontestable proof" that plague epidemics took place almost twice as often at times of maxima than at minima. He was so convinced of the link between Sun and disease that he wrote at least two papers urging all hospitals to install an armour-plated room to protect certain patients against the harmful effects of solar and cosmic EM radiation.

Chizhevsky also studied the question of overall mortality figures, using data going back more than 100 years from different parts of the country. Again, the coincidence appears striking, suggesting that although numerous factors other than the Sun obviously influence overall death rates, it is as though the solar output variation were a constant outline of general trends. Chizhevsky himself repeatedly insisted that the Sun did not cause anything specific, but simply made certain events on Earth more likely to happen.

Our only serious criticism of Chizhevsky is that he often fails to describe his experiments and research in detail. However, this applies to a great deal of other work done before the post-war information explosion forced reporting standards up, and in Chizhevsky's defence it must be said that his main thesis has been supported by many other Soviet researchers, who regard him today as the "father of heliobiology." This is the name they give to the study of possible effects of the solar output on biological systems, and the latest reports indicate a growing interest in the subject.[12]

Probably inspired by N. A. Shults's large-scale study at Sochi in the Crimea, which revealed an almost exact correlation between solar activity and white blood cell counts in normal people, other Soviet researchers have tracked down links between Sun and animal behaviour, Sun and disease, and to complete the picture, animal behaviour and disease. It is now known, for instance, that Siberian water rats will proliferate at certain points in the solar cycle, and since it is also known that plague epidemics are caused chiefly by an excess of rats, it is now possible to plan ahead of the next rat invasion, thereby possibly preventing an epidemic. The solar cycle also remains obstinately locked on to those of scarlet fever, malaria and polio outbreaks, and especially those of heart disease incidence, to such an extent that many Soviet doctors now co-operate with astronomers so that special care can be taken of patients when the Sun is unusually active.[13]

The day may come when every hospital will have its resident astronomer. Meanwhile, Chizhevsky's call for radiation-free rooms in hospitals has been taken a stage further with the advent of "speleotherapy," or the use of underground caves for curative purposes. This unusual form of treatment has an unusual history.

During World War II, inhabitants of the region between the Ruhr towns of Wuppertal and Hagen used to shelter from air raids in a deep cave in the Ennepe valley. After the war, a local doctor, Karl H. Spannagel, noticed that his asthma patients kept telling him how much easier they had found it to breathe while hiding in the cave. Lacking medicine for his post-war bronchial asthma patients, the doctor told them to go back to the cave for a few days. Results were so encouraging that other asthma sufferers soon flocked there, and a small sanatorium was eventually built on the spot. Other cave-hospitals have been opened in Hungary and Romania, while in the old salt mine of Wieliczka in southern Poland there is now a specially equipped sanatorium 600 feet below ground through which 3,500 patients had passed by 1976.[14]

In this chapter, we have been mainly concerned with the influence of solar energies on living systems. There can no longer be any doubt that such influences do exist, and that they can be significant. We also saw, in Chapter 2, that the solar output is not constant, suggesting that the Sun is influenced in turn by more distant sources of energy. Commenting on theories of planetary influences on the Sun, Dr. James E. Vaux, executive director of the Foundation for the Study of Cycles, has remarked that "until adequate mechanisms are at least theorized, they technically remain in the area of numerology."[15]

Such mechanisms are difficult even to theorize, since for all we have learned in the first two decades of the Space Age about the invisible forces in our solar system, we still know next to nothing about other more distant sources of energy compared with which our huge powerful Sun is barely significant. We have already mentioned the quasars and pulsars, possible sources of energy on an unimaginable scale that is very probably cyclic in nature, like the energy of the Sun and presumably of other stars.

Two Israeli astrophysicists, Dror Sadeh and Meir Meidav, have already recorded what they believe to be gravitational seismic waves originating from the pulsar known as CP 1133. Analysing the signal, they found periods that seem to fit in with those of seismic activity on the Moon as recorded by the equipment left there by Apollo 12.[16] There is still much debate as to the existence of superlong gravity waves, tachyons (hypothetical particles travelling faster than light), and other unknown waves and energies, which are proving just as elusive as the ether we mentioned in Chapter 1. However, if such particles or waves do exist, and if they were quadrupolar (with four poles instead of two, like known EM waves), it could be that when the planets of our solar system are arranged in certain geometric patterns, they can resonate more effectively to forces of inconceivably remote origin, thereby locking us on to cycles generated by sources we cannot yet hope to identify or measure, let alone understand.

For the time being, therefore, we must be content not with numerology but with numerical correlations, as we try to observe what is happening around us and note apparent coincidences. We have been quite successful in a century of scrutiny of the sunspot cycle, although all we still know for certain about a sunspot is that it is visible because it is slightly cooler than its surroundings. Even so, we are making good use of the sunspot periodicity as we discover more and more terrestrial events that seem to be related to it. We could perhaps be making more use than we are of our knowledge of the planets; we now have a good idea of their mass density, size and orbital periods, and thanks to such data we can compare their periodic motions with the fluctuating output of the Sun and hopefully record further coincidence and correlations.

We return, with more physical evidence, to the subject of the importance of cycles in human affairs in Chapter 10. In the following three chapters we again leave the mysteries of outer space and turn to those of inner space—the universe beneath our skins. We have already tried to relate man to his ecosphere, or local geophysical environment, from a new point of view (in Chapters 3 and 4), that of an interconnected electromagnetic system of which he is an integral part. We now look for further connections that are cyclic in nature; for it, as we have seen, there are countless cycles to be found in nature, so there are also cycles in man and in all living systems.

It is more than likely, therefore, that such "biorhythms" are to some extent driven by external forces.

Chapter 7

BIORHYTHM

MANY PEOPLE find it difficult to get up in the morning. It seems the most unnatural thing to do at the time, as if some internal clock of our own were struggling to unlock its cycle from those of day, night and the wretched alarm clock. Others find it equally difficult to get to bed at night: the day is not long enough for all that has to be done. Some of us simply cannot fit ourselves into a 24-hour schedule without regular use of pills.

One enterprising fellow, Donald R. Erskine of Philadelphia, has gone a stage further. He refuses to compromise with nature, and lives according to his own rhythms, going to bed about an hour and a quarter later every night. His rearrangement of the night-day cycle does not seem to have done him any harm, although he has kept it up for more than 20 years.*

However, for practical reasons most of us have to adapt to the 24-hour clock, and we have divided our days into roughly equal 8-hour periods of work, recreation and sleep. It follows therefore that our most common biological rhythm, or biorhythm, should be a 24-hour one, and so it is—but not exactly. Hence the term *circadian* has had to be coined (Latin *circa*, about, and *dies*, day) to describe the period of *about* 24 hours that is neither exactly a solar day of 24.0 nor a lunar day of 24.8 hours. Cycles shorter and longer than circadian are known respectively as *infradian* and *ultradian*.

---

* Quoted by Luce (see ref. 4).

Since Earth turns on its axis once every 24 hours, it is natural that some inputs to our ecosphere will have the same period, and possible that biological systems will respond to them. Within the circadian period lies the next most obvious of all biorhythms, the diurnal, or 12-hour day-night periods of alternating light and darkness. This rhythm is also called *photoperiodic*, since the amount of light reaching Earth varies according to the Earth's daily rotation, giving a "photoperiod" very near to 24 hours. One easily visible effect of this photoperiodicity is the way certain plants regularly open their leaves in the morning and close them again in the evening.

It was always generally assumed that the photoperiod is the main trigger, or what the Germans call the *Zeitgeber* (time-giver) of circadian rhythms; yet it cannot be the only one, for it has been shown that animals confined to a lightproof cage will continue to display circadian rhythms, as will their offspring born in total darkness and never exposed to any light at all. There are even some living creatures, such as the transparent fish in Mammoth Cave, Kentucky, that spend their whole lives in darkness. Clearly, there is more to the subject of biorhythm than was suspected until recently.

To many, the word biorhythm brings to mind the alleged cycles of 23, 28 and 33 days that govern our emotional, physical and intellectual lives. These are often spoken of as if they have been discovered, like a new continent, and proved to exist. This is far from true. We shall examine these "Fliess" ultradian biorhythms shortly, but it is important to bear in mind that these, if they exist, are only three of an unknown number of other infradian, circadian and ultradian cycles that must affect the workings of our bodies and minds, perhaps to some extent even driving them.

After much research with insects and animals, Professor C. S. Pittendrigh of Stanford University has come to believe that circadian rhythms must be instinctive even to the behaviour of molecules as they adapt to surroundings dominated by the 24/25-hour periods of Sun and Moon.[1] They certainly dominate many of man's most vital processes: temperature, respiration rate, blood pressure and flow, excretion of minerals, metabolic levels, urination, pulse count and many more. Our work capacity and creativity come and go in circadian cycles, and it has been well established that we are actually stronger and weaker at certain times.

Infradian cycles include those of hunger—most of us certainly get hungry more than once a day!—levels of sleep and wakefulness, and even periods of dominant breathing through alternate nostrils, a fact of which we are usually unaware unless we have a cold. Such infradian and circadian cycles are much easier to identify and observe than the ultradian ones, which may extend over days, years, decades, or a whole lifetime. It takes at least eleven years, for instance, to see if there is any biorhythm that corresponds to the solar cycle, while another eleven years are needed to confirm the discovery. However, thanks to long-term projects such as Dr. Tromp's blood sedimentation rate tests at Leiden, which have been going on continuously since 1955, we are able to examine at least some extraterrestrial effects on some of our internal parts, and compare variations in such effects with those of the various Piccardi tests in Leiden and elsewhere.

Another approach to the study of body rhythms is to take human volunteers and see what happens when they are deprived of normal external stimuli. In recent years, men and women have found themselves in extraordinary situations in the name of science. They have been locked into underground bunkers, they have lived for months in caves, camped on islands far to the north where the Sun never sets, and been suspended in darkened and soundproofed water tanks, though tank-tests have more to do with studies of sensory deprivation and psychedelia than of biorhythm.

When we are shut off from the familiar world, some of our internal clocks go out of control, while others keep good, even perfect, time. Man is the most adaptable of creatures, yet we all adapt in different ways. A Frenchman who spent 63 days in a cavern came out under the impression he had been there just over half that time. Another Frenchman, who stuck it out in a cave under the Alps for six months, showed an interesting combination of steady circadian rhythms and more erratic activity periods of up to twice the length of a normal day. Prisoners kept in solitary confinement have survived the most appalling conditions: the eminent Soviet astronomer Nikolai Kozyrev, whom we mentioned in Chapter 5, spent eleven years (1937-1948) in jail, being locked in a cell for more than a year with a man who had gone insane. Dr. Kozyrev kept going by meditating on the nature of the Universe and especially on "what Time and the passing of Time really are," and after his release was able to develop his ideas into a fascinating new theory of its properties.[2]

In a well-known experiment carried out in 1955, Dr. Mary Lobban of Cambridge University not only deprived her subjects of photoperiodic stimuli, but added an ingenious twist to their predicament. She left separate groups of students on islands within the Arctic Circle in permanent sunlight, providing members of two groups with special Rolex watches fixed to run three hours fast or slow per day, so that 24 watch-hours were actually 21 for one group and 27 for the other. After 42 (real) days, it was found that heart rate and blood pressure rhythms easily adapted to the artificial day—nobody knew if his watch was telling the right time or not—whereas excretion of water, chloride and potassium (ingredients of urine) obstinately held to a 24-hour pattern, especially potassium. Dr. Lobban concluded after this follow-up of an earlier test that "it is now impossible to assume that all diurnal rhythms in man are controlled by one common mechanism."[3]

Such experiments, of which there have been many, naturally lead to argument as to whether our body clocks are endogenous or exogenous, that is, regulated from within or from without. At least some regulation must be imposed from within: some scientists see the prevalence of circadian rhythms as an indication that living systems have been forced throughout their evolution to adapt to the environment, those unable to do so being eliminated according to the precepts of natural selection and the survival of the most adaptable. But this cannot be the whole story: some body rhythms are clearly at the mercy of Sun and Moon, and presumably all other exogenous forces as well.

The planets have their days, just as do the Sun and Moon, and these are close to, but not exactly, 24 hours in length. Planet-days and their influence on Earth events were first studied in detail by Michael Gauquelin, whose work will be mentioned later, and although the question of planet-driven biorhythm is a new and controversial one, it deserves a hearing.

Biologists are prepared to admit that social conditions may be a factor in the regulation of body clocks, but they seem reluctant to look further afield—further outwards into space, that is. Gay Gaer Luce, whose invaluable popular book on biorhythm lists some sixty pages of references, has next to nothing to say about external forces other than Sun and Moon. She does speculate, however, that since man has a tendency to lengthen his circadian activity when in isolation (or "free running") rather than shorten it, such longer periods might reflect a time structure "compressed to 24 hours by social conditioning throughout one's life," as she puts it. She also mentions that since we are always unshielded from gravitational fields, and almost always from most electromagnetic fields, such longer rhythms may reflect "other synchronizers," and she concludes her book by pointing out—rather late in the day, perhaps—that the time structure of our bodies is only partly within us. For, she says, "we are open systems, unable to detach ourselves from the beats of this nature of which we are part." And, as we must have made clear by now, much of this nature, possibly the greater part of it, is extraterrestrial.[4]

Ups and Downs

We all have our good and bad days. Sometimes we feel suddenly elated or depressed for no apparent reason, and then our disposition returns to normal. Creative work may involve more perspiration than inspiration, but there are days when the artist simply cannot create: however hard he perspires and racks his brain, the ideas refuse to flow. Yet on other days the poems, paintings or music seem to create themselves with the minimum of help from him. It would be useful if we all had wall charts telling us exactly when to expect good and bad days: we could plan important projects and appointments for good days well in advance and keep the bad ones free.

Some believe this is already possible, by using the so-called Fliessian method of charting our life cycles, which were calculated independently early in this century by a Berlin doctor named Wilhelm Fliess and a Viennese psychology professor, Hermann Swoboda. In 1897, Swoboda began to study biorhythms after noting that his own parents' activities seemed to fall into rhythmic cycles.

Poring over his notes, he thought he could see definite periods of 23 and 28 days emerging, and he designed a slide rule to help people locate their "critical days," which according to him were days on which the lines of each cycle crossed midpoint on the horizontal scale.

At about the same time, Dr. Fliess arrived at similar conclusions by very different means. Qualified as an ear, nose and throat specialist, Fliess developed some very odd ideas. He seems to have hit on his biorhythm cycle theory by studying the mucous lining of the nose, which he thought to contain "genital cells." These, like other human cells, were male and female, and each cell-sex had its own cycle. Fliess carried out two operations on the nose of Sigmund Freud, who is said to have been influenced by Fliess's ideas on one of his main obsessions—the bisexual nature of man.

The third founding father of pop biorhythm was an Innsbruck engineer named Alfred Teltscher, discoverer of the third cycle, that of intellectual performance. This completed the system of three cycles: the 23-day physical, the 28-day emotional, and the 33-day intellectual. These are said to start all together at the moment of birth, and to run regularly throughout our lives. If this is so, they should all reach mid-point simultaneously and start again a few months after our 58th birthdays, giving us surely the most "critical" day of our lives!

The quantity of published evidence for the existence of these cycles is impressive at first sight, but the quality of much of it is less impressive. Were it all true, it should now be possible to predict when we are most likely to fight our wives, fall off a bus, write great poetry or just feel terrible. To make things easier for us, all kinds of gadgets are now on the market, from Swiss-made watches to Japanese mini-computers, none of which is necessary for anybody numerate enough to add up how many days he has lived and divide the total by 23, 28 and 33. (Don't forget the leap years.)

United Airlines have supposedly halved accident rates among ground staff by locating employees' critical days with a computer and advising them to take extra care on such days. Swissair, it is said, has had no accidents at all on flights for which pilots' biorhythms were studied beforehand. In Japan, where pop biorhythm seems to have become a way of life, the Ohmi Railway Co. (which operates buses and taxis, not trains) claims astonishing success after four years of picking critical days for its 700 drivers. Again, a warning note to the driver seems to do the trick. Japanese insurance salesmen are even said to study their victims' charts in order to be able to catch them on days when their sales-resistance is low, while some Japanese motorists fly little flags on their vehicles to warn the world at large that they are having a bad day and are liable to have an accident at any moment.[5]

There are, however, several major objections to the claims of pop biorhythm supporters. As far as accident prevention is concerned, simple suggestion should be enough to make a man drive more carefully, whether or not the day is a "risk" day. If a taxi driver is told he is likely to have an accident next Friday, he will take extra care. (True, he *may* get nervous and have a crash.) Another objection concerns the fact that in an average month, everybody has at least six critical days. Nobody therefore can ever be more than a few days away from an allegedly critical day, so when we read that so-and-so died suddenly "only a few days from a critical day" we can hardly be surprised. Maybe Clark Gable, Judy Garland and Marilyn Monroe all died on critical days, but countless other celebrities must have died on non-critical days. We do not seem to hear so much about them. If it could be shown that a statistically significant percentage of *all* deaths occurs on critical days, the Fliess school of biorhythm will begin to look more plausible, but as far as we know, no pop biorhythm promoter has taken the trouble to do this. A statistical survey on the scale of that carried out by the Gauquelins (see Chapter 10) is called for here.

Two more serious objections to Fliessian biorhythm are that it uses too few cycles, and the three it does use cannot possibly be as regular as claimed for everybody. There are so many cyclic forces at work on the human body that it would be very strange if we converted them all into three neat cycles identical for every single man or woman. No two human beings are identical, so why should their cycles coincide? Moreover, cycles such as those of sunspots may have an average period of 11.1 years, but very few are ever exactly that length. This applies to countless other cycles that have been established far more scientifically than the three claimed by the Fliess school, so why should these be exceptions to the rule?

A final and more personal objection to pop biorhythm is that it simply does not work: at least, not for everybody. One of us (GLP) plotted his own charts for almost a year and marked days of special interest as they passed. Not one good or bad day ever fell where it should have fallen according to Fliessian lore, and nothing special ever happened on a critical day.

Yet there is certainly a baby to be found in the murky bathwater of Fliessian pop biorhythm. There may well be cycles of 23, 28 and 33 days in human activity, but if they have anything in common with the other known cycles of the body, they will be subject to variations in phase, amplitude and period, also to phase-locking or "bioentrainment" which we discuss in the following chapter.

The work of Dr. Rex Hersey is sometimes cited in support of the claims of the Fliess school. When examined more closely, however, it is found to go directly against those claims. Dr. Hersey, of the University of Pennsylvania, spent a whole year (1929-30) studying emotional cycles among industrial workers in a number of different factories. He then picked out a group of 25 apparently normal and contented workers of average intelligence and examined them carefully for 13 weeks. He wanted to find out if the "steady and regular" type of worker was really as steady as he seemed, and whether management staff were wise in treating all their men as if they were identical robots.

Hersey found that his test subjects did indeed have emotional cycles, though they varied for each individual— from 16 to 63 days—averaging at around 35 days, a whole week longer than the Fliessian 28-day emotional cycle! This is particularly interesting since his 25 subjects were picked for their apparent constancy of mood.[6]

We are surprised that little follow-up of Hersey's work seems to have been done. Comparative studies of emotional cycles in people of different backgrounds, races and nationalities might be of value in a number of ways. It would be interesting to see, for instance, if there is any statistical evidence for such popular conceptions as the excitability of Latins or the imperturbability of Japanese and Slavs.

One of the few recent attempts to check the validity of Fliessian biorhythm was made by three researchers from Wyoming University and state hospital. They studied a group of hospitalized psychotics, evaluating them for work, sleep and general well-being, and found that "the critical days hypothesis was not shown to be a meaningful concept" having "no predictive value beyond chance." However, one of the researchers admitted having noted a relation between subjective ratings of his own "ups and downs" and the hypothetical biorhythm of the Fliess school, followers of which could well criticize the Wyoming test on grounds of the very small sample used, and of the special type of subjects involved. Who knows, perhaps people lose their biorhythms when they become psychotic?[7]

Of the three claimed cycles of 23, 28 and 33 days, that of 28 days is most likely to be near the mark, simply because of the influence of the Moon. Correlations between Moon and man's behaviour are so well known that we have omitted mentioning them in any detail. But which Moon? Or rather, which "month"? As we mentioned in Chapter 1, there are at least five different months ranging in length from 27.2 days (draconic) to 29.5 days (synodic), not counting yet another month of about 30.4 days, which is one twelfth, or a subharmonic of one year. There is no natural month of exactly 28 days, and nor are there any obvious natural cycles of 23 or 33 days with which those of the body could be entrained.

The only way to settle the argument over biorhythms is to measure them instrumentally, and a start has been made in this direction by Dr. Leonard J. Ravitz, using methods developed by Professor H. S. Burr of Yale which we discuss in later chapters. Ravitz has found that there are "electric tides" in what Burr called the L-field, or "electrodynamic" life field, which he claimed all living beings possess. The most pronounced cycles to emerge from Ravitz's studies are of 14 to 17 days, 28 to 29 days, also diurnal, seasonal and semiannual. A lesser cycle of seven to eight days was also noted. Since Ravitz has described in great detail how these L-field measurements reflect a wide variety of emotional states, it is rather amazing that no follower of Fliess seems to have thought of checking his claims against Ravitz's instrumentally measured findings, which seem to indicate that at least one of the Fliessian rhythms may be quite near the mark.[8]

Anybody can monitor his own biorhythms, and we strongly urge everybody to try to do so. We recommend a scoring method similar to that used by Hersey, who asked his subjects to rate their own daily emotional level by a single entry on a scorecard: three points if you feel elated, two for happy, and one if you feel just about right. Score zero when you feel no special emotion, minus one for slightly off-colour, minus two for sad or depressed, and minus three for being in really bad shape. As the days pass, you link up the points on the graph of your personal emotional cycle chart (music paper is ideal, taking the middle line of the stave as zero) and wait for some pattern to emerge. At first there will probably be violent ups and downs, until you learn to distinguish moods induced by external events from those that seem to originate from within, which are the moods that are relevant to accurate biorhythm plotting.

This is much harder to do than it may sound. The tendency is for people to give it up when they go through a bad patch, which is just when the data would be most useful to them. Nevertheless, this is the most practical way in which accurate data about individual biorhythms can be acquired, and anybody with access to a suitable voltmeter can check his subjective impressions with electrical field readings. When there is enough data of this type, we should have a far more precise understanding of the nature of the human temperament than we have at present. The really serious chart-plotter can keep three separate charts for his emotional, intellectual and physical states, and a good tip for beginners is to plot the curve of sexual urges, the easiest of all biorhythms to identify.

Chronopharmacology, or Take Two Pills at 8 a.m. Precisely

The importance of knowing rather more than we do about biorhythms in general is illustrated by a remarkable experiment in which two groups of laboratory rats were given similar doses of x-rays at different times, one group during the night and the other exactly twelve hours later. The entire group that was radiated at night died, while the entire daytime group survived![9] It has naturally been impossible to repeat such an experiment with humans, although to some extent something similar may be happening all the time. If rat biorhythms are so marked that inputs into their body chemistry can kill them when wrongly timed, it is possible that the same general principle may apply to human beings.

According to traditional Chinese medicine, a subject rapidly gaining respectability in the West, the most vital body rhythm of all is that of the flow of *Ch'i*, or vital energy, around the twelve main meridians at regular intervals. The cycle, according to tradition, begins at 3 a.m. (local time) in the lung meridian and continues, at two-hourly intervals, to make up a 24-hour period as follows:

0300—0500 Lung	1500—1700 Bladder
0500—0700 Large intestine	1700—1900 Kidney
0700—0900 Stomach	1900—2100 Pericardium
0900—1100 Spleen	2100—2300 Triple warmer
1100—1300 Heart	2300—0100 Gall bladder
1300—1500 Small intestine	0100—0300 Liver

Then, at 3 a.m., the cycle starts again. According to this scheme, each organic deficiency should be treated according to the time when the *Ch'i* is concentrated in the right meridian. This would mean calling your local acupuncturist out at 3 a.m. to treat you for an attack of asthma, for example. Western practitioners may be doing their best to forget this part of Chinese tradition! (We return to the subject of acupuncture in more detail in Chapter 11.)

Experts in the use of herbal medicines have long known that doses have different effects if taken at different times of the day. The Tibetan herbal compound Padma-lax, for example, which is marketed in Europe as a laxative, can actually have just the opposite effect if taken at the wrong time, increasing constipation instead of relieving it. Many other herbal preparations are believed to have maximal effects when taken only at specific times of day or night.

Not only herbal remedies follow the body's natural rhythms. Dr. Georg Bentze of Oslo, Norway, says: "It is well known among doctors that *all* medicines have different effects depending on what time of day or night they are taken. Frankly, doctors never tell their patients this, because they have enough trouble as it is prescribing pills. Who wants more trouble?"[10] Doctors can argue with some reason that it would not be practical, given the already precarious state of medical services in many western countries, to introduce a time factor into their prescriptions which, on the whole, tend to solve their patients' problems. However, the fact remains that our bodies would appreciate more attention to their daily rhythms, to put it mildly, and this has led to the new field of "chronopharmacology."

"What is a safe dose at one phase of the circadian rhythm may not be safe at another time," says Gay Gaer Luce. "Timing may tip the balance between survival and death." This might sound like an exaggeration without basis in fact, but coming from one who has researched her subject so thoroughly it should not be dismissed. There is a mass of evidence indicating that both animals and humans respond very differently to a wide range of stimuli according to the time of their occurrence. Dr. Franz Halberg, who has produced a time-map of "vulnerability rhythms," has made a number of discoveries with animals that seem to have alarming implications for humans. He has found, for instance, that a group of mice seemed to ignore a noise as loud as that of a jet engine by day, but when the same noise was inflicted upon them at night it proved fatal: the animals suffered convulsions and dropped dead. He has also found that if mice are given the equivalent in human terms of a quart of vodka early in the morning, this is enough to kill 60 percent of them, whereas only 12 percent died when given the same does at the onset of their rest period.

Some vaccinations are known to be more effective at certain times of the day than at others. The impact of many injected drugs, including insulin and heroin, also varies during the day. Numerous cases of death due to "overdose of drugs" might not have occurred if the dose had been administered at some body-time other than the fatal hour. It is not unreasonable to suppose that everything we eat or drink will have a different impact upon our bodies according to when ingested. In many cases, differences in effect may not be biologically significant, but in other cases they may be striking. Very little is known about the time factor in the art of healing. It is common practice, says Gay Gaer Luce, for critical operations to be arranged at times when the surgeon is likely to be on his best form. But, she asks, what about the patient's best time?[11]

In Chapter 2, we mentioned the subject of resonance, the process by which energy is exchanged between two otherwise unconnected bodies by means of synchronous vibration. This principle does not apply only to tuning forks and piano strings, or planetary orbits and solar tides, but it also involves people. If our bodies' internal clocks are even partially driven by external stimuli, which seems likely, then we must respond to variations in such stimuli. Therefore, we might ask, what is going to happen when an external cycle falls into step (phase) with one of our natural biorhythms? This is the subject of our next chapter.

Chapter 8

BIOENTRAINMENT

Homo electromagneticus is as much part of the EM environment as a household appliance is part of a domestic circuit. Perhaps the comparison should end there, for man is far more complex than a refrigerator and is self-sustaining to some extent, running on his own bioelectric batteries. Moreover, he cannot be switched off, at least until death, and perhaps not even then. Throughout his life, at least, he is permanently situated within the fields, tides and currents of the universe and subjected to their influences.

As we have seen in previous chapters, a large number of natural phenomena come in cycles, including the EM waves that find their way into our immediate environment. We have seen how sensitive the biosystem can be to external EM effects, and how our bodies have natural rhythms that lie remarkably close to those of nature. The question now arises: what happens when a "free-running" biological cycle meets an external input in the right frequency range and of the right amplitude? The answer is *bioentrainment*, a newly coined word used to describe the process by which living beings become locked on to cycles, rhythms or waves from outside their body limits.

The very origins of life may have been generated by a process of entrainment. In 1953 Stanley Miller seized popular imagination when he managed to synthesize amino acids, the basic building blocks of proteins and therefore of all life, by simulating prehistoric Earth atmosphere conditions in a sealed glass jar through which he passed an electric arc, or artificial flash of lightning. Later, Leslie Orgel produced adenine, one of the components of the DNA molecule, by artificial methods again planned to recreate the Precambrian era in the test tube. These were exciting experiments, though neither Miller nor Orgel has yet been able to grow a little green man. But how did Nature manage to create life, unaided by modern laboratory equipment, some 600 million years ago?

It has been suggested that there must have been an asymmetrical environmental force in action when the first molecules of life were created, and that the most likely form of such forces would be some form of EM field. We could assume such fields to have had a role in organizing or structuring the geobiochemical environment, converting the entropic chaos of a pre-Creation "organic soup" in the oceans and atmospheres into the first primitive organized forms of life. Can it be coincidence that the most fundamental component of a living being, the DNA molecule, is in the form of a spiral or helix (a fact only discovered in very recent times), which is the way the life energies have been represented in mystic tradition for thousands of years? And what force other than an EM one would have such organizing ability? Future histories of the world may begin with the phrase *"In the beginning was the Field . . ."*[1]

There is already evidence for an influence of our invisible EM environment throughout every stage of life. A Soviet scientist has claimed that variations in the geomagnetic field (GMF) can affect the formation of the central nervous system of an embryo, and thereby lead to the birth of a future schizophrenic. GMF variations are known to be capable of entraining our hearts, and it is significant that in the artificial-day experiment mentioned in the previous chapter, the heart-beat was one of the first human functions to adapt to (or become entrained by?) the unfamiliar Arctic environment of perpetual light. Dr. Yuri Kholodov, a Soviet neurophysiologist and pioneer in the study of EM field and biosystem interactions, is convinced that EM fields act directly on the central nervous system, bypassing the known sense organs. We are, therefore, as Piccardi repeatedly insisted after a long lifetime of research and direct observation, *permanently* under the influence of the natural forces of space. Moreover, he pointed out, the action of EM radiations and fields is total, in that it affects the whole mass of an organism and evokes a response in any structural element capable of responding to such stimuli.[2,3,4]

Without going into too much detail, we can point to numerous parts of the human body that offer a striking resemblance to parts of radio sets. There are nerve groupings that resemble coils, ganglia that remind us of triodes, parts of the body where AC seems to be rectified into DC and, according to Dr. R. O. Becker, we also seem to have developed a bio-transistor in our nervous systems. Itzhak Bentov, an Israeli scientist working in the U.S., claims to have identified five separate resonating systems in the human body: in the heart and aorta, the skull, the third and lateral ventricles, the sensory cortex, and in each hemisphere of the brain, where pulsating magnetic fields of opposite polarities are set up, these being very sensitive to environmental fields and providing a possible mechanism by which the brain picks up information from the environment through "resonant feedback."[5]

To return to our biological clocks and *Zeitgebers*; a growing band of biologists, largely inspired by the research of Professor Frank Brown of Northwestern University in Evanston, have come to believe that the regulation of the clock must be at least in part of external origin. To prove that the rhythmic systems of living beings were wholly independent of the rhythmic environment, Brown has said, one would have to assume the complete absence of any periodic input into the organism, which is clearly impossible. Natural geophysical fields, presumably modulated by extraterrestrial fields, must have been present throughout the whole evolutionary development of life, on Earth and perhaps elsewhere. Surely it is reasonable to suppose, then, that the organism has always relied to some extent, perhaps totally, on an external rhythm in order to set the pace of its internal *Zeitgeber* or body clock? Or must we fall back on "chance coincidence" to explain the close similarity between so many physiological and natural external cycles?

These are merely speculations: now let us get back to the laboratories where attempts have been made to gather the evidence we need for better theories. There are two obvious ways of seeking to establish the link between external rhythmic sources and the triggering of an internal body clock. One is to shield out such potential external triggers as we can (bearing in mind that we can never hope to shield them all), such as radio waves and the Earth's static magnetic field. The other is to take one of the suspected exogenous agents, simulate it artificially, and see what biological effect it has in the laboratory. As recently as 1969, W. Ross Adey, whose research we shall discuss shortly, was lamenting the fact that very little work of this type had ever been done on either animals or man.

Adey, a space biologist with the Brain Research Institute at UCLA, is one of a group of scientists interested in the effects of EM fields on biosystems in spaceflight conditions, when men are deprived of their normal environment and subjected to heavy particle concentrations and fields stronger than those to be found on Earth, once they have passed through our protective layer of atmosphere (or, perhaps, even closer to home, as suggested by the fate of the Soyuz XI crew).

An interesting early experiment in body-time artificial distortion was carried out at the Max-Planck Institute for Physiology in Germany. Scientists fitted out a bunker-apartment in the institute's subterranean laboratory as a Faraday cage, in such a way that the test subjects living there for the experiment were unaware that the walls and ceiling of their temporary home were lined with hidden metal shielding. Thus they were isolated from almost all known forms of electric and magnetic field effects. A control group spent the same time underground, but without the shielded conditions.

Both groups began to develop longer circadian periods than normal, the unshielded subjects averaging 24.84 hours and the shielded group 25.26 hours. The latter also showed more irregularities than their colleagues, notably in periods of urination. This is especially interesting in view of the fact that in Mary Lobban's Arctic experiment, urination was the one rhythm that stuck faithfully to the 24-hour cycle even as other rhythms ran off the rails.

The Max-Planck scientists then tried the effects of a strong artificial electric field on a group shielded from all other fields. This immediately had the effect of shortening the average "day" to just under 24 hours. Some volunteers even seemed to have previously erratic body functions "corrected" by the artificial treatment.

We should not read too much into the results of these tests, for social conditions obviously played their part in altering physical rhythms. This aspect of the test was studied by neuroscientist Dr. Ernst Pöppel, who did not believe the social factor could be wholly responsible for the alterations in circadian rhythms. Moreover, whenever people are involved in any scientific test, there are so many individual personality features to be taken into account that generalizations are unwise. The fact, for instance, that astronauts on the whole have survived their ordeals in space without damage to their bodies should not be taken to mean that anybody can go to the Moon and come back feeling fit and relaxed! Astronauts are hand-picked for their combination of a number of physical and mental qualities and most of us, if blasted out of the atmosphere, would probably either die of shock or suffer severe spacesickness.

Even so, we must think it is reasonable to conclude from the Max-Planck underground experiments that human bodies *do* respond both to the natural rhythms of Earth and to a sudden change of them, and that the indications are definitely towards a lengthening of circadian rhythms rather than a shortening of them under shielded conditions. This in turn implies a degree of entrainment from external forces. If our body clocks were wholly internal, they should keep ticking away regardless of changes in the environment that we cannot perceive consciously.[6]

Other early work in this field has indicated that reaction times of humans can be significantly affected by very weak artificial fields, both electric and magnetic. With such experiments in mind, Adey and his colleagues set out to see whether or not low-level electric fields could have any effect on patterns of electrical activity in the brain, and hence on behaviour. This was before the "biofeedback revolution" of the 1960s which we mention in the following chapter, so Adey was working in the dark at first.

He did not know which brain structures, if any, would show effects resulting from the presence of the electric fields, nor did he know what kind of changes to expect in the EEG. In his first tests, with monkeys, he placed electrodes in their brains and looked for any differences in the EEG pattern on a trial and error basis when the artificial fields were on and off. He used the standard "press the lever for a food reward" routine, because this gives a good idea of how monkey brains react to simple situations.

Using fields of only 2.8 volts, less than the voltage from a couple of torch batteries, at frequencies of 7 and 10 Hz., he held daily four-hour sessions with the fields switched on, and similar sessions without them. Results with the 10 Hz. field were inconclusive, although some changes were noted; but under the 7 Hz. field conditions, a consistent difference in response time was noted in all the animals' brains. The shift was towards faster reaction time, and Adey found evidence of the EEG rhythm being driven by the applied field, thereby indicating that "biological transduction" of electromagnetic energy into the brain is possible and can produce measurable reactions. It is also interesting that effects were noticed on one biofrequency but not on another very close to it.

In another series of tests, this time with cats, Adey used VHF fields of 147 MHz., in the radio range, modulating them at lower frequencies in the brainwave range, 1 to 25 Hz. Using another standard test, he found that cats irradiated with waves modulated at biofrequencies differed considerably from the control group in performance rate and accuracy of reinforced patterns, and also tended to forget what they had been taught.[7]

Adey's results, published in 1973, were among the first of their kind to appear outside Eastern Europe, where such work has been going on intensively for some time, including experiments with humans as well as animals. As early as 1966, Dr. Kholodov published a book on the influences of EM fields on the central nervous system, in which he discusses the effects of UHF, VHF and DC fields, providing considerable evidence that changes in the functional condition of the CNS can be brought about by such fields. In tests involving fish, for example, he was able to establish what he calls a "positive electrodefensive conditioned reflex" by applying VHF and gamma radiation in a magnetic field. In other words, he was able to scare the living daylights out of the fish simply by beaming EM waves at them. He does not say what he has been able to do to people, though a more recent book of his bears the ominous title *Man in a Magnetic Web*.

An interesting feature of current Soviet research in this area is the remarkably large number of master's and doctoral dissertations now being devoted to interactions between all types of EM fields and living systems. Kholodov's own master's dissertation dates from 1959, and dealt with the subject of the action of magnetic fields on animals. It is unfortunate that the difficulties of obtaining up-to-date information on research from Eastern Europe have added to the overall problem of keeping up with the information explosion. The U.S. Department of Defense has translated a large number of recent Soviet works of great potential interest to scientists. These are normally classified for ten years and then declassified, by which time they are probably out of date. Meanwhile, there is an urgent need for a general review of recent Soviet discoveries in medicine, physics, biology and astronomy, all of which have a direct bearing on the question of man's interaction with his environment.

Above all, there is a need for more long-term studies of these interactions. The implications of some of the few that have been carried out are quite alarming. One American experiment, involving several generations of mice bred and confined in a weak artificial electromagnetic field, showed that the animals' motor activity gradually decreased to the point where they lay on their backs with their feet in the air, a typical posture of defeat among animals. One hopes nothing analogous is happening, without our knowledge, to us.[8]

One intriguing subject which we offer to any young scientist looking for a theme for his dissertation or thesis is the question of whether air ions, which we mentioned at the end of Chapter 4, are carriers of the effects of modulated EM fields into biological systems. This possibility was raised by Dr. E. Stanton Maxey and James B. Beal in a 1975 conference paper, in which they recalled a 1963 prediction of Dr. Tromp's that control of air ions would one day be routine, especially in submarines and spacecraft. They point out that atmospherics, or sferics (EM radiation produced by lightning and other natural phenomena) have magnetic components that penetrate aircraft cabins and are known to affect brain rhythms; artificially produced sferics have been able to put human brain waves directly into a dominant theta rhythm, which is usually characteristic of dreaming or reverie states. Is it coincidence that waves of 7 and 10 Hz. are characteristic of geomagnetic field fluctuations, ELF and VLF waves produced in the Earth-ionosphere cavity (Schumann resonances), and the natural, relaxed rhythms of the human brain?[9]

We have already briefly mentioned acupuncture, and will have more to say in later chapters about this form of bioenergy therapy. One of its traditional features is the network of points, or *loci*, dotted across the skin and mostly linked to the *meridians*, channels that apparently convey the stimulus from the needle to the appropriate organ. There is some doubt as to the true nature of these meridians, as we shall see, but the loci undoubtedly exist. Or, to be precise, there are points on the skin where a detectable change in electrical skin resistance exists, and they happen to coincide with what the Chinese call acupuncture points. (Hill has built a small acupoint detector that emits a sound when making contact with one of them.)

When the Sun sends us an extra burst of radiation and causes a geomagnetic storm, our immediate EM environment is shaken up considerably, and some Soviet researchers believe that this directly affects the electrical acupoint potential on our skin. Could it be that the acupoints, forming a circuit of least resistance, are the points most responsive to the incoming electromotive force, which then in effect acts repeatedly like the electro-mechanical stimulation of the acupuncturist's needle, stimulating the organs to which the acupoints are linked? Soviet geophysicist Dr. Aleksandr P. Dubrov, of the Institute of Earth Physics in Moscow, considers it quite possible that the electromotive force interacts with the constant electrical currents that flow through the neurons, thereby affecting the central and peripheral nervous systems, neural networks that regulate our living organism.

Thus a direct line of influence from sunspot to human central nervous system has been suggested, and if its existence is not yet proved, it is beginning to look possible. In a recent book, Dubrov cites no fewer than 360 Soviet books and papers, most of them dating from the 1960s and 1970s, of which a great many offer evidence for an influence of the Earth's magnetic field (and therefore the Sun) on an enormous range of human physiological activities. All life is indeed beginning to appear far more closely bound to cosmic forces than we had ever imagined.[10]

Moon and Hamster

Biological clocks and rhythms are widespread throughout nature in plants, animals and micro-organisms as well as in man. Professor Frank Brown believes that most physiological processes probably indicate their presence to some degree. He sees this as an indication that such rhythms must be of very ancient origin, "long prior to when living creatures left their ancestral marine habitat." There is now much evidence, a good deal of it arising from Professor Brown's own extensive research, for the existence of a basic Sun/Moon timing system reflected in the body clocks of marine organisms, which are of all living creatures the most directly exposed to the effects of solar day and ocean tide. Brown is best known for his oyster experiment, in which he showed that these creatures will continue to respond to the influence of the Moon even when they have been shielded from its rays and moved hundreds of miles in longitude from their home beds, modifying their circadian rhythms to those of a non-existent inland ocean tide.

Another important experiment of his involved a group of hamsters. He wondered, after establishing that oysters respond to phases of the Moon even when they cannot see it, whether the behaviour of land organisms might have a similar lunar periodic component despite the fact, as he put it, that "no useful role of a lunar periodic component is ordinarily obvious for the non-marine species." He knew of experiments suggesting that some animals have a lunar day component of 24.8 hours in addition to their normal diurnal rhythm, so he decided to see whether the Sun and Moon could be observed to entrain them over a long period. His hamster experiment began in June 1964 and lasted two years, the animals' activity being recorded continuously by standard methods. Their compartment was confined to a room dimly lit by a 6-watt lamp switched on at 6 a.m. and off at 6 p.m., setting up a precise artificial diurnal rhythm of day and night.

The hamsters' activity was clearly synchronized with the lighting schedule, trailing off to zero before the light came on and beginning just before the following dark period. Yet when the data for the whole period were analysed, another rhythm stood out clearly. Brown expected that if any influence from the Moon had managed to find its way into the test laboratory, its effects would move gradually across the solar day cycle at the mean rate of about 50 minutes a day; and this is exactly what he found. Throughout the two-year test period, highest daily activity values occurred four days after full Moon and a day or two after new Moon. The hamsters *were* responding to the influence of the invisible Moon after all, even over 24 months! This experiment confirmed his earlier findings of similar lunar entrainment with fiddler crabs, and seems to settle the matter conclusively. We can see now that it is not just a question of living systems responding to day and night through the presence or absence of visible light: the invisible part of the spectrum also plays its part, to which the organism may respond unconsciously.[11]

Phase-Locking and Lunaception

Where there is life, there is oscillation. This can vary from periods of minutes to years. What happens when one oscillator entrains another, that is, develops a stable phase relationship with it?

The mechanical theory of *phase-locking* was first developed by the Dutch scientist Christiaan Huygens in 1674, when he managed to make two pendulums, which are mechanical oscillators, swing together in time, as one entrained the other through vibrations transmitted by the wall on which they hung. Huygens cannot have imagined the implications his discovery was to have in the study of biosystems. It now seems clear that it is natural for any living thing to be entrained, and conversely that it is unnatural to be disentrained. More precisely, it may be natural to be entrained *naturally*, and not natural to be entrained artificially. This question becomes an important one when we find that disentrainment and unnatural entrainment both tend to lead to ill effects, some of which are quite definitely fatal. So it may be worthwhile to look briefly at some such entrainment effects and their probable causes.

One example of disentrainment with which many are familiar is that known as jet lag. When jet passengers fly long distances from east to west or vice versa, thereby crossing several time zones, their bodies become disoriented with respect to new local time. It may take up to a week for the body to become re-entrained to the new diurnal cycle, in which time, if he is a businessman or politician, he may have had plenty of opportunities to make wrong decisions. A scientist travelling from Germany to the U.S. once refused to do any work on the days following his arrival on the pretext that "my soul has not yet arrived."

He had good reason to feel this way. He may have known, for instance, that experiments have shown insects and animals to have their life spans shortened by a factor of up to six percent—corresponding to about four years for the average human—when forced to live under simulated jet lag conditions. Airline pilots and crews usually give the impression of being the most stable, efficient and responsible of men and women, but it is a fact (not often publicized for obvious reasons) that they suffer more than the average from digestion problems, lack of appetite, eye irritation and insomnia. Stewardesses also find their menstrual periods going haywire. (It may be significant that aircrews inevitably must spend a great deal of time inside sealed metal structures that function as Faraday cages.)[12]

Dr. Ernst Pöppel has discovered that people react surprisingly well to brief distortions of the diurnal cycle. Tests carried out on subjects kept awake all night showed that many of their motor functions kept to their normal rhythms with only a slight shift in phase and amplitude. However, after his experiences at the Max-Planck Institute which we have already mentioned, he feels that whereas some people actually benefit spiritually from periods of voluntary desynchronization with natural rhythms, anybody with neurotic tendencies is likely to become more internally desynchronized.[13]

Jet lag is nothing compared with space lag, and it is interesting to note differences in U.S. and Soviet politics towards artificial days for astronauts. The latter prefer to keep on a rigid Earth wake-sleep cycle, while Americans are more adventurous and are quite willing to work in space for a "day" of anywhere from three to 48 hours! A final tip for jet-setters: experiments have shown that aircrews travelling from east to west tend to adjust more rapidly than when flying from west to east. By travelling round the world in this way, the first day is lengthened instead of shortened, and numerous tests have shown that our bodies are happier after an artificially long day than after a short one.

Jet lag is disentrainment of circadian rhythms: more potentially dangerous, perhaps, is accidental entrainment of our shorter infradian rhythms, and this can happen in a number of ways. There have been cases of people going into trances in cinemas where projection speed was too slow, thereby feeding spectators' eyes with photic flicker in the 7 to 15 Hz. range. Brain entrainment can even provoke epileptic fits: a Manchester neurologist described three cases in *The Lancet*[14] in which female patients had suffered grand mal seizures while adjusting the controls of flickering TV sets. Such a link between photic flicker and epilepsy was known to the Romans, who used to rotate a potter's wheel in front of slaves' eyes to see if they were liable to fits. (Can there be *anything* that was not known to the Romans, Greeks or Chinese?) This is bioentrainment in its simplest form, and it may be happening to us more than we realize in less obvious ways.

Strobe lights flashing at 10 Hz., the dominant brainwave frequency of a man in a slightly dissociated state, can also cause petit and grand mal epileptic seizures in patients with weak central nervous systems, and it is a matter of conjecture what effects they have on frequenters of discothèques (who might be interested in re-reading our section on infrasound in Chapter 4!). Many road and air disasters have almost certainly been caused by entrainment between the driver's or pilot's brain and the Sun. Long straight roads lined with evenly spaced trees, as often found in France, are potential death traps: if the Sun is at such an angle that it shines through the tree trunks, it will pulse light at passing drivers at a frequency determined by speed of car, width or gap between trees, and position of driver. If he is unlucky, a driver moving at a steady speed may have his brain change gear into a dominant alpha state (around 10 Hz.), with almost inevitably disastrous consequences. Ironically, it may be that drivers who stick faithfully to speed limits near "black spot" areas are those that suffer most.

As for air crashes, the accident rate among helicopter pilots is all too well known. A factor not often discussed is the undeniable fact that many helicopter pilots are subject to flashing sunlight through the whirling rotors above their heads, which theoretically puts them in the same potential position as the careful French Sunday driver who suddenly drives off the *route nationale* into a tree, killing himself and family and subsequently baffling the police, who are assured that he was the most cautious driver anybody knew.

We have heard of at least one case in which brain-Sun entrainment by photic flicker has been named as the most probable cause of a serious road accident. It is a very real possibility, and if drivers bear it in mind in future when travelling on tree-lined roads at sunrise or sunset, more accidents might be avoided. The best action in such a situation is to accelerate and slow down in alternating bursts, to avoid the constant speed that is a major factor in the syndrome.

Bioentrainment must not be thought of as a danger in itself. On the contrary, our very lives depend on it. For example, our hearts are mechanical pumps, and to function properly all the cells in them must beat in unison. Normally, they do this under the guidance of the vagus nerve, which provides the electrical stimulation that signals the beginning of each contraction—the P-wave on the EKG. When the heart cells fall out of step, or disentrain, a situation known as fibrillation, the heart begins to function erratically. This can be fatal unless corrected within seconds by shocking it with an electric current and re-entraining the cells. Bioentrainment here can mean the difference between life and death. It is when one of the cycles to which we are entrained fails, or when a strange cycle intrudes, that we are in trouble.

We have mentioned just a few of many possible examples of ways in which biological rhythms are entrained or disentrained by periodic environmental stimuli. Now we come to the most important examples of all, the human female's menstrual and fertility cycles, about which surprisingly little is known even by many women, let alone men. It is hard to say which is worse off, the unwanted child or the childless mother who wants one but cannot, or thinks she cannot, conceive one. Each is a potential social misfit, the one unloved and the other frustrated.

The word *menses* comes unaltered in spelling from the Latin for months, or monthly periods, and must originally have referred to synodic lunar months of 29.53 days. The fact that some women still have regular periods of about this duration suggests that ovulation or the fertility cycle may originally have been entrained to that of the Moon. If this is so, something has certainly gone wrong, for menses can be anywhere from 16 to 75 days in length, and can also be highly irregular from one month to the next.

Clear correlations between Moon and fertility are to be found in many living beings. The ingenious behaviour of the Californian grunion fish (*Leuresthes tenuis*) is well known. It lays its eggs in the sea at the very moment of peak high tide, and lets the waves deposit them on the sand, where they wait to be hatched upon immersion during the next high tide. This is family planning of a very precise order, and it is not confined to grunion. Dr. Bob Johannes, a marine biologist who has studied the natural history of the Palau islands, has found that more than 50 local varieties of fish also have their spawning cycles entrained to that of the Moon. Fish, he found, spawn like clockwork, and the Moon is the pendulum. In this respect, he has stated that one uneducated local fisherman taught him more than the entire scientific literature on the subject.[15]

Even land-based creatures are also locked on to lunar cycles. Johannes was able to observe and film the remarkable sight of fiddler crabs staggering out of the bushes exactly at full Moon to lay their eggs in the sea. Entrainment to lunar rhythms has also been found in fresh water fish and in the land beetle *Calandra granaria*. It has also been positively identified in at least one species of mammal, the wildebeest of Africa. Here, the interesting discovery was made that the animals become sensitive to the lunar cycle only when the cycle of climate falls below a certain threshold. In this connection, an important discovery was reported in 1977 by scientists from the University of Utah, who were able to make rodents reproduce out of season simply by feeding them fresh grass for two weeks. Obviously, an animal thinks of its food more than it thinks of the Moon, but it may still remain aware of the lunar cycle background, a fact that may also apply to humans, as we shall see in Chapter 10.[16]

Could not humans once have displayed an entrainment to lunar rhythms as precise as that of the grunion? Fish and many land creatures spend their lives unshielded from the rays of the Moon, whereas we have lost touch with our cosmic environment to the extent that many city dwellers probably never see the Moon at all. Perhaps, long ago, when we lived in closer communion with nature, our body cycles were entrained with those of her clocks and pendulums.

American biophysicist Edmond Dewan firmly believes this. Ovulation, he thinks, may originally have taken place at full Moon, a time when primates would have been less occupied with defence and would have had more time to mate in safety. Noting that chickens will lay eggs at night if a light is left on in their hutch, Dewan wondered if an artificial Moon in the form of an ordinary light bulb could have a similar regulating effect on the fertility cycle of women. Light, he reasoned, could affect the ovaries and testes by a direct route from the retina along the central nervous system and through the brain to the pituitary gland, where the hormone LH is normally released during ovulation.

Accordingly, Dewan and Dr. John Rock took a group of 19 women at the Rock Reproductive Clinic in Roxbury, Massachusetts, who were suffering from menstrual irregularity, and made them sleep with the light on during the 14th, 15th, 16th and 17th nights of their cycles, counting the onset of menses as day one. Dewan hoped that ovulation would be triggered at the right time when the light entrained a process leading to the release of the LH in a single pulse. "If it could be triggered," he thought, "a woman's cycle could be made perfectly regular. The rhythm method would make sense."

It seems to have worked. Dewan's simple experiment, costing nothing except for the presumably bigger electricity bill, enabled most of the women to regularize their cycles *at once*. Several of them later became pregnant after failing to do so beforehand. A control test was run, with lights left on two or three days before day 14, and no effects were noted. Dewan concluded that "possibly such entrainment by artificial light can assure a firmly predictable natural rhythm method of birth control."[17]

It sounds too good to be true, and if repeatable on a large scale the implications would be staggering. If all menstrual irregularities could be normalized simply by sleeping with an electric "full Moon" switched on for four nights a month, there goes the market for the pill and the loop. The multinational drug lobby may not welcome this idea, but they might take notice of the success claimed by Louise Lacey in her book *Lunaception*.* Unfortunately, however, neither Dewan nor anybody else to our knowledge has yet carried out a full-scale investigation of such methods, and Dewan's original report on his test was far from definitive. If ever an experiment called for repetition, this surely was one.

---

* New York: Warner, 1976.

Further evidence for a link between light and fertility comes from the discovery that girls are now reaching menarche, or menstruating for the first time, at increasingly earlier ages than in the past. (Menarche is another word with lunar associations, from the Greek *men* for Moon and *arkhe*, beginning.) This is a well known fact, and it is often ascribed to better nutrition, although this theory does not stand up to all the facts.

A team of Pakistani doctors has suggested that girls menstruate earlier because they spend more time exposed to artificial light than their ancestors did.[18] Tests with rats have shown that prolonged exposure to light brings earlier onset of puberty, and as for humans, a short time spent studying the legendary girls of Ipanema and other Brazilian urban beaches will suffice to establish that girls, well nourished or not, will mature precociously to an astonishing degree if they spend almost every daylight hour on the beach for about nine months of the year.

However, it is also known that girls born blind reach menarche earlier than normal, which seems paradoxical. If light provokes early menstruation, one might think lack of light would inhibit it? The key to this puzzle is found in the ever-surprising pineal gland. Tests with animals have shown that the pineal needs to be triggered by light in order to start up its function as an inhibitor in the growth of the gonads. No light—no inhibition!

Girls are certainly exposed to more light today than they were before domestic electricity was installed. Here is a perfect example of mass environmental feedback, as the ecosphere that we modify modifies us in return.

Finally, we turn from the subject of entrainment by natural sources to the much more disturbing area of deliberate man-made entrainment.

Electrosleep and Other Horrors

In Chapter 3 we mentioned the Russian electrosleep machine, and suggested it might one day replace the sleeping pill. One of us (Hill) was able to examine one of the Electroson range of sleep therapy machines exhibited at the 1976 Soviet Trade Fair in Copenhagen, where he was told that a good deal of research in this area is going on in the Soviet Union. Versions of some of their machines are now on sale in the U.S., though we have no reports on their clinical use as yet, though we do know that the U.S. Department of Defense is sitting on a large pile of literature on the subject that it has translated.

Electrosleep machines may be a taste of things to come in tomorrow's world, both good and bad. They work by placing electrodes on the surface of the scalp and applying low-frequency "noise" or high-frequency signals modulated by low signals within the known biofrequency range. In this way, sleep can be induced in human subjects—whether they want to go to sleep or not.

A British patent has been granted to a U.S. company, Bio-electronics Inc. of New London, New Hampshire, for a new electro-anaesthesia technique in which the input is a radio frequency carrier modulated by selected frequency components from the subject's own EEG signals.[19] Electrodes are attached to the temples, and current does not exceed 5 milliamps. When the phase of the applied signal has the correct relationship to the natural electric signals generated by the brain, the apparent electrical activity drops to the level associated with deep sleep. Signal frequency can be varied so as to produce local anaesthesia in arm or leg without loss of consciousness. Potential benefits to patients seem considerable.

We wish the same could be said about a new development known as electrodeless electrosleep, or *radiosleep*. As far as we know, this has not yet been developed for clinical use, and western scientists are generally sceptical, if not thoroughly alarmed, about the whole idea. The method involves a UHF generator (modulated by electrosleep-like pulses) and consequently all the unknown factors of biological effects of microwaves that may be inflicted on patients.

This has not kept people from working on the idea. We have seen one report from Dr. Rentsch of a Leipzig, East Germany, neurological clinic, on work involving the possibility of magneto-inductive transmission of stimuli to the brain. Rentsch points out that although the efficiency of the energy transfer by this method is extremely low, the amount of energy is still sufficient for electrosleep, which can be induced without a costly primary energy source. As for the influence of such impulses on the brain, experiments have shown that uncommonly long and deep sleep could be induced. *Without electrodes, and at a distance.*

In other words, the East Germans can now put anybody to sleep merely by beaming radio waves at them. The spy story writers seem to have missed this one.

The centre in the west for this type of research is the University of Graz, Austria, where an international symposium on electrotherapeutic sleep has already been held, and where the *Journal of the International Society for Electrosleep and Electroanaesthesia* is published. Reading through such few reports as we have been able to get hold of dealing with research in this area, we come across references to Pavlov's conditioning of the reflexes by inhibition-causing "unusual stimuli," to "altering the normal condition of the organism," and other phrases that would seem more in place in Kesey's *One Flew Over the Cuckoo's Nest*.

We are not exaggerating. Whatever the potential benefits of various forms of electrical entrainment of our natural rhythms, the dangers are considerable. Each of us has first-hand evidence from reliable witnesses of the misuse of electrotherapy. One particularly distressing example must suffice here:

In some U.S. states, it is still possible to commit a person to a mental hospital without his consent and without medical justification. This happened not long ago to someone personally known to one of us. The man had intended to divorce his wife. There being no alimony in the state in which the couple lived, the woman decided to intern her husband. One night she obtained the necessary signatures of two doctors, had the man forced into a straitjacket and driven over the state line. Then, he was given systematic electroshock treatment at a private sanatorium to the extent that today he *is* mentally deficient, parts of his memory having been virtually wiped clean.

Similar methods are known to be widely used in the Soviet Union, though personal accounts of them are rare for the simple reason that victims are dosed with phenobarbital beforehand and remember nothing of their ordeal, the purpose of which is to "alter the normal condition of the organism" of those who persist in thinking thoughts that are not permitted. As for more direct forms of torture by electric shock, for purposes of extracting information from prisoners, an old favourite is the U.S. Army field telephone, which has been in regular use in Brazil at least since 1969. Unfortunately neither East nor West, nor the "third world" can be proud of its record in the misuse of electricity.

Biological effects, as we now know, can result from EM influences in our natural and artificial environments, yet it would be wrong to give the impression that there is nothing we can do about "EM pollution." EM fields can be reduced, and in some cases eliminated altogether, by Faraday cage-type shielding, which is no harder to build than a fallout shelter and may prove just as necessary. There are many ways in which the buildings we live and work in can be made "clean" of acoustic and EM pollution, and we hope architects will include some geobiology in their reading lists in the future.

In Chapter 4 we mentioned that some of the effects of negative atmospheric electricity can be offset by taking ionic salts as recommended by Dr. Caymaz. It is now also quite easy to buy small negative ionizers to regulate the ionic content of office and home microclimates. We would like to see these more widely used in public areas as well as private.

We have also seen that a knowledge of human biorhythms can be used to advantage in reducing the effects of exposure to EM waves. Radio and radar operators and x-ray technicians on shift work could perhaps have work scheduled for times when their organism is at its most resistant. (Remember the rats who died from radiation at night!)

Recent research from the Soviet Union indicates that certain natural substances can diminish the effects of harmful radiation by increasing the non-specific resistance of the organism. These include the well known *Panax ginseng* root and the "Siberian ginseng" *Eleutherococcus senticosus max.*, also known as Russian root, for which remarkable claims (as yet unverified in the West) have been made by Dr. I. Brekhman of the Institute for Biologically Active Substances in Vladivostok, Siberia. He has reported that a single dose of *Eleutherococcus* is enough to prolong the lifespan of rats exposed to x-ray radiation.*

---

* In 1977, the authors were able to obtain a piece of this root and arrange for laboratory testing of it in a London laboratory.

Finally, as we see in the following chapter, it is now possible for us to overcome the effects of bioentrainment by learning to entrain ourselves, that is, learning control over our own internal rhythms and vital processes, both physical and mental. The scientific information explosion of the Space Age has fortunately now been matched by rapid progress in such techniques as biofeedback, transcendental mediation and what the Soviets call "psychic self-regulation," through which a number of new ways have been found in which we can resist and overcome the invisible forces around us.

Chapter 9

BIOFEEDBACK—SELF-ENTRAINMENT

A NEW human activity needs a new word, and since *biofeedback* is not yet in all dictionaries we can define it provisionally as "the process of acquiring information about our body processes of which we may not normally be consciously aware, thereby learning how to control such processes through feedback of information."

We obtain such bioinformation every time we take our temperature or feel our pulse, actions that require no more than an ordinary clinical thermometer for the one and an ordinary finger for the other. But more complex instruments are needed if we are to become consciously aware of many other body processes, such as brain waves, muscle potentials or tensions, skin resistance and temperature, or blood pressure.

The yogi has supposedly been able to control such functions for centuries, and anybody can now do the same with the initial help of an electroencephalograph (EEG) connected to the brain, an electrocardiograph (EKG) to the heart, or an electromyograph (EMG) to the muscles, coupled with some kind of feedback device so that the subject can actually see what is being recorded.

The Sanskrit words *yoga* and *yuga* mean *union* and *yoke*, and yoga has come to mean two things: a state of union, or entrainment, with absolute reality, and the system of exercises one must do to attain such a state. As a yoke locks two oxen in step with each other, so the yogi seeks to lock his mind on to a broader reality. Thanks to Space Age technology, this can now to at least some degree be done without lengthy training, and one of the achievements of the biofeedback movement has been to bring yogi and scientist together in the laboratory, as we shall see.

Before we can improve the quality of our state of consciousness, we must add to the quantity of external and internal impressions that constitute it. We cannot very well control our bodily functions unless we are aware of them, so the first link in the biofeedback chain is awareness.

When a bright light shines into our eyes, we shut them. Why? Because our conscious nervous system, the system we use when we interact with our environment wilfully, and respond to conscious stimuli, has sent information from eye to brain, where after a flash of computing, messages are sent to the appropriate muscles—in this case those of the eyelids—to take defensive action. This process is called a feedback loop. However, while the nerve channels are carrying these messages from eye to brain and back to eyelid, something else is happening. Our pupils are contracting in an attempt to shut out the bright light. This is not done consciously, but is directed by the autonomous—or autonomic—nervous system, the network that also handles such functions as heart rate, skin temperature and brain wave rhythm, all of which are usually regulated without our conscious mind becoming involved. The neuroregulatory centres used by the autonomous nervous system are mostly located in the lower brain stem, quite separate from the higher centres we use for conscious problem-solving. Thus we have two separate systems lying almost side by side and controlling the same organ, in this example, the eye.

Normally, our bodies look after themselves. There is no need for us to be consciously aware that our pupils are contracting. We know that closing our eyes is enough to solve the problem of the bright light. We need not worry all the time what our blood pressure, heart beat, or brain-wave rhythms are. But now and then our self-regulation system goes wrong: blood pressure shoots up, the brain suffers a spasm, or the heart breaks down. For centuries, doctors could attack such malfunctions only with medicine, surgery, or simple suggestion in the case of minor ailments. ("Take two of these and call me in the morning!")

Tales were told of holy men in the East who could control their own hearts, brains and blood, even stopping the heartbeat altogether or spending hours underground without breathing. But those were travellers' tales . . .

Western scientists no longer laugh at such tales (if they are up to date with the scientific literature), because they are true. Moreover, they are no longer confined to the mysterious East. Indeed, published evidence for the control of body by mind in the West dates back at least to 1885,[1] though it is only recently that scientists have been ready to confirm the feats of the yogi with reliable instruments. Prior to the 1960s, psychologists concentrated on man's observable behaviour rather than on the inner workings of his mind. Adepts of the behaviourist faith headed by B. F. Skinner even managed to persuade themselves that the mind was a "spurious entity," in the master's own words. Barbara Brown, a leading figure in the biofeedback revolution, has likened such utterances to "the ecclesiastical mandates of the fifteenth century."[2]

Revolution is the right word to use with regard to the emergence of biofeedback on the scientific scene. It was one of the quickest revolutions ever. The Bio-Feedback Research Society (BFRS) held its first meeting in 1969, bringing together a number of scientists including Elmer and Alyce Green, Joe Kamiya, Thomas Mulholland and Barbara Brown, some of whom had been repeating each other's research without knowing it. One of its revolutionary aspects was the breaking-down of boundaries between disciplines, which had reached the point where, in Dr. Brown's words, "the branches of biological science were all sliced from the mother tree and transplanted in distant nurseries.[3] As for the scientists, "their specialities had proliferated like rabbits, and like rabbits each marked the territory not to be defiled by talking across borders."[4]

But the real revolution was that psychologists and physiologists joined forces to help people train themselves to control themselves rather than to impose control on their external behaviour. Such a change in direction called for a new word, and biofeedback (now usually spelt without a hyphen) is here to stay.

Biosignals

Living bodies emit a variety of electrical signals, and with the right equipment we can follow the flow of the electric currents within us as easily as we can hear the heart beat through a stethoscope. The heart, in fact, emits some of the strongest of our electrical pulses: about once every second a wave of polarization sweeps across it from one side to the other, creating a dipole field that can be detected with a galvanometer or a voltmeter. The modern science of electrocardiography is based on this electrical information from the most vital of our organs.

The brain also emits signals, about ten times weaker than those of the heart. Their frequency ranges from less than one cycle per second to several thousand, although little is yet known about high frequency components of brainwaves, since we do not have the instruments to measure them. Whereas heart signals are relatively simple, brain potentials revealed by electroencephalography are very complex; several dominant rhythms are mixed together with a welter of "noise" from spontaneous firings, and much brain electricity is in fact dissipated and lost before we can catch it and record it.

The four dominant brain waveforms, and their associated states, are:

delta   0.5 to 3 Hz. Deep sleep, higher states of awareness.
theta   4   to 7 Hz. Reverie, dreaming states.
alpha   8   to 13 Hz. Passive, blank state.
beta   14   to 30 Hz. Thinking, active state.

These are by no means all the brain's rhythms, nor are the limits of each group exactly defined, but this is the rough classification normally used today.

Next, in addition to the heart and brain signals, we have muscle waves of up to a hundred kilocycles or more in frequency. And since any surface warmer than its surroundings gives off thermal radiation, the human body is also a source of infrared radiation.

Now that we know about some of the signals emitted by the body, the question arises: what use is this information? Do the various signals correlate with certain definite mental or physical states? They do indeed, and a knowledge of them can be used to bring much benefit in a number of ways.

Much research in the pre-BFRS age was concentrated on the alpha rhythms of the brain, in which an EEG machine was used to feed information on such rhythms back to the subject in the form of flashing lights, wavering tones, or mild electric shocks. It was found, independently, by Kamiya (a psychologist) and Brown (a physiologist) among others that people could be taught quite quickly to keep a light flashing by somehow making their alpha rhythms dominate the others. (The brain emits more than one rhythm at a time: most EEG tests measure output from at least four different areas, and when we speak of an alpha state we mean one in which there is more alpha than any other frequency to be seen on the EEG chart. It does not mean that the whole brain has suddenly "gone alpha.")[5]

It must have seemed like Instant Yoga, and while it is unlikely that any short cuts to *nirvana* (ultimate bliss) exist, biofeedback can certainly help people take the first steps more quickly. We can immediately see visible proof of what was previously a matter of empirical judgment. It is difficult to convey the sense of self-assurance gained in this way to somebody who has not tried it. To watch a pointer moving or a light flashing in response to your own *mental* activity is a most rewarding experience. Barbara Brown was so awestruck by the sight of 20 billion brain cells chattering away on the EEG that she had "the feeling of suddenly having discovered where haloes came from."

The brain gets most of the publicity, but biofeedback research is interested in information from the whole body, the heart, muscles, skin, blood and even single cells. We must be able to listen to anything that any part of our body may be trying to tell us.

Mind over Body

Muscles. The training of voluntary muscle control has come a long way since J. H. Bair found people could be taught to waggle their ears in 1901, perhaps the first biofeedback experiment to be written up in a scientific journal.[6] It is now established that the mind can make the muscles relax far more than was ever thought possible. Many forms of chronic pain due to muscle tension, such as migraine headaches, are proving amenable to self-healing through biofeedback.

Every time we order a muscle to contract or extend itself, a burst of EM radiation is emitted that can be picked up on an electromyograph, amplified, and fed back to the patient, who can then teach himself to control his muscles in much the same way that he regulates his alpha brainwaves—by will power. This type of treatment, if more widely available, would save a lot of doctors' time. The only part of the process that calls for expert help is the exact placing of the electrodes: biofeedback is no use if you are feeding yourself information about the wrong target.

One prominent biofeedback researcher, John Basmajian, has made the astonishing claim that voluntary control can now be exercised over a *single cell*.[7] If this is so, it is hard to imagine the limits of what may be possible in the near future. There are already signs that people with muscular deficiencies can be helped by biofeedback methods. One ingenious researcher, Dr. Edmond Dewan, has even managed to "write" a word in a kind of brain morse code printed out through a computer, short and long bursts of alpha waves being used for the dots and dashes. This is a laborious process, but it may lead to new hope for the totally disabled who cannot communicate any other way.[8]

Heart. The rate at which the heart pumps blood during normal activity varies a good deal, though tachycardia, or chronic speeding up of heartbeats, can be dangerous. EKG feedback can now be used to warn a patient when his heart is exceeding the speed limit set by his doctor, by flashing a red light, for instance, when it is going too fast or a green when the beat is too slow. Elmer Green has reported that a patient can be taught in three months to lower the heartbeat from 110 to less than 70 beats per minute.

One way to control the heart beat may be through music. According to Barbara Brown, the heart rate can synchronize with musical rhythms. This fact does not seem to be generally known, but one of us (Hill) has been able to verify it personally in tests with a Swedish lady, Mrs. Brite Kinch, a former air hostess with SAS. On her visits to Latin America in the 1940s, Mrs. Kinch had found the rhythms of the local music strangely compelling, and she discovered that her heartbeat would change to keep time with it. She thought this might interest science, so she visited a cardiologist for testing, and eventually was written up in the press as the "stewardess with the dancing heart." Her doctors, however, were not interested. "Don't you know," they told her, "that the heart is nothing but a mechanical pump?"

Hill examined Mrs. Kinch in 1975 and found her heart could still dance as well as pump. He arranged his equipment so that he could record radio music on a tape recorder and monitor her EKG electrical heart signals at the same time, while Mrs. Kinch could only hear the music, and not the beats of her heart. After a thirty-minute control run to register her normal EKG rhythm, he switched on the radio and tuned it to a music programme. As soon as he did so, a strange phenomenon took place.

For a few seconds after the music began, her EKG rhythm disappeared for a few cycles, as if her heart had "skipped a beat." When its rhythm was re-established, its main peak (the so-called QRS complex) was now in time with the music! He switched to other stations, and was able to test Mrs. Kinch's dancing heart with Latin, light popular and classical pieces. Each time, it promptly joined in, keeping perfect time. But when heavy rock or jazz were played, it did not; and Hill learned that she does not like either type of music. Her heart would respond only to pieces that pleased her. Could the heart really be the seat of emotion, as lovers have suspected since time immemorial? It seems so. Five other subjects were tested for "musical entrainment" with a variety of musical stimuli, but none displayed Mrs. Kinch's ability.

She insists that everyone can do this, though we have so far failed to verify her claim. We have also failed to interest doctors and even some biofeedback researchers in this discovery, so we mention it here in case anybody is interested. We all like to relax to our favourite music; and it would be the most pleasant form of therapy imaginable for a cardiac patient to relax to carefully selected rhythmic music, and perhaps to allow his suffering heart to be regulated by it.

In 1971, J. D. Taylor, a surgeon at the North Middlesex Hospital in London, announced that he had managed to bring three patients out of comas by playing nonstop pop music on the BBC Radio One programme at them. Two were victims of road accidents in which they had suffered serious brain damage; one had been in coma for almost nine weeks, the other two for twelve days. All three recovered consciousness within two days of what Mr. Taylor calls "Radio-One-Therapy." He suggests that people who go to visit unconscious patients in hospitals should try talking to them in future, instead of staring at them in an awed silence![9]

Although many scientists have been and are both music

lovers and talented musicians, it is only recently that the links between music and the human brain have been studied in detail. In a radio discussion of a collection of essays entitled *Music and the Brain* (Heinemann, 1977) a group of psychologists and musicians agreed that man's ignorance of what music really is and does is matched only by similar ignorance of the full extent of the brain's power and qualities.

The conductor Herbert von Karajan has had his pulse beat recorded while piloting a jet plane and while conducting Beethoven's *Leonora No. 3 Overture*. For some inexplicable reason, his pulse was far higher and more erratic while conducting than while flying, even when carrying out a particularly difficult manoeuvre. During the overture, his pulse rate also seemed to leap up when nothing especially tense or dramatic was going on in the orchestral score. The same variations were also noted when the tape of his performance was played back to him. We hope that this experiment, included in the book mentioned above, will be repeated. The prospects for research into interactions between music and the human body seem promising. For the time being, however, let us return to self-entrainment without the use of external stimuli.

Blood Pressure. Hypertension, or high blood pressure, is one of the curses of the twentieth century, and it is an open question whether the millions spent on researching it are matched by the profits of drug companies that market products which fail to cure it. It is essentially a muscular problem, and anybody who has learned to control his muscles can ease his blood pressure. Indeed, in some cases it may be that biofeedback is the *only known cure* with no possible side-effects, and very little expense, for a problem for which there is still much argument as to the real cause. We know it is caused by tension in the muscles lining the arteries, but what causes that? Since it has been shown that the mind, plus the appropriate machinery, can ease this tension, is it not likely that the mind also causes it in the first place?

Skin. By using a device called a thermocouple, variations in skin temperature much finer than those shown on an ordinary thermometer can be read, studied, and controlled by the patient. Some types of headache can be eased by voluntary control of body temperature distribution. If the hands are made warmer, less blood will flow to the head, thereby cooling it and soothing the pain. It is quite amazing to watch a skilled biofeedback student literally *think* himself hot or cold to order.

The skin's electrical resistance can also be controlled. This is closely related to certain anxiety states and overall degree of relaxation, as has been well known for some time through polygraph ("lie detector") work. As Barbara Brown points out, it is a sad comment on our sense of priorities that all to emerge so far from the study of our skin is a way to detect lying. And a method only 73 percent accurate, according to highest estimates.

When skin resistance is high, relaxation will be increased, and when it is low, anxiety states are likely to arise. One of the most easily learned biofeedback procedures is one in which a loud tone is generated for the anxious state and a lower tone for the more relaxed state. The psychologist and psychiatrist C. G. Jung almost invented biofeedback in 1904, when he showed that electrical skin activity responded to emotional stimuli. But that was the age of Pavlov and his wretched slobbering dogs, whose "conditioned learning" fascinated psychologists to the point where they forgot about mere *people* and their age-old problems, not to mention the "psyche," and embarked upon half a century of playing with rats pressing levers in boxes. Fortunately, people and their inner torments are coming back into fashion, thanks to a considerable extent to the biofeedback revolutionaries.

Healing by Self-Regulation. It is well established that the use of biofeedback training methods can help ease minor ailments, but what about major ones? There is little to report as yet, but many grounds for optimism. One is that it has been shown in controlled tests that people can learn to control two related body functions at once: not only that, but they are actually better at doing this than they are at controlling each one separately.[10] When we learn to control *all* our body functions *all* the time, we shall make it much harder for many types of disease to set in. This is theoretically quite possible.

An ingenious new development in this context is a device whereby sufferers from stomach disorders can monitor the level of acidity at the spot affected, and learn to control it. Instead of lying on his back all day with a milk drip rammed down his gullet, a patient with a serious ulcer can now treat himself using nothing more than a small gadget and his mind.

Another cause for optimism regarding healing through biofeedback concerns the capability for self-healing that the body already has, and the possibilities for strengthening this self-defence, or "immune" system. Despite all advances in medicine, some of the most impressive instances of healing take place when the patient simply "makes up his mind" to get better, and does. Take the case of Niki Lauda.

Motor racing is a dangerous sport, and serious accidents at high speeds happen all too often. What is really surprising is the number of drivers who almost kill themselves and recover in record time after the most hideous smash-ups. In 1976, world champion Niki Lauda suffered severe external and internal injuries in a crash during the German Grand Prix, being dragged from the blazing wreck in which he had been trapped for almost a minute. It was thought certain at the time that his season, if not his career, was over, and quite possible that he would die in hospital.

Six weeks later, Lauda was back in his Ferrari, and he ended the season within inches of retaining his championship title. How did he do it? In a TV interview, he said that at one point, after he had regained consciousness in hospital, he had felt he was dying but had simply resolved not to. His will to live was strengthened considerably by a Catholic priest, who administered the last rites (while Lauda was fully conscious) without even asking how he felt!

Not everybody has the motivation or the determination of a Niki Lauda. Many people with "incurable" diseases resign themselves to them and die obediently, as if not wishing to offend the doctors. Yet the Lauda type will recover, because he *knows* he can, even if the doctors know he cannot. It is difficult to explain this repeated fact (the similar cases of Stirling Moss and Graham Hill come to mind) along Pavlovian-Skinnerian lines of stimulus-and-response. The stimulus is clear enough, but how and where does the response originate? Cases such as that of Lauda remind us how we still underestimate the power of the human mind over the human body, even a body apparently fatally damaged.

On the other side of the coin, the will to live can be matched by the will not to live. Barbara Brown reports a dramatic example in which a prisoner underwent a difficult operation shortly before his release was due, making a full and rapid recovery only to find that he faced another jail sentence. He collapsed at once, dying.[11] There have been many reports about the custom of certain Eskimos, whose old people simply "decide to die," and wander off into the snow when their time has come. Nomadic tribes in parts of the middle East will leave the old behind to die when they move to fresh pastures. Life, as well as disease, is subject to spontaneous remission—to borrow the phrase used by science to explain "miracle cures," though it explains nothing and cannot be repeated to order.

Biofeedback researchers are coming to believe that clues to the mystery of self-healing may turn up on the EEG charts. If the process by which we activate our defences can be made conscious, it will be registered by the brain and we will be able to record and measure it with instruments. Then it is only a question of finding out how to reproduce this brain-change pattern to order. It would be very interesting to have seen exactly what went on in Lauda's brain waves when he took the decision not to die.

Mind over Consciousness

As we have seen, we are now able to exercise conscious control over our internal functions by reading our internal signals and responding to them. This is a great breakthrough in itself, but there is even better news to come, as we find out not so much what the mind can do as how we can influence it to do even more. Research in this area offers tantalizing glimpses of the possible.

A number of physiological investigations have been made of various kinds of meditation, from the "relaxed awareness" of Japanese Zen Buddhism[12] to the "hypometabolic state" of Transcendental Meditation, or TM.[13] It has been shown that yogis and trained meditators can exercise a remarkable degree of control over their brainwave patterns, in some instances making a regular progression from fast beta activity through alpha to slow theta and delta as they achieve deep meditation. These studies have helped establish a link between brain rhythms and states of consciousness, while with TM subjects it has been claimed that states of mind can be learned which affect oxygen consumption, electrical skin resistance, arterial blood pressure and respiration rate, all in accordance with depth of the learned meditation states. (This can take some time.)

The rapid growth of Maharishi Mahesh Yogi's TM movement has led to the setting up of institutes devoted to the scientific study of meditation and its effects. The scientists involved are not ageing mystics, but include the pick of the crop: men like R. K. Wallace, Herbert Benson and Lawrence Domash, and others who would be as much at home at MIT as they are at MIU—Maharishi International University. Domash, a physicist, suggests that certain states of consciousness are directly measurable, claiming there is a crude analogy between quantum mechanical states, as in chemical and physical systems, and "pure" states of the brain. He compares the transition from one level of consciousness to another with the way water is transformed from its unordered liquid state into ordered crystal formations, or ice.[14]

Some feats by psychics and yogis in the laboratories have caused consternation, especially those of Swami Rama and Jack Schwarz. At the Menninger Foundation lab in Topeka, Swami Rama astonished scientists by stopping his heart on demand for 17 seconds, during which period his EKG shot up to 300 beats per minute representing fluttering in the atrium. The Swami was also able to speed up his heart beat from 66 to 92 and back again, and to create a temperature difference of up to 11° (F) between two points *on the same hand*. He even offered to stop his heart for three or four minutes, but researchers Elmer and Alyce Green thought it wiser not to accept. (Though they are unlikely to be surprised by anything a yogi can do: on a recent trip to India, they observed one who spent seven and a half *hours* in an apparently airtight box producing continuous alpha brain rhythm.)

Swami Rama's views on mind and body control are just as interesting as his ability to control them. "All of the body is in the mind," he says, "but not all of the mind is in the body." The physical body is an energy structure, part of a larger and less dense energy structure in which ten different energies, or *pranas*, are involved. The whole planet, he says, is surrounded by a "field of mind," and with sufficient training, we can learn to control some of the various *pranas* outside our bodies as well as those inside.

"Every piece of the body, every cell, you can control, because it's all in the mind," Elmer Green reports him as saying.[15]

Such views are not new to psychical researchers, but scientists are more likely to listen to them when they come from a man who has just stopped his heart while hooked up to an EKG! They make an interesting comparison with those of Jack Schwarz, a Dutchman who has given research scientists some nervous moments. He once rammed a needle straight through his biceps without batting an eyelid in front of fifty doctors. He can also control the flow of his blood to an astonishing degree, making it stop flowing from a scratch on his skin on demand. Schwarz insists that all the cellular material in the body can be controlled by the mind, and he has given evidence for his claim by removing all traces of skin punctures in as little as 24 hours. (Similar claims have been made by witnesses of "psychic surgery" in the Philippines; and Playfair has also observed this phenomenon in Brazil.[16] It is impressive to see such claims being confirmed under more orthodox conditions.)

Swami Rama had other surprises in store during his Menninger tests. He obligingly produced any of the four dominant brain states to order, even managing to hang on to his delta rhythm while remaining responsive to external stimuli at the same time. This was generally thought to be impossible, since delta is normally dominant only in deep sleep or advanced meditation states. Before the biofeedback age, with its visible printouts of body information, few scientists would have believed it could be done.

The Swami upset another traditional belief by insisting that his moments of *satori*, or peak meditative experience, did not coincide with a dominant alpha rhythm, as most researchers thought they should. In fact, *satori* did not coincide with a theta or delta state either. To add to the confusion, during periods he indicated as particularly significant his EEG showed nothing but beta activity, which is supposed to indicate the state of trivial conscious behaviour, though it could be that the EEG was producing high frequency waves too fast to measure. Clearly, the yogi is still a step ahead of science.

Can there be more subtle essences in the brain that cannot yet be measured by our present instruments? It seems likely, for the brain is known to have components of high frequency, and the fact that the Swami Rama showed only relatively high frequency activity at *satori* moments suggests that present measurement techniques might be extended to higher frequency ranges. Not so long ago there were thought to be only three subatomic particles: electron, proton and neutron; today several hundred have been found. Brain rhythm research may be today where atomic physics was in the nineteenth century. It is not likely that an organ consisting of several billion cells will produce only low frequency activity, stopping at the end of the beta range, at about 30 Hz.

A start has been made to explore the higher frequencies of the brain waves. In 1965, U.S. Patent 3,195,533 was awarded to Victor Fischer for his "Hyfreeg" (*hyperfrequency electroencephalograph*), which can measure up to 150 Hz. Its inventor claims that it provides a much clearer graph than normal EEGs, especially those of brain aberrations such as during epileptic seizures or "brainquakes." Fischer also claims he can induce sleep by applying low voltage pulses at about 80 Hz. to eyeball and occipital areas.

A Polish scientist, Jan Trabka, has made studies of animal brains using a computer to sort out the high frequency brain waves from the general background noise. He has identified waves of up to 500 Hz.[17] Itzhak Bentov, after studying 156 subjects using meditation techniques, has reported the existence of brain waves of up to 12,000 cycles per second. The current record for HF brainwave detection, however, is claimed by Soviet scientist B. M. Nudel, who has reported the existence of brain vibrations of up to 100 kHz.[18]

For all we know, the brain may have a spectrum of its own as broad as that of electromagnetic radiation—all the way up to 10^{23}Hz. or thereabouts. "Asking half a dozen electrodes to tell us what the brain's ten billion cells are up to is tantamount to representing the opinions of two or three people as the results of a national poll," says Marilyn Ferguson.[19]

New Maps of the Mind

Once upon a time, there were only five states of consciousness. Man was either awake, asleep, half way between the two, unconscious, or dead. Today, things are much more complicated: more than twenty different mental states have been identified or suggested, and some progress has been made in mapping out what are known collectively as altered states of consciousness, or ASC. These range from those of meditation and creativity to psychedelia and cosmic, or total, awareness.

Much useful research has been done on the ecstatic and meditative states of yoga and Transcendental Meditation (TM). It has been conclusively shown, for instance, that a distinct decrease in temperature accompanies the onset of certain types of meditation. An easy (though very costly) way of verifying this is by using thermographic equipment, on which you can actually *see* temperatures altering on a screen like that of a TV set.

It now seems possible that biofeedback methods may help us find biological bases for such unexplained phenomena as telepathy, clairvoyance and psychometry. During a series of tests with the well known psychic Peter Hurkos, Barbara Brown noted a clear correlation between the state of his brainwave patterns and the accuracy of his psychometry. (This is the art of obtaining information about an object and its associations simply by touching or holding it.) When a certain double spiking pattern appeared on part of Hurkos' EEG chart, he would reach 90 percent accuracy. On another occasion, when some magicians had come along in the hope of debunking his supposedly paranormal abilities, his EEG remained normal and his abilities failed him completely. This offers evidence for the claim often repeated by psychics and mediums that they cannot perform in the presence of hostile or sceptical onlookers. This claim has always been met with sarcastic laughter from sceptics, but there is now evidence in support of it. (And still none against it!)

Lack of feedback can be as intriguing as an abundance of it. In a test with Jack Schwarz, who can apparently make himself totally immune to pain, Dr. Brown found he could keep his heart rate, breathing and brain waves absolutely steady while having a cigarette stubbed out on his skin! There was not even a burn mark. Schwarz also invited her to stick a skewer through his forearm, but she refused.[20]

In a more conventional experiment of this type, researchers in England found that a hypnotized subject's skin could be pierced by a needle without causing pain or bleeding; measuring the electrical skin resistance on his hand, they found it to be 70 percent below normal. There is, they say, a relationship between depth of trance and variation in skin resistance.[21]

Successes so far in linking visible measurements to specific mental states gives us hope that one day we will be able to draw up detailed charts of the mind, and trace its various connections with the body. "For the first time," says Barbara Brown, "the mental self can communicate intelligently with the physical self," and she believes that biofeedback has a major role to play in a new era of mental and physical health, though she laments the fact that the number of immediate uses of it far exceeds present capacities of medicine, biology and psychology to explore them. When properly used, she is convinced that it could transform society. The more we learn to control our own minds and bodies, the more harmonious interpersonal and international relations will become.[22]

On the lighter side, there is even the possibility that biofeedback can be taught to children through toys. The ever-inventive Dr. Brown has come up with a racing car set that works on brainpower. Competitors have to wire themselves up to an EEG and generate alpha to make the cars move! It may be a while, though, before Junior Biofeedback Kits are on the market, for the whole subject is far more complex than we have been able to indicate in this chapter.

Mind over Metabolism

We should not omit mention here of biofeedback research in the Soviet Union, where they prefer the term "physical self-regulation" (PSR). During Hill's visit to the All-Union Conference on PSR in Alma-Ata, Kazakhstan, in 1976, he was able to meet and talk with Dr. A. S. Romen, a leading figure in Soviet PSR research who is also an accomplished yoga practitioner. In 1965 he went to India to study the phenomenon of metabolic control as practised in yoga, and on his return home he made a film, *Who are the Indian Yogis?*, in which he shows, among other things, a man skewering his forearm with a long metal spike. No blood appears, although the spike goes in one side of the arm and comes out on the other.

There was another surprise in store for Hill. After seeing the film, he learned that the skewered arm belonged to Dr. Romen himself. "These abilities are not confined to Indian yogis," said Dr. Romen, offering to teach Hill the method in ten days. A woman conference delegate claimed she had learned to stop the flow of blood from a wound a few days after starting Romen's training, which is based on the creation of a state of inner calm and detachment during progressive relaxation, much as originally taught in Europe by Dr. Johannes Schultz and later known as "autogenic training."

Soviet PSR is not confined to the laboratory, but is being put to practical use. Dr. Romen's latest book is entitled *Psychical Self-Regulation and Sport*, and it seems that Soviet athletes are being encouraged to learn metabolic control, to help them release the extra energy required to make the difference between a bronze and a gold medal—often only a matter of hundredths of seconds.

In Hill's contribution to the conference, he pointed out the close resemblance between biofeedback control of events inside the body ("ins") with the presumed nonmotor outputs or manifestations known to parapsychologists as psychokinesis (PK) or "outs." This includes the ability to move objects at a distance, such as pendulums and compass needles, and it has been demonstrated by Swami Rama, who regards such *siddhis* (paranormal events) as concomitants of learned control over inner functions. The idea was received with sympathy by the Soviet scientists at Alma-Ata, who, like their American counterparts, back up studies of yoga with modern instrumental techniques. (For a most unusual and largely overlooked study of the relation between yoga and PK, see the article in *Fate*, July 1959, by Dr. Mayne Reid Coe, Jr., who taught himself to move objects at a distance by what he assumed to be control of his body electricity. Dr. Coe's claims, if repeated, could do much to explain a number of supposedly paranormal phenomena.)

The Soviet PSR researchers have confirmed the main results of western work on the control of internal organs and body functions, and some idea of the amount of work they have done can be had from the bulk of the *Proceedings* of the three PSR conferences held in Alma-Ata to date. These average 500 pages each and are largely devoted to clinically documented studies on everything from treatment of alcoholism to acute clinical disorders. Dr. Romen is a co-founder of the International Society for Psychical Self-Regulation, and we hope to hear more from him.

Mind over Death

Earlier, we mentioned the case of Niki Lauda and his recovery from the brink of death. Even more remarkable evidence for the ability of man to save his life by controlling his body comes from the experience of Gustavo Zerbino, a medical student who was one of the survivors of the 1972 air crash in the Andes. This event has received much publicity due to the fact that the survivors survived only because they resorted to cannibalism: less widely publicized is the fact that seventeen days after the crash, 13 of the original 29 survivors were killed in a sudden avalanche of snow that buried them alive.

Zerbino was also buried alive, but instead of struggling desperately with his muscles, he used his head and reasoned that he had about a minute's supply of oxygen left in the small space around his body where its heat had begun to melt the snow. He spent this minute allowing his body heat to melt more snow, thus making more air available. Then his hand found a pocket of air, and after ten minutes in the snow he escaped. For the next eight weeks, before rescue came, Zerbino and his fellow survivors practised a form of self-regulation of body temperature, and we understand from a report published in the Danish newspaper *Politiken* (27 February 1977) that subsequent medical examinations showed profound changes in their bodies. They were, doctors said, "supernaturally healthy." Like Dr. Romen and Jack Schwarz, Zerbino can now pierce himself with a needle and control the bleeding at will.

Mind over Field?

It is seldom that an anatomist and a philosopher join forces to write a scientific paper, as did Professor Harold Burr and F. S. C. Northrop of Yale in 1935 when they announced their "electrodynamic theory of life," to be mentioned later in this book in more detail. Briefly, their theory is that all living organisms possess an organizing electric field: a pure voltage field independent of variations in local skin conductivity or ionic currents flowing in the tissues. The "L-fields," measured in millivolts, range from DC to slow AC components and occasional sudden shifts, and have been observed over long periods of time to reflect both internal changes and external fluctuations.[23]

Leonard Ravitz, whom we mentioned in Chapter 7, has

found that people can learn to control their bioelectric fields to the point where they can guess what their field readings are at a given time. This discovery seems to have been ignored even by biofeedback researchers, which is strange in view of the remarkable claims made by Ravitz concerning correlations between field readings and states of mind, especially in mentally disturbed subjects.[24]

Biofeedback has enabled science to pursue the "ghost in the machine," the spirit or mind in the mechanistic physical body, and to come tantalizingly close to catching it. Yet though we have clarified some aspects of the mind-body dualism, we still cannot *demonstrate* that memory is more than electrochemical storage, or that thinking is not confined to the summation of all electrical potentials of individual neurons. The physiological records of the Zen monk and the well trained western meditator may appear to be identical, but does this mean they are experiencing identical states? At this stage, such questions are endless and unanswerable. The future for biofeedback research, however, looks promising. It has won quick victories in the control of physical functions by the trained mind, and there is every reason to expect an eventual solution to the mystery of mind itself.

Brain Control

Nothing but good, it seems, can come from properly guided self-entrainment. But control of other people's minds and brains is another matter, and in this sensitive area it is becoming difficult to sort out the science fact from the science fiction. We will do our best to keep to the former.

The name of Phineas Gage goes down in history as that of the world's first, and perhaps only, do-it-yourself brain surgeon. In 1848, while working on a railway construction gang, he accidentally drove a tamping rod 41 inches long and an inch and a quarter in diameter clean through his head. It entered his left cheek and came out through the top of his skull. A brave local doctor pulled it out, and Gage lived a further 13 years, though his personality and behaviour altered considerably, and for the worse. More recently, methods of interfering with the brain have become more elegant, though the frequency with which unsuccessful cases of brain surgery are reported, with side effects too dreadful to think about, indicates that our knowledge of this subject has advanced too little since Gage's pioneer experiment.

To many, the very idea of tampering with the brain, except perhaps to remove a tumour, is a repugnant one. The Portuguese neuro-surgeon Egas Moniz faced a storm of criticism, even from his colleagues, when he first removed a human frontal lobe in 1935. True, he later won a Nobel Prize, but he also ended his life a hemiplegic after one of his ungrateful lobotomized patients had shot him in the spine. Controversy has also surrounded the use of electroshock therapy ever since Ugo Cerletti introduced it after World War II to treat depressed and schizophrenic patients. As for psychosurgery—surgical intervention in the brain to treat mental disorders (not to be confused with "psychic surgery"!)—this has been a very touchy subject ever since 1891, when Gottlieb Burckhardt of Switzerland removed parts of the cerebral cortex in order to calm patients suffering from vivid hallucinations. In 1967 there was a major scandal in the U.S. after a letter had appeared in the *Journal of the American Medical Association* suggesting the use of psychosurgery to combat urban violence, the aim being "to pinpoint, diagnose *and treat* those people with low violence thresholds before they contribute to further tragedies." (Our italics.) *Ebony* magazine promptly headlined a story: "New Threat to Blacks: Brain Surgery to Control Behavior."

Brain functions can be drastically changed even without surgery. Dr. José Delgado, a Yale brain researcher, has attracted much publicity for some of his more dramatic demonstrations of what can be done by placing electrodes in the brains of animals and beaming stimulating signals at them. In this way, he has put a monkey to sleep in 30 seconds, "frozen" a cat stiff as it lapped milk from a saucer, and even stopped a charging bull in its tracks. (More precisely, he made it run round in circles rather than gore its human target.)

Could any of this kind of thing be done to people? Is it already being done?

A surprising number of people believe that somebody—usually their least favourite intelligence agency—is controlling them through electrodes implanted in their brains. We have had personal contact with two such people, both of whom appear totally sane, though one of them has repeatedly been certified as paranoid, which only means that he displays symptoms similar to those previously classed as associated with a still little understood condition. (After interviewing one of them in a Copenhagen sidewalk café, we found, on playing the tape back later, that his voice faded from the tape after a few seconds, though the machine seemed to be in perfect order.) If any reader thinks he is being electronically entrained, we suggest he have his skull x-rayed, to reveal implanted electrodes, and also have his teeth checked. Miniature radio receivers exist that can be fitted into the mouth, and it is even theoretically possible to do this without the patient's knowledge. If the history of mankind is anything to go by, any invention will sooner or later be put to evil purposes.

On a more cheerful note, it seems that electrical brain stimulation can be of value in relieving pain, epileptic seizures, and spasticity. And more than that: in a somewhat unusual experiment, Robert Heath of Tulane University enlisted the aid of a young prostitute to help him treat a male patient who, among other problems, was unable to feel attracted to women. The girl lay on the bed beside the patient, Dr. Heath applied septal stimulation to his brain, and there seems to have been a happy ending. Professor Elliot S. Valenstein, a neuro-scientist at the University of Michigan, has written: "If brain stimulation could be demonstrated to be safe and beneficial, it is conceivable that it would come to be regarded as routine as the use of tranquillizers and anti-depressants." This statement implies that it is not always safe or beneficial.[25]

Borders between fact and fiction melt away when we come to the Stimoceiver; a tiny device that weighs only 70 grams, and can be fixed to the head and concealed by a hairpiece. The idea is to set up a two-way radio link between the brain of the patient and a computer, so that when a certain brainwave pattern appears, the computer emits the appropriate stimulus and the patient possibly avoids an epileptic fit. Epileptics have already been trained to avoid brain frequencies characteristic of their particular form of attack, and the advantage of the Stimoceiver is that it allows full-time monitoring over long periods without the presence of a doctor.[26] However, as Professor Valenstein points out, the problem in this kind of situation is still what to monitor and where to stimulate? As we mentioned in Chapter 8, stimulation of the brain can have distressing and even dangerous effects, so it is not to be undertaken without caution.

We all experience some form of brain stimulation and entrainment every time a light is switched on. Certain types of neon tube produce flickering that causes immediate distress and irritation, and as we have said before, the whole question of electromagnetic pollution still has to be examined in more detail. Little is known about the extent to which we can develop tolerance towards attempted entrainment of the brain, let alone immunity. Unless we all go around in suits of copper-plated armour, we are almost defenceless against the entraining forces most of the time. This makes further research in this area a matter of some importance.

We come now to the subject of entrainment from extra-terrestrial sources. We have seen that man responds to the cosmic environment, both consciously and unconsciously, in many more ways than was once thought possible. We also know that the cosmos is constantly providing us with external stimuli in the form of electromagnetic and gravitational inputs. Theoretically, therefore, we could respond to such stimuli, and it now only remains to be seen whether such responses are of any biological significance. This is the subject of the following chapter.

PART FOUR

BEYOND THE FRONTIERS

Chapter 10

BABY AND BATHWATER

IN 1975, a group of 186 scientists, including a handful of Nobel laureates, signed a statement condemning astrology. "It is simply a mistake to imagine that the forces exerted by stars and planets at the moment of birth can in any way shape our futures," they said. "We believe that the time has come to challenge directly and forcefully the pretentious claims of astrological charlatans."[1]

One scientist who particularly objects to astrology is astronomer Bart Bok, who complains that since the gravitational forces exerted at a baby's birth by the doctor present "far outweigh the celestial forces," it is not possible that planetary influences on a new-born baby can be anything but insignificant and infinitesimal. This particular point has been dealt with amusingly by José Feola, who took the trouble to calculate the forces concerned and came up with:

Gravitational force exercised on baby by	*(in newtons)*
Earth	9.80
Sun	5.90×10^{-8}
Moon	3.36×10^{-5}
Jupiter	3.60×10^{-7}
Mars	1.86×10^{-8}
Doctor (70 kg.)	4.76×10^{-10}

From this it can be seen that whatever Professor Bok thinks, four of the celestial bodies do in fact exercise more gravitational pull on a baby than an average-size doctor at a distance of one metre.[2] Not that such force is thought important in itself by astrologers, as Feola points out. Their point is that, given the state of motion of all the celestial bodies, no two babies can ever be born under identical configurations. We are all unique, with our own finger print and our own gravitational or EM "ether print."

There is reason to suppose that planets do exercise some influence on Earth's inhabitants. In 1967, H. Prescott Sleeper Jr. of the Ryan Aeronautical Co. noted that since light travels round the Earth in about an eighth of a second, the resonant cavity between Earth's surface and the ionosphere will have a dominant frequency of about 8 Hz., the same as that of the lower end of the human alpha brain waveband. He suggested that the geomagnetic field might provide a fine tuning mechanism for this frequency selection which would set up a direct chain of influence from planet to human brain via the Sun, assuming there to be a link between planetary position and solar flare output, for which there is plenty of evidence. Now, if our brains beat against a natural background of resonance vibrations, they must respond to variations in them. Admittedly, such background influences are very weak compared with the more obvious stimuli with which our brains must cope, but they are still there, and it is not unreasonable to suppose that sudden geomagnetic field fluctuations, which we know to be closely linked to solar events, may be responsible for highly significant changes in human emotion and behaviour.[3]

Since this is not a book about astrology, we have no space for the claims of astrological charlatans, nor for those of scientists who denounce something they do not appear to have studied. We are concerned with the interactions of outer and inner space, and whether this is called cosmobiology, heliobiology, biometeorology, astrology or whatever, we suspect that all we are really trying to do is rediscover a lost science rather than invent a new one.

There have been times when men knew more than we give them credit for today. The Babylonians had a more accurate system of astronomy at their disposal than Renaissance Europe had before 1600. They had tables with which they could predict eclipses much more precisely than we could until we invented the telescope. It is less than a century since we developed the technology that would make it possible to construct a building cosmically aligned as accurately as the pyramids. These, whatever their main purpose, were very likely used as astronomical observatories and calculators, producing results of an accuracy that could not be equalled in Europe before the development of automatic telescope alignments.

How are we to explain the apparent present-day astronomical knowledge of the Dogon tribe of Mali? Without encyclopedias or telescopes, it seems that they have known for centuries about such things as elliptical orbits, rings of Saturn, moons of Jupiter and the turning of Earth on its own axis. They have long known about the circulation of the blood, that there is "red and white blood," and that the body absorbs oxygen. They even compare the Milky Way with circulating blood. But their really astonishing knowledge concerns the star Sirius, about which they seem to know far more than they possibly could by normal means. They know that Sirius is orbited by Sirius B (totally invisible from Earth without a telescope), that its orbit is elliptical, with Sirius as one of the foci, and that its period of revolution is just under 50 years. They know Sirius B is small and heavy, which it is, and have made as yet unverified statements which, if proved, would add credibility to their claim that life on Earth descends from amphibious creatures from the Sirius system who came here in spacecraft. These include descriptions of a third star in the Sirius system, larger than Sirius B, with an orbital period of 32 years, and also a planet. The source of the knowledge of the "primitive" Dogon, who, it has been suggested, may be descendants of the Argonauts, is indeed a mystery.[4]

Another modern mystery is that of the origins of acupuncture, an ancient practice that has survived thousands of years and is now being rediscovered by western scientists. The traditional invisible points on the body where acupuncturists place their needles can now be located with EM instruments. But how did the ancient Chinese find them? Not all, surely, by sticking spears in their feet by mistake? Is it reasonable to suppose that some two thousand spots a fraction of a millimetre across can be discovered by chance? We can assume either that the Chinese located them psychically, or that they had instrumentation. If they invented rockets 2,000 years before Pearl Harbour, could they not also have invented some form of voltmeter? Or were they actually able to *see* the human aura, energy body, and *chakra* centres? We may never know, but the fact is that the knowledge was there, and we do not know how it got there.

How much knowledge of the true nature of man and his relationship with the solar system and cosmos did the ancients really have? Did we live in caves for 90,000 years to become civilized overnight? We have lost the entire body of knowledge that existed before the means, or perhaps the desire, to record it also came into being. There will be no *Collected Works of Hermes Trismegistos* in paperback editions. But rather than lament our loss of ancient knowledge, we can try to rediscover it. If they learned it, so can we. This also applies, amazingly enough, to the knowledge of such recent "ancients" as Kepler and Newton, two of the most influential men of all time, vast amounts of whose writings have yet to be published. Even more amazing is the quantity of published writings that are never read: when Arthur Koestler waded through Copernicus' *Book of the Revolutions of the Heavenly Spheres* he had a surprise. Everybody knows perfectly well that Copernicus reduced the number of "epicycles" in the Ptolemaic system. Yet everybody, including an Astronomer Royal, is wrong. Copernicus actually increased the number of them, from 40 to 48! This is a fine point of scholarship, perhaps, but it goes to show how easily a myth can be transformed into "fact."[5]*

Everybody also knows that the great Johannes Kepler, the man who formulated the laws of planetary motion, was a violent opponent of astrology. Wrong again. He had a strange love-hate relationship with what he called "the step-daughter of astronomy," although he was not above doing some part-time astrology himself, for money, and seems to have been quite good at prediction. He was convinced that "some useful and sacred knowledge" lay hidden beyond the lunatic fringe of prophets and soothsayers. "The belief in the effect of the constellations derives in the first place from experience, which is so convincing that it can be denied only by people who have not examined it," he claimed. Which goes, we suspect, for many of the scientists referred to at the beginning of this chapter.

---

* Galileo, for instance, never said "E pur si muove." The original Sherlock Holmes never said "Elementary, my dear Watson," and Bogart did not say "Play it again, Sam."

(Though certainly not for Professor Bok himself, who has studied it extensively.)

"That the sky does something to man is obvious," Kepler says, "but what it does specifically remains hidden." Rejecting the general silliness of stargazers was fair enough up to a point, he thought, but he warned fellow scientists not to "throw out the baby with the bathwater." With enough diligent searching, the truth could be found, just as the persistent hen will find the golden corn in the dung-heap. We would like to know what experience convinced Kepler of the reality of extraterrestrial influences, but since *only one* of his numerous works is currently available in English, we shall have to wait. In the meantime, we can take a look at the work of one man who, after many years of hunting for the golden corn, has produced statistical evidence of a physical relationship between sky and man.

Michel Gauquelin was born in Paris in 1928. He showed a precocious interest in astrology, for reasons he himself cannot explain, and by the age of ten he had learned how to cast horoscopes. Five years later, he was cutting classes at the Lycée Charlemagne to browse in Monsieur Chacornac's bookshop, which specialized in astrology, and before he was seventeen he had read a hundred books on the subject. Despite the gaps in his formal education, Gauquelin got through the Sorbonne with qualifications in both psychology and statistics.

At about the time John Nelson was putting his telescope together on the RCA Manhattan rooftop and pointing it at the sunspots, Gauquelin and his wife Françoise embarked upon a stupendous programme of data gathering, which has resulted in sixteen published volumes. They were determined to find out once and for all if people's birthdays really had any effect on their futures. Was there such a thing as an Aries personality? Did more people die under the influence of Saturn than chance should dictate? Was there anything, in fact, to the fundamental precepts of astrology?

Initial results, based on about 27,000 birth dates, were totally negative. There was no evidence for an Aries personality, or a fatal influence of Saturn. There was, however, something else as surprising as anything ever claimed by astrologers: an evident relation between a baby's future profession and its time of birth—not with respect to the solar year or day but in the *planetary* day. More famous doctors were apparently born just after the rise or culmination of Mars or Saturn than should be by chance. Great soldiers-to-be showed a preference for Mars or Jupiter in similar positions. In some cases, there was only one chance in five million that such clusterings were due to chance alone. Yet something even more remarkable emerged: such highly significant results only seemed to apply to groups of people who were successful in their work. For instance, 1,458 scientists who had never won any prizes, published more than average material, or discovered anything new, did *not* tend to have been born at any specific time during the Mars, Jupiter or Saturn day, whereas a larger group of distinguished scientists did. The chance of this correlation being due to chance alone amounted to one in 300,000. (Odds of a mere 200 to one are accepted in most branches of science as being highly significant.)

Gauquelin has made many other discoveries, some of which seem to support traditional astrology, while others do not. He has shown, for example, that two placements of some planets in the sky—culmination (directly overhead) and rising or setting (on the horizon)—are apparently far more significant than any other positions with regard to the subsequent professions of babies born at such moments. This was not new to astrologers, but it was to everybody else. Although Gauquelin did establish a link between planetary-day position and future profession, he failed to find any evidence that being born under any particular sign of the zodiac was in itself of any significance at all.

One of his strangest discoveries makes an interesting comparison with a statement of Kepler's, to the effect that babies are born at a precise time that corresponds to that of one of its ancestors' births. For reasons unknown, Kepler named the baby's grandfather or uncle as the relevant ancestor rather than the mother, whereas Gauquelin has found a more plausible and direct "planetary heredity" to exist between baby and one or both parents. After a statistical analysis of no fewer than 160,000 planetary horoscope comparisons, he felt confident enough to state that children tend to be born after the rise or culmination of the Moon, Venus, Mars, Jupiter or Saturn *if* the same circumstances held for one of their parents. There is only one chance in 500,000 that this correlation is purely fortuitous.

Michel Gauquelin has put forward some fascinating ideas concerning the possible role of the baby, rather than a planet, timing its own birth. "Is it possible," he asks, "that throughout the entire labour procedure, it has some invisible contact with the planetary signals?" Nobody knows the work of Takata and Piccardi better than Gauquelin, who has done more than anybody to publicize them in his own books and papers; and in view of their discoveries it would be unwise to underestimate planetary effects on anything involving a liquid, let alone such a highly liquid system as a human maternal womb.

Gauquelin is a scientist, who has arrived at his conclusions after carrying out a vast amount of original research. He has also produced a coherent and provocative hypothesis that goes far towards explaining some of those conclusions. It is based on a paradoxical finding: although the Sun is far more powerful in every way than any planet, no solar influence has been found to correlate with birth patterns, whereas numerous planetary effects have been identified and studied. However, the solar output is not constant. It follows a cycle of spot production and associated flare activity, which means that Earth receives periodic variations in its solar input, as we outlined in Chapter 2. It is well established that there is a close correlation between solar activity and geomagnetic field activity. It is also well established that the latter affects living systems in a number of ways, as we have repeatedly mentioned. Moreover, as the work of Nelson was enough to indicate in itself, there are good reasons to suppose that the planets influence the periodicity of the solar output. Thus we seem to have a possible direct chain of cause and effect, from planet through the "sunspot junction" to Earth. Now, if it could be shown that geomagnetic field variations are reflected in birth patterns, we will have established that mechanism of which Dr. Vaux lamented the absence as we mentioned in Chapter 6.

Gauquelin seems to have done this. He took 16,000 birth dates and paired them with the corresponding records of geomagnetic activity. (This has been measured precisely since 1884 according to "International Magnetic Character" indices, in units of 0 to 2.0, a day with an index of more than 1.0 being rated as disturbed.) He found an overwhelmingly significant correlation: children born on magnetically disturbed days will have 2.5 times as many planetary hereditary similarities with a parent than those born on quiet days. The odds against this being due to chance are a million to one.

It only remained to find a way out of the paradox mentioned above: why the position of the planets seemed to affect the time of birth while the Sun did not. Gauquelin theorizes that "the Sun is the motor and the solar field is the medium," and the solar field will be agitated by those bodies exercising the strongest force on it, namely, the Moon, Mars, Jupiter, Saturn and Venus. (Gauquelin does not regard the effects of the more distant planets as likely sources of significant influences.) For the purpose of this theory, it does not matter whether the planets have any *direct* influence on sunspots or not. The net effect is the same if it is the solar field that bathes Earth which alters in quality because of the planets. As far as the baby is concerned, a magnetically active day seems to increase its "planetary sensitivity" and greatly increase the probability of its popping out in accordance with a planetary day position similar to that of one of its parents.

Gauquelin admits he is baffled as to how a baby about to be born can distinguish among all the various radiations surrounding it and respond only to one of the weakest. We can only suggest that planetary configuration receptivity may be programmed into the genes, so that the baby will instinctively feel when the time has come to be born. Countless other questions are raised by Gauquelin's hypothesis, and we join him in calling for more research.[6]

The important thing is that he has established that a planetary influence exists in human affairs, which has never been shown before on the same scale. His work suggests that Kepler knew what he was talking about in insisting that the kernel of corn was there in the dung heap, waiting to be discovered, and that Professor Bok and his colleagues are doing just what Kepler told us not to do—throwing out the baby along with the bathwater.

Gauquelin has described his research methods in detail, in both scientific papers and popular books, and has also published his source material. No scientist has yet managed to shake the evidence on which he has built his main hypotheses, though many, notably the Belgian "Comité Para" group, have tried their best. French academician Jean Rostand has made a last-ditch defence to the effect that if Gauquelin has proved astrology by statistics, then he no longer believes in statistics! Other scientists, such as Professor Hans Eysenck, have been encouraged by Gauquelin's work to carry out their own research, to be mentioned shortly.

Yet Gauquelin's work is often misunderstood. There is no point in carrying out mini-surveys of ninety actors and expressing surprise that they were not *all* born with Jupiter rising, as a British TV programme did in 1976, seeking to suggest that they had undone the Gauquelins' years of work involving hundreds of *thousands* of birth date correlations at one fell swoop. The point they missed is that the Gauquelin planetary heredity effect is a slight one. It only becomes apparent when large samples are taken. They also overlooked the fact that the Gauquelin effect applies only to actors (or doctors, sportsmen, etc.) who are outstanding in their careers. In the birth charts of ordinary, undistinguished professionals, it does not appear. We cannot help comparing the Gauquelin effect with the female fertility cycle: in so far as the latter may once have been entrained to that of the Moon. Perhaps in ancient times, when our relationship with nature was less obstructed, we responded more sensitively to subtle planetary influences, and all it needed then was a little prompting from the magnetic field for babies to be born on cue. Such babies were claiming their full heritage of ability, and went on to become better at their professions than the disoriented baby who missed the planetary cue. Perhaps, today, babies *not* born under their parents' planetary configuration are undergoing a form of unnatural birth, an interruption of a natural process. It may be that we now need a boost from the magnetic field to bring us back into line and restore our sensitivities to what they once were. All this is of course pure speculation, but when a scientist produces controversial data, all we can do is speculate on them and try to repeat them.

A British researcher interested in the planet-baby question is sociologist Joe Cooper, whose initial findings are of great interest in that they go against some of Gauquelin's negative solar findings, while fully supporting his evidence for planetary effects. Working with Dr. Alan Smithers of the University of Bradford, Cooper studied some 35,000 birthdates and announced that he had found the link Gauquelin had failed to find: an apparent statistical correlation between certain professions and preferred times or seasons of birth. The clusters he found, and considered to go beyond chance expectation, were:

Soldiers	mid Summer to late Autumn
Doctors	early Summer to mid Autumn
Artists	late Winter to late Spring
Musicians	late Autumn to mid Spring

The divisions he chose closely follow those of the zodiac, early Spring corresponding to Aries, and so on. Studying British data covering a hundred years, he was able to produce a chart of "expected birth frequencies" with which to compare his tables of professionals' births. This confirmed Ellsworth Huntington's earlier discovery that births tend to peak in late Spring and decline in late Autumn, a rhythm that Huntington believed to have been established in primitive man because it favoured the survival of young babies to have them live their first weeks in warmer weather.[7]

Looking at our professions, Cooper found no seasonal correlations for clergymen, cricketers or poets, though significant deviations from the expected random frequency turned up for novelists who seem to prefer to be born in late Summer. "For some occupations at least," he says, "there is a relationship to the time of year a person is born." Without making any claims for astrology, he adds that "a possible explanation might be that astrological beliefs grew up as rationalizations of intuitively recognized patterns of behaviour, which could not otherwise be accounted for." However, he enjoys pointing out that "on the basis of birth sign, astrologers might have predicted that Hitler and Lenin would have been placid, easy-going and conservative; that Molotov, Kosygin and Eichmann would have been cuddlesome and sympathetic."[8]

In 1976, Cooper repeated an experiment described in C. G. Jung's often quoted (and misunderstood) treatise on what he called "synchronicity," in which he collected birth data on 400 married couples and analysed them for evidence of truth in the astrological tradition of attraction between opposites: persons born with Sun or Moon in the ascendant. Jung's results seemed impressive, though the statistical expert he consulted to evaluate the significance of the results had some trouble arriving at a figure of 1:1,500 for the probability of the pairings being due to chance. To which Jung observed that if he only had one chance in 1,500 in making a telephone connection, he would prefer to write a letter![9]

Cooper collected birth data for 200 married couples and looked for the traditional "favourable" planetary aspects, finding a deviation of 27 from the expected random pairing, there being one chance in 100 for this result. "There is an obvious need for replication," Cooper admits, "yet this is an example of a simple collection of data to test one of many available astrological hypotheses which are generally ignored by conventional wisdom."[10]

Cooper works his way through any set of birth date records he can lay his hands on to test such hypotheses, and as the results pile up, they consistently come out well above chance level. Thus he has found that TV newsreaders are more likely to have been born under Taurus or Scorpio than Pisces or Cancer, and that telephone operators are definitely Aries, Taurus, Leo or Aquarius types. As for Australian Army officers, their births follow the pattern of their northern hemisphere counterparts, also peaking in mid to late Summer.

"I do believe that signs may exhibit common physical characteristics," Cooper says, and to prove his point he has persuaded a TV producer to seat people according to their zodiac signs in the front row of a studio audience and have him guess their sign characteristics. "Aquarians shine out," he says, "as does a row of eyebrowed Scorpios!" We wonder how many professional astrologers would care to demonstrate astrology so publicly. On a more serious note, Cooper believes that an accumulation of consistent evidence for patterns in birthdates may be marginally useful in vocational guidance. It could save parents much distress if they could be told that their children had a statistically better chance of success in some professions than in others, and to be told this not after the child has left (or been thrown out of) college or university, but the minute he/she is born. It should also be helpful to undecided teenagers to know in which areas they are statistically most likely to succeed, for again and again evidence turns up indicating that it does make a difference when you are born. Here are just three of numerous recent discoveries in this context:

In 1973, three British doctors looked for correlations between season of birth and mental disorder. They did a very thorough job, using not a random sample but all cases of persons born between 1921 and 1955 in England and Wales who were admitted to psychiatric beds in 1970 and 1971, a total of more than 28,000. The survey revealed that 9 percent more schizophrenics and 7 percent more manic depressives had been born in the first three months of the year than chance would predict, allowing for known overall seasonal birth variation. Yet figures for sufferers from all other psychoses, neuroses and personality disorders showed variations of only 1 percent from expected frequencies. This study fully supported the results of an earlier project carried out in Sweden, and its authors conclude that there is "evidence for a real association between season of birth and functional psychosis."[11]

Nature, one of the world's leading general science magazines, has allowed a spot of astrology to creep into its pages, with a brief but intriguing survey that uncovered strange correlations between signs of the zodiac and scientific specialities, also between different types of specialist and even different subtypes of one scientific speciality! Molecular biologists, for instance, seem to be born under Aries far more often than any other sign;* indeed, fewer are born under the very next sign, Taurus, than any other. On the other hand, taxonomists cluster under the sign of Cancer and tend to avoid Scorpio. Young biologists wondering which field to specialize in might find this report more helpful to them than tossing a coin or consulting the *I Ching*.[12]

Some readers might have thought *Nature* was joking, but shortly after the above paper was published, staff writer John Gribbin took a census of his office colleagues, 23 in all, and found that eight out of the 16 science journalists among them had been born under Pisces—fifty percent! Of the seven non-journalists, only one was a Piscean. No staff member was born under Aries, Taurus or Gemini, while the only two Leos happened to be the editor and deputy editor. Gribbin does not mention, perhaps understandably, whether the eight Piscean journalists were more distinguished than the others, as might be expected according to Gauquelin's findings.[13]

---

* Co-author Hill, an Aries biophysicist, is in good company.

Several correlations have been noted between season of birth and overall intellectual and physical development. A 1929 Moscow survey showed that children with high IQs tended to have been born in the first half of the year. A study of 45,000 American college freshpersons showed that those conceived in the warm season grew taller and fatter, though those conceived in the winter had a 60 percent better chance of matriculating. Girls conceived in winter also reached menarche earlier. Studies of distinguished American scientists and of British entries in the *Encyclopaedia Britannica* revealed that the highest percentage of the eminent was conceived in April, the lowest in September. English statistics show that more babies with defective brains are born between October and March, with a peak in December. It seems almost as if both the intellectually gifted and the physically retarded tend to be born in winter, whereas both healthier and stupider babies prefer the summer.

An analysis of 15,000 birth dates from England and Holland showed that schizophrenics had a significantly higher birth rate in February and March (though this did not apply to epileptics or sufferers from other mental disorders), with a minimum in June and July. Data from Australia showed exactly the reverse effect, and such correlations have been confirmed independently often enough to look fairly plausible. One such confirmation comes from the Psychiatric Institute of Columbus, Ohio, where a study of birth dates of mentally defective children born between 1913 and 1948 revealed that significantly more schizophrenics had been born in January to March than should be expected according to U.S. standard monthly birth distribution rates.

We are not suggesting that all these correlations are due to the Sun and planets. There may be causes much nearer home. The Columbus researchers, for instance, thought it might be significant that the third month after conception (June to August in the case of the schizophrenics) was the period when organization of the future child's cerebral cortex was taking place. It was also the hottest time of the year, when mothers might be expected to eat less protein, and the combination of high temperature and protein deficiency might have affected the future functioning of the baby's intellect. This is plausible in view of the fact that the hotter the summer, the more mental defectives tend to be born in the following January/March period, which is also the time when overall abnormalities in pregnancies can be expected.[14]

Assuming this theory to be true, it does not mean that all other correlations with birth dates can be explained away so easily, as many rationalists, materialists and humanists (new and old) would probably hope. Nobody has yet managed to advance a physical theory to explain the effects discovered by Gauquelin without taking extraterrestrial forces into account.

Mention must be made here of the work of John Addey, a former president of the (British) Astrological Association and one of the few who have tried to reconcile astrology and science. Lamenting the fact that astrology has become "ossified into rigid formulas and doctrines from which the illuminating principles have largely been lost," Addey has set out to recover "the pure Pythagorean principles of number symbolism as they relate to the microcosm and the dynamic interrelationships of its parts and principles." Some of his most interesting original research, which deserves repetition, has been into the apparent tendencies of certain diseases to fluctuate not so much from year to year but over long periods apparently independently of supposed immediate causes. He suggests that some whole generations are more prone to certain diseases than others.[15]

There are also intriguing indications that both dates and places of birth of certain types of person tend to cluster from time to time. In *Science Since 1500*,* H. T. Pledge included a number of maps showing remarkable birth-clusterings of eminent scientists in south-east England and the Netherlands in the seventeenth century, in northern Italy up to about 1700, and in Switzerland and south-east France in the eighteenth century. More recently, a disproportionate amount of scientific and literary talent seems to have originated from the Lake Constance region, while over a long period the city of Edinburgh has produced far more than a fair share of eminent scientists. Pledge points out that such clusterings cannot always be ascribed to wealth or population density.

---

* London: H. M. Stationery Office, 1966.

Astrological Birth Control?

It is now clear that there may be a relation between the exact time of birth and the positions of certain planets. Could a study of the planets therefore enable us to control birth, to some extent? Some extraordinary claims have been made for what has been called astrological birth control (ABC), and in view of the shortcomings of most existing methods of family planning, any new approach deserves a fair hearing, however inherently improbable it may seem.

We have seen that there is reason to suppose that the female fertility cycle was once entrained to the lunar synodic month of 29.53 days. The precision with which fish and crabs spawn under the full Moon suggests that higher forms of life may once have responded to nature's clocks with equal regularity, their rhythms becoming desynchronized as new and artificial cycles intervened to throw them off course. The work of Dewan and Rock suggests that such rhythms can still be re-entrained by providing artificial "moonlight."

Most of the claims for ABC originate from a Czech gynaecologist and psychiatrist, Dr. Eugen Jonas, who is convinced that every normal woman has about four days of fertility a month that can be predicted exactly on the sole basis of her time and place of birth. He also claims that women can choose whether to have a boy or a girl, and even pick the right day for the beginning of a successful pregnancy.

It sounds very easy to locate a woman's four-day ABC period. According to Jonas, all that is needed is the angle between Sun and Moon (on a geocentric horoscope) at the moment of her birth. This angle comes round once every 29.53 days (29 days, 12 hours and 44 minutes is the exact figure) and each time it recurs, it indicates the fourth day of what Jonas calls the "cosmic fertility period." To complicate matters, it seems that this is in addition to the mid-cycle fertility period located by the Ogino-Knaus method, and the two periods may well not coincide, in fact they cannot always coincide for any woman whose period is not exactly synchronized with that of the lunar (synodic) month.

Accurate information on Jonas's work is hard to come by, much of what has appeared in English being based on very dubious sources. Some reports speak of tests involving

30,000 Czech women to determine their cosmic fertility period resulting in 97.7 percent accuracy. If true, this is indeed sensational, but we have yet to unearth a properly written report on any such test—and not without trying! Let us try and disentangle fact from fiction regarding ABC.

Jonas set up his Astra Birth Control Research Centre in the town of Nitra, and according to a typed statement signed by Jonas of which we have a copy, studies have been done on 10,000 women volunteers, the research falling into four categories:

1. Calculation of "sterile days" on which conception cannot take place;
2. Calculation of days on which the birth of a boy or a girl, as desired, can be planned;
3. Calculation of "days of increased susceptibility to conception in the case of women for whose sterility no medical reason can be given";
4. Calculation of days on which conception of a healthy child can be expected.

The report states that 1,252 women took part in a yearlong test under the first category, of which 1,224 (97 percent) confirmed the method to be reliable. Nothing is said about categories two, three or four, and this amounts to virtually all the first-hand information we have regarding the Astra method.

To summarize Dr. Jonas's main conclusions, as we understand them:

1. The ability of a mature woman to conceive tends to occur under exactly that phase of the Moon (i.e., Earth-Sun aspect) prevailing when she was born;
2. The viability of an embryo is influenced to a great extent by the positions of certain celestial bodies at the time of conception;
3. The sex of the child depends on whether at this birth time the Moon is in a "positive" or "negative" field of the ecliptic, or angular position in the plane on which Earth orbits the Sun.*

Jonas seems to be claiming that the time of a woman's maximum fertility in each synodic lunar month is the point

---

* According to astrological tradition, positive (fire and air) signs alternate with negative (earth and water) signs. These are given at the end of this chapter.

at which the Moon reaches exactly the same relationship with the Sun as it had at the time of her birth.

But in terms of astronomical co-ordinates, all popularizers of Jonas and his ABC method that we have read have assumed he meant a relationship on the Sun-Moon ecliptic longitude as measured against the fixed-star background. This makes no sense at all. It would be like saying there should be a total eclipse of the Moon every time the Sun and Moon are 180° apart with Earth between them, which is obviously not the case, since the plane of the Moon's orbit around the Earth is inclined to the Earth-Sun ecliptic by about 23.5°. Even the builders of Stonehenge must have known this, hence the elaborate computer they built to predict eclipses. Since the Moon is usually either above or below the Sun-Earth ecliptic, another angle is involved. This is called the *declination* or *latitude*, and when this is taken into account it is plain that *exact* Sun-Moon relationships do not repeat every month, but over a much longer period. This crucial point seems to have eluded Jonas and his followers in the English-speaking world. Several popular magazines run advertisements to "compute your fertility cycle" for a fee. Before readers part with their money, we suggest they ask how many angles are being computed. If the latitude/declination of the Moon is not included, their money will probably be wasted.

A group of California scientists who work under the name Biodynamics Research[16] has begun a serious and systematic examination of Jonas's claims with the help of a small grant from the Point Foundation. BR has failed to confirm any of them as stated by Jonas, though it has found indications of some lunar and solar effects that deserve further study, also of a possible secondary fertility cycle synchronized with the phase of the Moon. It can do little useful work without further funding, and has been trying to interest U.S. and world health authorities in its work in the search for a reliable, universal and easily taught method of birth control that can hardly raise objections on any grounds, religious or medical.

Should they fail, there is another most remarkable method to fall back on, which may prove harder to research. This is simply "mental" birth control, as practised by what two journalists from the German magazine *Stern* (August 1972) described as "the most harmonious society we have ever seen."

The Muria, a tribe living in Central India and numbering about 200,000, was visited some time ago by a British missionary named Verrier Elwin, who planned to convert them to Catholicism. Unlike most missionaries, however, Elwin was so impressed by the happy and peaceful way of life enjoyed by the Muria that he decided to learn from them instead of trying to teach them anything, and he eventually wrote a book, *The Muria and their Ghotul*, followed by others, on the tribe's social organization. This, to say the least, is highly unusual.

The secret of their success is an extraordinary institution called a *ghotul*, or children's dormitory. Here, boys and girls come to live—and sleep—together at a very early age, and stay there until they get married. The kids are happy and hard-working; crime, prostitution and homosexuality are said to be unknown among them. Most astonishingly of all, so apparently is premarital pregnancy. Elwin reckoned that only 4 percent of Muria girls became pregnant before marriage, despite many years of uninhibited sex in the *ghotul* and the complete absence of any conventional form of birth control.

How they seem to defy the laws of nature as we understand them is quite simple. When a girl enters the *ghotul*, a ceremony is held at which she asks the local god to keep her from becoming pregnant until marriage, and that is that. A girl is never obliged to make love against her will, and it has been speculated that the Muria youngsters develop an instinctive awareness of their fertility cycles. If this is so, it would imply that all women might once have been able to do the same.[17]

Elwin's account of Muria mating habits makes an interesting comparison with those of the anthropologist Bronislaw Malinowski as described in *The Sexual Life of Savages in North-Western Melanesia*. The Trobriand islands he studied knew no form of contraception, and did not believe there was any connection between sexual intercourse and pregnancy! Unmarried women were allowed to mate as they liked, and premarital pregnancy, though higher than among the Muria, seems to have been less common than one might expect. Western women who wonder what the secret of these peoples may be are recommended to look at the method of natural family planning developed by Dr. John Billings and approved by the Roman Catholic church, in which women are taught to identify their own fertility cycles by studying the production rhythms of their cervical mucus.[18]

Women who have trouble locating their ovulation periods can try two safe and easy methods at home. One is the temperature method, which involves no more than taking the temperature every morning and noting when it rises, which it does after the onset of ovulation. Since variations of less than one degree C may be involved, this must be done with care. Another possible method, though not yet established, is that used by Professor Burr (to be discussed in Chapter 12) involving measurements of the voltage in the quasi-electrostatic field of the body, which he found to rise sharply during ovulation. Burr's method, involving daily readings on a millivoltmeter, has yet to be properly validated or shown to apply to women in general, but it is worth looking into, and instructions can be had from the Aquarian Research Foundation.[19]

Astrological Death Control

In addition to his astrological birth control work, Dr. Eugen Jonas has put forward a theory that there are cosmic forces at work in determining the exact time and type of death we can all expect. He presented a paper on this bizarre topic at the IV International Biophysics Congress in Moscow in August 1972. Taking a sample of forty deceased persons, he obtained details of times of birth and death, also of manner of death, and divided his sample into "vitality groups" based on life expectancy at 0-30, 30-60 and over 60 years of age. He then claimed he could predict the day any person was going to die to within half a lunar cycle, apparently using no more than the Sun/Moon position at the subject's birth.

As unfortunately is his custom, Jonas gave much less detail than we would wish, and we mention him again here only because we have found better evidence supporting the popularly supposed age-old link between the lunar cycle and human behaviour, giving us such words as *lunatic* and *moonstruck*. Hard evidence for this link gathered by modern statistical methods has begun to appear in reputable scientific journals. To summarize just one recent report:

In 1972, two researchers published a report in the *American Journal of Psychology* on their study of 15 years' data on homicides in Dade County, Florida, and 13 years for the region of Ohio that includes the city of Cleveland.

Grand totals for the periods were 1,887 for the Florida county and 2,008 for Ohio. Next, they divided the lunar synodic month into thirty periods of almost one day and plotted each homicide on a chart for each state county, according to the date of the lunar month it was committed. For both areas, major peaks in crime coincided very closely with full Moon, with secondary peaks just after new Moon. The probability of this being due solely to chance was one in 700 (Florida) and one in 3,000 (Ohio). Studies of the two sets of data showed similar trends, and closer study revealed something even more intriguing: a variation according to latitude, suggesting that while the Moon affects us all, it affects those of us in certain latitudes more than those in others.[20]

The Presidential Death Cycle

In 1840, W. H. Harrison was elected president of the United States, and in the following year he died of pneumonia. Ever since, every president without exception elected or re-elected at 20-year intervals has died in office:

Lincoln	1860	Assassinated	1865
Garfield	1880	Assassinated	1881
McKinley	1900	Assassinated	1901
Harding	1920	Died	1923
Roosevelt	1940	Died	1944
Kennedy	1960	Assassinated	1963

Thus we apparently have a cycle of U.S. presidents elected at 20-year intervals dying in office. Is this a genuine cycle or just a string of coincidences? To be a true cycle, it must have predictive value and it must also be caused by something. We cannot say with certainty that *all* presidents elected every 20 years will die in office. Or can we? But first, the apparent coincidences . . .

Of the 27 presidents *not* elected in a 20-year interval since 1840, only one has died in office, while 26 presidents did not, and neither did those elected in 1800 and 1820. The "cycle" therefore begins in 1840, and there is one chance in 2,500 according to the Fisher test of Exact Probability that the ensuing string of deaths was due only to chance. This, we must point out, is not to rule out chance altogether.

The number of coincidences arising from the deaths of Lincoln and Kennedy is endless. Both were shot in the head from behind on a Friday with their wives beside them, and both were succeeded by Southerners named Johnson who were born exactly 100 years apart, as were the supposed assassins, Booth and Oswald, both of whose full names contain 15 letters and both of whom were shot with a single bullet in a confined space while physically handicapped, dying about two hours later. Both Lincoln and Kennedy seemed to predict their own deaths. Lincoln had a secretary named Kennedy who advised him not to go to the theatre on the fateful night; Kennedy had a secretary called Lincoln who advised him against a trip to Dallas. These are just a few of the Lincoln/Kennedy assassination coincidences first noted, we believe, by author Jim Bishop shortly after Kennedy's death.*

Coincidence, or cyclic forces at work? As we have said, to be useful a cycle must have predictive value. We must be able to say with reasonable certainty that something will happen because it always has. For instance, we can say that the Sun will rise tomorrow unless there is a total eclipse, and we know when to expect the next eclipse. We are virtually certain that if we mix two parts of hydrogen with one of oxygen under the right chemical conditions, we will have water. Yet we cannot be certain that every U.S. president elected at 20-year intervals is going to die in office.

Nor can we be certain that they will not! Psychologist Stanley Krippner points out that power in the U.S. is represented by one man to a far greater extent than in many other countries; the U.S. chief executive is one of the most powerful men in the world. He speculates that a climate of resentment against the archetypal father figure could build to a peak every 20 years; the fact that a president is killed, or dies in office, at such intervals possibly providing a release for the hostility of other potential assassins. "As a result, they may not centre their hate upon another father-image-authority figure for approximately two decades."

While there may not be a true permanent cycle of presidential deaths, there may be a 20-year cycle of aggression among malcontents which has been relieved in similar ways

---

* See chapter 7 of Edward Dewey's *Cycles*. (New York: Manor Books, 1973).

seven times in a row. Krippner admits his hypothesis is highly speculative, though he points out that it may help explain "why the U.S. has such a high casualty rate among its chief executives."[21]

There is also an astrological hypothesis to account for the "cycle." David Williams, president of the American Federation of Astrologers (and one of at least three to predict well in advance that Kennedy would not survive his term), has pointed out that all seven presidents in the "cycle" died when their terms of office included a Jupiter-Saturn conjunction in an Earth sign (Virgo, Capricorn or Taurus) of the zodiac. Williams also predicted that the president elected in 1980 would break the "cycle" and survive.

Conjunctions of Jupiter and Saturn, the two largest planets, repeat at intervals of just under 20 years. They are of special interest because their combined mass is more than eleven times that of all other known planets put together. (In Chapter 5 we mentioned the possible influence of such a conjunction on the rearrangement of the California coastline.) The sight of these two celestial giants in line with each other, or only a degree or two apart, is said to be awe-inspiring; Kepler saw it in 1603 and found it worth recording. It has even been suggested that the Star of Bethlehem was in fact Jupiter plus Saturn, which would have been in conjunction in the year 7 B.C., now thought the most probable year of birth of Jesus.[22]

Jupiter and Saturn will be within two degrees of conjunction from November 1980 to March 1981, which should give us all time to admire the spectacle. Over the next couple of years, several other outer planets will line up with each other—but not all at once, as Gribbin and Plagemann implied in the original edition of *The Jupiter Effect:* in fact the astrological action will go on until 1988/9, with a rare triple conjunction of Uranus, Neptune and Saturn. Arthur Prieditis, whose interesting piece of earthquake prediction we mentioned in Chapter 5, has assured us that "world-shaking political upheavals of the first magnitude" are in store for us early in the 1980s.[23]

One way and another, the 1980s promise to be an exciting decade. But unfortunately the mystery of the Presidential Death Cycle will remain unsolved until a cause-and-effect mechanism can be discovered. We mention it here to show how difficult it can be to decide whether some events are truly cyclic or just arranged by chance in a series that closely resembles a cycle. It may be that Jupiter-Saturn conjunctions prompt Americans to take pot shots at their presidents, and when combined with solar maximum periods also cause earthquakes. Or it may not. This must be borne in mind when assessing the work of one of the most remarkable scientists of this century, whom we have already mentioned in Chapters 4 and 6.

The Father of Heliobiology

Professor Aleksandr Leonidovich Chizhevsky (1897-1964) was an interdisciplinary scientist of great versatility. A professor at the Moscow Faculty of Medicine, he was also a Fellow of its Archaeological Institute, an assistant at the astronomical observatory and a collaborator at the Institute of Biological Physics. He was also a talented musician, playing both violin and piano, a successful painter, and a poet whose work was praised by Mayakovsky. (One of his early poems is a sonnet to the Sun.) He also found time to make detailed studies of geography and history.

In addition to his pioneering work with air ions, he did a vast amount of original research into cyclic phenomena of all kinds, and before he was 30 he had published several papers in Russian, French and German on cycles linked to one of his main interests: the sunspots.

He became interested at a very early age in both cycles and sunspots, making his first observations of the Sun in June 1915. At that time, a large group of spots crossed the central meridian of the Sun, the aurora borealis was unusually strong in several places (this is invariably an index of solar activity), and magnetic storms became strong enough to disrupt radio and telephone communications. Chizhevsky also noted that the toughest battles of the Great War were being fought in that year, and he wondered if there could be any overall connection between solar and human activity. He devoted much of the rest of his career to the search for such connections.

In 1917, he observed that the Bolshevik revolution took place close to an unusual burst of solar activity, as had the abortive uprising of 1905. By 1922, he had drawn up an extraordinary chart which, he claimed, showed that a period of no fewer than 2,400 years of "mass movements," including *all* major wars, battles and uprisings recorded in the histories of all peoples, revealed not only regular cycles, but cycles in phase with that of the Sun. He had, he decided, hit upon a universal cycle of historical events. Periods of mass movement would rise and fall with regularity even in nations that had no contact with each other. This suggested that some external factor was causing the cycles, and the most likely such factor would be the Sun, or more precisely the forces that cause the solar cycle.

"We must assume," he said (in 1926), "that there exists a powerful factor outside our globe which governs the development of events in human societies and synchronizes them with the Sun's activity; and thus we must also assume that the electrical energy of the Sun is the superterrestrial factor which influences historical processes."

Chizhevsky found there to be an average of nine mass-movement cycles every hundred years of just over eleven years each: exactly the average length of the solar cycle. Not only were the two cycles in phase with each other, but over and over again a peak year of popular unrest would coincide with, or come very close to, a year of solar activity maxima. The French revolutions of 1789, 1830 and 1848, the commune of 1870, and the two Russian uprisings of 1905 and 1917 all took place near times of solar maxima. (As did the outbreak of World War II, the Communist take-overs of many of the East European countries, the Soviet invasion of Czechoslovakia and the worldwide period of student unrest in 1968.)

Having identified his universal cycle of events, Chizhevsky went even further. He also found that each individual cycle could be divided into four parts, exactly matching the progress of each solar cycle, corresponding to minimum, increase, maximum and decline of mass excitability. Nearly 80 percent of all major historical events in modern times took place in parts two and three, and only five percent in part one. Chizhevsky divides the "excitability" cycle into lengths of three, two, three and three years, and we must assume he meant this to be the *ratio* of each sub-cycle rather than the length in years, since he was perfectly well aware that solar cycles were not always 11.1 years in length. (In fact, he knew more about sunspots than the authors of several recent textbooks: we read again and again that "sunspots were discovered by Galileo after he had invented the telescope in 1610," a statement that contains at least three errors. Isolated sunspot observations with the naked eye were in fact made and recorded in Arabian, Chinese, Russian and Armenian chronicles and also in German public records. A collection of 45 early Chinese observations was published in 1889, and Chinese sunspot records certainly date back to 28 B.C.; Chizhevsky included all this data in his calculations, and felt justified in extrapolating to fill the gaps.)

The four subdivisions of the "historiometric cycle," according to Chizhevsky, are as follows:

In the first period, the masses are peaceful and tolerant, but also lacking in unity and generally indifferent to political matters, slow to resist and quick to capitulate. In the second, the masses begin to unite, new ideas emerge, new leaders appear, and alliances are forged between nations and groups. The immediate solution of a predominant question becomes necessary. In the third period, of maximum excitability, nations are aroused both to their greatest achievements and their greatest madness. The excited masses respond willingly to their leaders. This is a time of war, uprising, persecution and emigration. Finally, in the fourth period, the masses become exhausted and enervated, slipping back easily into the first part of the next cycle. (Sunspot maximum falls in the middle of part three.)

Chizhevsky found correlations with the sunspot cycle in the most unlikely places: in the emigration of Jews to South America and Norwegians to the U.S.A., in the ups and downs of Whig and Tory governments in nineteenth-century Britain, in executions under the Lynch law in the U.S.A. between 1889 and 1923, and in strikes and political terrorism in Russia around the turn of the century. We have already mentioned his numerous correlations between solar activity and the incidence of epidemics; and he repeatedly pointed out that "psychic epidemics" were as regularly distributed as those of disease.

No theory has any value unless predictions can be made using it as a basis, and in a paper written in 1926 or earlier (there is much repetition in Chizhevsky's published papers), Chizhevsky predicted "human activity of the highest historical importance" for the period 1927-1929, which would change the political map of the world. In this period, Salazar took over Portugal to begin the longest dictatorship of modern history, Chiang Kai-shek captured Peking and set up his Nationalist government, Italy elected an all-Fascist parliament thereby paving the way for

Mussolini, Hitler was well on the way to the top, while in Chizhevsky's homeland Stalin threw out Trotsky, thus establishing himself firmly in control. Altogether, it was a remarkably good period for dictators. It was also a time of distress in Britain, with the collapse of the Baldwin government, and of the greatest economic disaster in history in the U.S. with the 1929 Wall Street crash. Some of these events could have been predicted in 1926, but surely not all of them?

Going over Chizhevsky's work forty years later, Edward Dewey made use of data not available to his predecessor in cycle-hunting, and discovered that there was actually a slight time-lag between peaks of mass excitability and sunspot indices, with the former peaking first. The sunspot cycles, in other words, turned *after* the cycles on Earth that they were previously thought to cause! How, Dewey wondered, could cause follow effect? Presumably, some other factor caused both solar and historical cycles, the Sun taking longer to respond. The actual position on Earth of events also seemed to be important; sunspot activity took place on an average of 14° latitude from the solar equator, while most cyclical events studied on Earth fell between 40° and 55°N latitude. If solar and terrestrial cycles had a common cause, Dewey reasoned that a time-lag according to latitude could be expected.

If, as is beginning to look more than probable, planetary forces are acting on the Sun and also on the solar field that envelops the Earth, we would expect the latter to be far more responsive than the former.

Numerous objections can be raised to Chizhevsky's theories, as they can to any theory involving the interpretation of statistics, yet his fundamental hypothesis remains unshaken and largely ignored, perhaps because of its implications, which are too alarming for some of us to face squarely. Nothing generates more heated discussion than mention of the dreaded word *predestination*.

Chizhevsky's hypothesis will take a long time to prove or disprove, since it involves matching of day-to-day events with at least one whole 11-year solar cycle. However, a few minutes' work is sufficient to provide some intriguing evidence that seems to support one of his most extraordinary claims; that political trends in Britain during the nineteenth century followed the rhythm of the sunspots. British politicians, especially those with marginal seats, may hope this proves as nonsensical as it sounds. Yet if we dig out the statistics on general elections since World War II, what do we find?

Wars invariably throw cycles other than those of nature off course, so if we take 1954 as base year (the first solar minimum after the end of the war), we find that the percentage of the British electorate voting Conservative rises after a minimum year (1954, 1964, 1976) and falls after a maximum (1957, 1968). Local elections held in 1977, as a prolonged period of solar minimum activity was ending, showed a tremendous swing to the right. Twice in succession, percentage votes for Conservative and Labour, as plotted on a chart, have crossed over very close to a turning point in the Sun cycle. It also appears that general elections have tended over the 1954-1976 period to be held about two years after a turn in the cycle. This, like the deaths of U.S. presidents every 20 years, may be pure coincidence. We shall have to wait and see. (A Conservative victory in a late 1978 general election would be interesting in this connection.) Without mentioning Chizhevsky, Iben Browning has claimed a similar sun link wth Republican voting trends in U.S. politics, and indeed, as can be seen on our own chart here, recent swings to left and right do show similarities in both British and U.S. elections.

Are we all slaves of the Sun?

Chizhevsky thought that up to a point the answer was yes, we are. Yet he stressed that solar forces do not compel us to do anything specific. They merely oblige us to do *something*, and this need not involve violence or destruction. He hoped that national leaders in the future would be able to channel the aroused energies of the masses at times of maximum excitability into such peaceful outlets as sport, scientific expeditions, collective creative art, or "the building of stupendous structures," as he put it. (The building of Brazil's new capital, Brasília, one of the most stupendous constructions of modern times, got started close to the 1957 solar maximum, by the way.)

Chizhevsky put forward enough challenging hypotheses to keep researchers busy for lifetimes. Yet at the height of his own career he found himself in trouble, for reasons not hard to imagine. According to Marxist dogma, human history is driven by class conflict based on the forces of

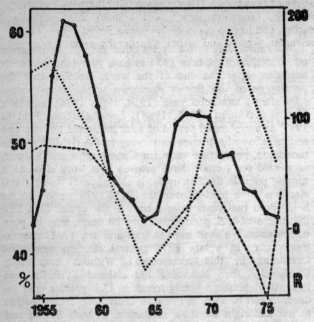

Fig. 10
Cycle or non-cycle? Chizhevsky claimed that Conservative voting patterns in nineteenth-century Britain followed the ups and downs of the solar cycle. Iben Browning has made the same claim for Republican voting trends at presidential elections. Examining the period from 1955 to 1975 (almost two complete solar cycles) we can see that there may be something to such claims. Thick line is the sunspot cycle (right vertical column), dotted and dashed lines respectively are U.S. Republican and British Conservative percentage (left col.) share of votes cast. Note that U.S. elections are held at fixed intervals, whereas British ones are not, also that data for the latter are based on estimates following by-election trends.

economics, and the change from imperialism/capitalism to socialism/communism is the inevitable result of struggle by the masses, and not the appearance of blobs on the face of the Sun or Jupiter-Saturn conjunctions.

We have been told on good authority that Lenin was in fact quite impressed by Chizhevsky's work, some of which he read shortly before his premature death in 1924 and spoke of favourably. His successor felt otherwise, evidently, and like Giordano Bruno and Galileo three centuries earlier, and like so many of the young Soviet Union's finest scientists, writers and artists, Chizhevsky eventually found himself to be holding the right views in the wrong time and place. In 1939 or thereabouts (just after a solar maximum), his brilliant career came to an end, and he was sent off to a labour camp. He was still in his early forties.

He was released and rehabilitated about five years before his death in 1964, and a number of books and papers of his were published from 1959 onwards, though many of these, perhaps all of them, were written before 1939. He has had much influence on young Soviet scientists, who call him the "father of heliobiology." In February 1968, the physics section of the Moscow Society of Naturalists held a meeting in his honour, the proceedings of which were published under the title *Sun, Electricity, Life*.[24] That he is known at all outside the Soviet Union is almost entirely due to Michel Gauquelin, who has frequently mentioned him in his own widely-read books.

The science of the Space Age (perhaps *cosmobiology* is the most apt word for it), has thus been foreseen and outlined by the sleepwalkers of the twentieth century: the successors to men like Kepler, Copernicus and Galileo who, though solidly grounded in science, made their important discoveries by unconventional, somnambulistic and even somewhat unscientific methods. Chizhevsky was perhaps the first modern cosmobiologist, a man with a vision of a new science combining the study of the Sun, electricity, life and air ions with that of cycles in natural events and in human affairs—a regularly fluctuating background of war and peace, excitability and passivity, epidemic and health. Cosmobiology is the study of the living Universe as a whole, and such a vast field cannot be examined in the necessary detail by one man. The specialists' dedication to single parts is as necessary as the cosmobiologist's breadth of vision, and it is to be hoped that in the science of the future, specialist and cosmobiologist will work side by side —or at least in neighbouring rooms—with occasional glances at the other to see what he is up to. There can be no more barriers between disciplines.

Astrologers will no doubt feel that they have been badly treated here, and that theirs is the science of the Space Age and always has been. We cannot agree. Astrology may once have been a true science based on the study of repeatably observable interactions between man and environment, combined with the correct interpretation of such study. Today, however, the position is such that, as one eminent astrologer has admitted, "astrology is almost as confused as the earthly chaos it is supposed to clarify."[25] It is interesting to note that the two men who have done most, it might seem, to produce evidence in support of astrological traditions—Michel Gauquelin and John Nelson —not only do not regard themselves as astrologers but even go to some lengths to dissociate their work from astrology.

Yet astrology cannot be dismissed out of hand. The sky does do something to man, as Kepler insisted, and we are gradually finding out what it does and how it does it. In the process, some of astrology's cherished tradition is blown to pieces, yet some of it remains intact: and although astrology as practised today can hardly be called a science, there are encouraging signs that rediscovery of its original scientific basis may be taking place.

Geoffrey Dean, the British scientist we mentioned in Chapter 2, spent four years in what is almost certainly the first systematic and critical survey of the entire literature on astrology, covering about 80 percent of that available in English and a fair amount in other European languages. Working with a team of fifty collaborators, both scientists and astrologers, Dr. Dean has concluded that "astrology works, but seldom in the way or to the extent that it is said to work." He has also provided a solid base for future research in what will perhaps come to be known as cosmobiology rather than astrology.[26]

Why, it might be asked, do astrologers not collaborate with scientists? With modern statistical techniques it should be very easy to test a good many of astrology's claims. Take two of the most fundamental of these claims: that there is a correlation between what sort of person you are and the sign of the zodiac under which you were born, and that persons born under the "water signs" (Cancer, Scorpio and Pisces) tend to be more emotional and neurotic than the rest of us. Surely an astrologer, a psychologist and a computer could all get together to prove or disprove these two claims at least once and for all?

They could, and at long last, they have. In 1977 considerable interest was aroused by the appearance of a paper co-authored by three men including Jeff Mayo, one of England's best-known professional astrologers, and one of the most eminent living phychologists, Professor Hans J. Eysenck of the University of London. Using a sample of 2,323 men and women, they set out to test the above-mentioned claims, and they found strong support for each of them.

Using a standard personality evaluation test devised by Eysenck himself in 1964, they found an extremely clear correlation between odd-numbered zodiac birth signs (Aries, Gemini, Leo, Libra, Sagittarius and Aquarius) and a tendency to extroversion, with an equally clear link between even-numbered signs (Taurus, Cancer, Virgo, Scorpio, Capricorn and Pisces) and introversion, precisely as stated by traditional astrological lore.

As for the water sign link with emotionality and neuroticism, the graph peaked precisely on cue for all three signs, with an equally high but unpredicted peak for Aries people, which apparently came as a surprise even to the astrologer, Mayo. He and Eysenck conclude that "the astrological hypotheses tested have not been disconfirmed" and that their results "are not to be derived from any theoretical considerations likely to be discovered in textbooks of psychology." The latter comment is of special interest in view of the number of standard works on the subject by Professor Eysenck himself.

The authors state modestly that their work "may present some difficulties to those who maintain the negative position vis-à-vis astrology," specifically mentioning a Bok group we referred to at the beginning of this chapter.[27]

Their work cries out for repetition, and now that evidence of the quality amassed by Gauquelin, Nelson and Dean is available to all, we cannot imagine that it will be long before more scientists and astrologers join forces to tackle some more of life's ancient mysteries. In the process, there is a good chance that they will rediscover the true basis of astrology.

In the following chapter, we return to a theme that has turned up briefly in earlier chapters, that of the ways in which we can make use of the invisible forces around us. For reasons of space, we concentrate on the uses of ultraviolet and various colours of laser light (which is not always invisible) and of the energies at work in acupuncture, since these are fields in which a good deal of interesting work has been going on in recent years.

Chapter 11

THE ELECTROMAGNETIC WEB

MUCH OF this book has been about possible biological effects of external inputs in the form of energy that is periodic or cyclic in nature. Yet not all cosmic inputs come into this category: often we are exposed to fields, waves or particles that are transient, or one-of-a-kind events. On the Sun, for instance, though the spots come in almost predictable cycles, each solar flare is an individual event, releasing a huge amount of EM energy in a very short time, increasing the background radiation count to far beyond normal, and then fading away.

In the laboratory, short pulses of EM radiation have been aimed at biological targets, for good or ill, in the cause of scientific research; and the study of this kind of once-only input and its biological consequences is just as intriguing and as important as that of the cyclic events we have mentioned.

Since the early 1960s it has been known that Soviet and East European scientists have placed more emphasis on this type of research than have their Western counterparts. Even so, there were many surprises in store for Scott Hill when he travelled to the Soviet republic of Kazakhstan in late 1976 to attend an interdisciplinary conference on the theme of biological systems subjected to short impulses of EM energy, including laser light. Much of the information in this and the following chapters appears here for the first time in English.

The oldest known form of manipulated biological energy is the oriental practice of acupuncture, through which much has been learned about man's response to external stimuli in the form of prickings with a needle. Achilles, we are told, met his death after a sharp-eyed bowman located his only vulnerable point by piercing his heel with an arrow. Achilles' heel may have been what the Chinese call a "forbidden" point, of which there are several marked on the traditional charts. If stimulated, they can injure or even kill, and whatever the truth of the Achilles legend, it is a fact that the Chinese have found a point on the lower leg —the *san yin chiao* on the spleen meridian—which must never be touched with the needle during a woman's pregnancy, or labour may be induced prematurely and an unwanted abortion result. And, as so often, we now find that the same set of circumstances can bring benefit as well as distress. Three Hong Kong doctors announced in 1976 that they had successfully induced drug-free labour in 21 out of 31 women by using nothing more than electro-stimulation. This is one of the recent additions to traditional hand-manipulated needles in the acupuncturist's repertoire of techniques, which also include the use of "sonopuncture," or ultrasonic waves, and now "laser-puncture," or low intensity coherent light beams from a laser.

There is no doubt, say the doctors, that uterine contractions can be initiated by stimulation of certain "acupoints" (acupuncture points, also known as "loci"). Although the mechanism by which such "acustimulation" actually induces labour is not yet known, they noted there was a delay of 20 minutes to three hours, suggesting that a hormone-related reaction, enzymatic transfer, or some other agent diffused in fluids of the body was involved. Uterine contractions were similar to those of normal labour with no foetal or maternal complications, and although this method of artificial labour-induction is hardly routine as yet, it is sufficiently developed to have had a textbook written about it.[1,2]

Acupuncture has always been notably successful in inducing anaesthesia. Nobody really knew why or how until it was discovered very recently that our pituitary glands contain a chemical called beta-endorphin, which can be released under suitable stimulation. This is a natural pain-killer that seems to have some of the properties of morphine, and has been hidden away inside our bodies all this time.[3] (See also pp. 302–03).

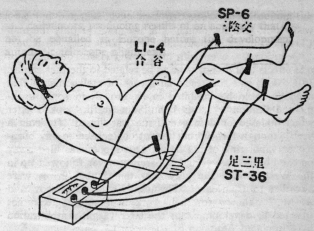

Fig. 11
Electric acupuncture used to induce drug-free labour. SP-6 is the San-Yin-Chiao "forbidden point." (Courtesy: American Journal of Chinese Medicine).

Medicine of the Future?

We return to traditional acupuncture later in this chapter, but first we want to describe some recent research into the influence on living systems of something that has been generally overlooked—light. Work has begun on biological effects of microwave, radar and radio frequency energy, but studies of the effects of light itself, visible and invisible, have been few. We know, for instance, that eye surgeons can use lasers, beams of high intensity coherent light concentrated into a very narrow band of the EM spectrum, to destroy cells selectively without damaging neighbouring tissue; and this technique is being used to repair detached retinas by "welding," bore tiny holes in tissue, and help perform other very delicate surgical tasks.

This type of laser-surgery involves strictly thermal effects, and in the West there is surprisingly little else that lasers have been used for in medical therapy. In the Soviet Union, however, research has been going on at least since 1965, using lasers of only 20 mW, or lasers with one *three-thousandth* of the power of a household 60 watt bulb, a power that cannot possibly warm up biological tissue by more than a fraction of a degree. Whatever the nature of the interaction between this highly concentrated but very weak light and the human skin, it cannot be only thermal. (The heat effect is about 0.001°C.) What then can it be?[4]

Occasionally, even with high power lasers, one hears of odd effects that do not seem to be related to thermal causes. In 1964, it was reported that a case of melanoma (a skin cancer) could be inhibited in its growth in some unexplained fashion by laser light. Skin grafts "took" better when diseased or damaged tissue was "lased" (to coin a verb), though this did not seem to be related to any burning or vaporizing of the tissue. This report was filed away on the library shelves and apparently never followed up in the West. It was the very year that Nobel Prizes were awarded to Charles H. Townes, Nikolai G. Basov and A. M. Prokhorov for their work in quantum electronics that led to development of the laser (*L*ight *A*mplification by *S*timulated *E*mission of *R*adiation).[5]

In the Soviet Union, however, no time was wasted in finding out what effects weak laser light could have on plants and tissues. Results were presented regularly at conferences that were unfortunately closed to Western scientists until 1976—and even then, only three attended: two Norwegian doctors and Hill.[6]

The medical implications of using laser bio-stimulation in healing human tissue are promising enough, though in the long run it may be in the area of food production that the laser has its greatest part to play. While in Alma-Ata, site of the Kazakh State University, Hill had a bite out of a tomato weighing 500 grammes, which had an unusually sweet taste. It also had an unusual history, for it was one of the local crop experimentally grown with the help of laser stimulation.

This agricultural innovation is not exclusive to the U.S.S.R. In 1970, *Nature* published a report from two Australian plant physiologists who had discovered that their cabbages seemed to pick up a little when red light from a small laser was beamed at them: even when only 3 milliwatt lasers were used from a distance of more than a quarter of a mile. They described their methods as calmly as if discussing the latest development in fertilizers, and they stirred memories in some minds (at least in ours) of the claims of radionics practitioners, who supposedly influence plant growth at a distance with their "black boxes."[7]

The Australians seem to have made their discovery independently of the Russians. At first, they used normal floodlights to irradiate fields of barley at night, finding that plants could be brought to maturity 3 to 4 weeks quicker than by standard growing methods. When they tried the laser, they found results were even better despite the relative weakness of the light. Curiously enough, they found that short exposures of about 100 seconds were more effective than exposures ten times as long. Soon, we suspect, every gardener's shed will house a mini-laser along with the spades and forks.

In the medical sector, remarkable claims have been made by Soviet and East European scientists for laser-induced cures, not only of body surface wounds, such as trophic ulcers or unclosing sores, but also of internal disorders. The laser is now in daily use in several Soviet hospitals, and the list of its uses grows steadily. Facial paralysis, for example, is a disease often impossible to cure, yet, as Hill was able to see for himself during a visit to a children's hospital in Alma-Ata, this disease is now responding to "laserpuncture" or "photopuncture": stimulation of traditional acupuncture points with laser beams, an intriguing blend of ancient and modern practices.

Soviet "laserpuncturists" use the electrical characteristics of the acupoints themselves to determine how much radiation to use, how often, and where. By comparing the minute voltages between points on the left and right sides of paralysed children's faces, they have noted a "biological semiconductor effect," or imbalance in voltage. This, they find, can be restored to normal in some cases by laser therapy. This is literally healing with light, since when voltages were normalized, the paralysis was found to be cured.[8]

The healing power of laser radiation has also been investigated by researchers at the Semmelweiss University Medical School in Hungary, who have studied the effects of daily doses of radiation from a gas laser emitting visible red light. Energy used to treat wounds on rabbits' ears was one joule, somewhat higher than that used by the Russians, and they found that this considerably accelerated the forming of new vessels. In another test, using rats, a similar improvement in wound healing was shown in the "lased" group over a control batch, and the Hungarians believe that the mechanism by which laser beams interact with biological tissue is a humoral one: that is, an influence on the enzymatic conditions, especially the activity of succinic and lactic acid dehydrogenase, acid phosphatase and nonspecific esterase in the wounded areas. The activity of all these substances irradiated by laser beams was higher than in the control group, thus supporting the enzymatic hypothesis.[9]

In West Germany, the "akup-las" laser acupuncture device manufactured by Messerschmidt-Bölkow-Blohm has been on the market since 1975, and 30,000 patients are said to have been treated with this for a number of ailments.[10] A number of Scandinavian doctors have also begun research into laserpuncture, though the method is as yet far from generally accepted. One Western acupuncturist of long experience has assured us privately that he was able to cook a steak with a device (manufacturer unknown— not the firm mentioned above) claimed to be an acupuncture laser. "I do not wish to cook my patients," he told us. Clearly, more research into laserpuncture is required, and perhaps tighter control of equipment put on the market.

Photoecology: Light and Life

Before we can understand how laser light can affect biological tissues, we must consider the broader context of how light and life are interrelated.

Visible light is only a small part of the electromagnetic spectrum, less than one octave of the range from direct current, or infinitely long waves, to wavelengths shorter than the diameter of a single atom. It serves a number of biological functions (in addition to enabling us to see where we are going), as we shall explain shortly. It may seem paradoxical to talk about "invisible light," like the "inaudible sound" we mentioned in Chapter 4, but the word "light" can be applied quite correctly to EM radiation to either side of the tiny band of light we can actually see unaided. Thus we have infrared (IR) and ultraviolet (UV) light, with wavelengths respectively longer than red and shorter than violet light.

Infrared rays are given off by the human body as heat, and have been used for diagnostic purposes by the firms that make "thermovision" equipment, which makes IR radiation visible on a TV screen, and has a number of interesting uses. In medicine, it is used to detect breast cancer in women; military uses include location of enemies in total darkness; builders use it to locate heat-leaks; while according to one catalogue in our possession, the British Home Office has one "for the detection of underground corpses."

Ultraviolet light can have notable biological effects. In small doses, it gives us a suntan, but in larger doses it is bacteriocidal and can even be lethal. The Sun emits large amounts of UV light, but fortunately the protective covering of our atmosphere prevents too much of it getting through to us.

Even the air we breathe is closely related to light, through the phenomenon of photosynthesis; by which sunlight is transformed into chemical energy enabling all plant life to grow, exchanging carbon dioxide—a poison to man —into life-giving breathable oxygen in the process.

In Chapter 3 we mentioned Gurvich's "mitogenetic" radiation, ultraweak light apparently emitted by dividing or dying cells. It now seems that a plant or animal can emit its own autoradiographic imprint on UV film, and this weak luminescence can be added to the long list of chemo- and bioluminescences that have been discovered in connection with bio-energetic processes in cells, cultures and chemical solutions.[11] Some bioluminescences, such as the light emitted by the firefly, can be seen with the naked eye, but most are much weaker and can be recorded only with special equipment.

Summarizing the effects of outside energy on biosystems, Polish scientist Przemyslaw Czerski points out that the living organism is a self-regulating system equipped with several interdependent homeostatic mechanisms. When the system absorbs radiant energy, it is disturbed, leading to the activation of these mechanisms and the setting up of a chain of biological events with immediate, indirect and delayed effects.[12]

The effect of an EM wave on body tissue should be proportional to the wave's thermal content, according to current theories. For instance, a strong microwave or a radar beam can "cook" tissue, even on the inside, in a very short time, and this process has been put to work in the "microwave oven" now on the market in many countries, that can boil a kettle in six seconds or produce a well-done steak in a few minutes.

Using this model of thermal effects, it has been calculated that nerve cell membranes cannot be thermally excited even at high field strengths at frequencies greater than 100 MHz. Yet if this were so, then many of the experiments we have described in this book (see, for instance, reference 7 to Chapter 8) could never have succeeded; and here is one more example:

In an experiment mentioned by Czerski, it was found that chick brains exposed *in vitro* to 147 MHz EM waves, amplitude-modulated at the biofrequencies of 9, 11, 16 and 20 Hz., led to an efflux of calcium from the brain. Calcium ions are vital to life processes, in fact no nerve cell could function without the proper amount of them. This experiment, like others we have mentioned, indicates that the purely thermal hypothesis mentioned above simply cannot hold up. Indeed, the whole literature of EM effects on biosystems is full of unexplained effects. To mention just one: it was observed in 1958 that in garlic root tips exposed to pulsed waves of 5 to 40 MHz, chromosomal and mitotic aberrations could be induced even when the duration of the pulses was only 15 to 30 milliseconds.[13]

Not only plants have been shown to suffer such damage from high-frequency radiation. Studies have also been done on chromosome damage to animal and human cells. In a 1971 report to the U.S. Department of Health, Education and Welfare, two scientists listed some alarming damaging genetic effects including breaks in and "partial de-spiraling" of the DNA helix in animal cells. Similar "electromagnetically-deteriorated" chromosomes were also detected in human white blood cells. Since the DNA double-helix is the carrier of the complex genetic code of life itself, here is a new worry for those who wish to impose a moratorium on genetic engineering.[14]

The mounting pile of evidence for the bioactive properties of weak EM waves, including light, points to the fact that the interaction mechanism with tissues cannot be only thermal in nature, as previously believed. It must be electronic-resonant, and here we return to the question of bioresonances which we touched on briefly in an earlier chapter.

Bioresonance and Bioresonators

The problem of biological resonance is one that has been shelved for a long time, and is overdue for re-examination. Most physicists and engineers deal, both experimentally and theoretically, only with dead matter, and many tend to ignore biological problems altogether, because they are just too complicated for the relatively simple models used in the physical sciences. This is partially understandable, since science must advance by reducing problems of great complexity to more simplified problems that one has a chance of solving within a human lifetime. However, it is not justifiable when it means that major and real problems (as opposed to mathematical abstractions) are all too often simply swept under the carpet, and this surely applies to the concept of bioresonance.

If the whole human body resonated at its characteristic wavelength, a wave about two metres long would be necessary. However, since our insides are made up of numerous channels, tubes, nerve bundles, air spaces, liquid-filled enclosures and cavities, we can expect our bioresonators to be of varying shapes and sizes. We can also expect that a *set* of different frequencies might bring about bioresonance, which could be electrical, magnetic or acoustic in nature, perhaps a combination of all three.

Could we have bioresonators even at cellular level, so small that until recently they have escaped detection altogether? One scientist who thinks so is Dr. Stuart Hameroff, a medical researcher from Tucson, Arizona, who is especially interested in a recently discovered element of the cytoplasm called a microtubule (MT).

Microtubules were something of a mystery until the 1960s. They were thought to be artefacts caused by imprecise staining techniques, but as electron microscopy became more accurate, it was found that they do indeed exist, though it is still not certain exactly what they are for. They are hollow cylindrical bodies about 270 ångströms in diameter (one ångström is one hundred-millionth of a centimetre), and are thought to play a structural role in the contractile mechanism of the cell membrane. They behave in an interesting way, moving around within the cell, dissolving themselves and reforming in another place, and Hameroff thinks that they may be actually guiding the DNA molecules to their "right" place in the double helix structure, serving as cellular pathfinders.

He has calculated that they may resonate at close to UV light frequencies (10^{16}Hz.) and serve as light fibre waveguides; as such they could be related to cell division and serve as a link between light and cell growth. Hameroff notes that the outer layer of skin is translucent and highly refractive, admitting UV light into the body. Once inside, he suggests, this could be resonantly amplified by MTs and then flow around the body. He even puts forward the idea that this internal circulation of UV light may be identified with the flow of *ch'i* energy traditionally associated with the meridians of acupuncture, which we will discuss later in this chapter.[15]

Some support for such unusual thinking comes from recent studies on the "third eye" of occult tradition, the pineal gland. Once regarded by anatomists as an obsolete body (in man), this is now being looked at both as a photoreceptor and as a neurochemical transducer. In Chapter 8, we mentioned its role as inhibitor of the gonads when triggered by light, and it now seems quite possible that light may penetrate deep into the human brain, and stimulate the pineal gland.[16]

It has even been suggested that UV light communication may be the *primary* regulatory process in cells, and that when disrupted, it could lead to such cell abnormalities as unchecked growth, as in cancer. This raises the question of possible detrimental effects of man-made chemical pollution of the ozone layer of our outer atmosphere as a result of the disturbing effect of freon, the gas we use every day to drive our harmless-looking domestic spray cans. Though scientists in official positions are constantly denying the suggestion, some believe that by disturbing the ozone balance, we may also be upsetting that of the UV light levels to which we are exposed every day.

Theoretical work along the lines of UV light bioregulation mechanisms has already been taken up by Fritz A. Popp, a biophysicist at the University of Marburg, who points out that since about ten million cells in the human body die every second, and have to be replaced by new cells created by mitosis (division of old cells), whatever the system that *regulates* this process must be very delicately balanced, and must "know" at any given time just how many dead cells need replacing. Could the necessary information carrier be an electromagnetic wave?

He has suggested that a specific energy transfer between a carcinogen and a bioreceptor is an important and perhaps primary step in the development of cancer. Energy transfer is the exchange of photons (or any other quanta) in any form. This cellular connection is most obvious in the case of cancer induction by radiation, where the direct interaction of photons and bioreceptor is decisive.

Radiation must have energies above approximately three electron volts (eV) to induce cancer, and Popp has pinpointed a specific set of resonances which might cause disturbance of absorption and re-emission of UV photons in this critical energy range. These resonances are active in polycyclic hydrocarbons, which in turn have been linked to carcinogenetic activity.

The part of the cell bearing special attention, says Popp, is that of the nucleic acids, which become conductive after excitation with energies above the critical 3 eV level. He feels it is presumptuous to regard the DNA and RNA macromolecules as in "stationary states" that are relatively stable, as is commonly assumed. On the contrary, he thinks, due to "weak quantization" (also a consequence of the interaction of the macromolecule with its cellular environment), the DNA helix may act as a *resonant circuit*, with the DNA as a coil and the membranes as capacitances connected in parallel as in a tank circuit.

If one looks at the DNA/cell energetic coupling in a certain way, according to Popp, the energy scheme resembles that of a four-level biolaser. He has even developed the mathematics of this model to a point where one might begin to believe that such a fantastic concept could actually work.

Cell growth regulated by modulation of sound and light waves? Cancer caused by incorrect information transfer? We cannot tell as yet, though we were intrigued to learn that Dr. Popp seems to have reached conclusions similar to those of Soviet scientists with whose work he was not familiar until Hill introduced him to it.[17]

Major breakthrough in cancer research, or red herring? Only more research into the mysteries of light-life interactions will settle the matter.

Theories Old and New

Professor Chizhevsky, whose work we mentioned in Chapters 4, 6 and 10, was aware, many years before astronomers knew anything about solar radiation in any detail, that there was more to sunlight than met the eye.

He suspected that it contained some unknown and biologically active component, which he named z-radiation. Today, now that more is known about cosmic rays, solar neutrinos and other high energy particles, we can possibly identify his "z-rays" with *nuclear* radiation from the Sun.

A nineteenth-century Danish doctor also had ideas before his time. In 1893, Dr. N. R. Finsen wrote his M.D. dissertation on "the influence of light on the human skin and illnesses thereof." He believed that blue and green light could heal a variety of ailments, and invented a "light bath" for this purpose. Today, the Finsen Institute in Copenhagen, Scandinavia's largest specialized cancer clinic, treats patients in its "light bath" section for a variety of skin problems, using an array of mercury and carbon-arc lamps, sunlamps and UV "black lights."[18]

It seems paradoxical that ultraviolet light can either retard or accelerate cell division according to the frequency and dosage used, and can even totally destroy the chromosomes of animals.[19] Dr. H. F. Blum of the (American) National Cancer Institute has suggested that hyperplasia induced by UV light may be related to cancer induction, and such ideas seem more plausible now that we know other ways in medicine in which extremely weak radiation can have very definite effects.[20] For example, a Russian 20 mW laser applied for two seconds, as in the treatment for facial paralysis described earlier, produces only 40 milli-joules of energy—one *three thousandth* of that provided by a 60 W reading lamp, in the same length of time.

Even this sounds like a massive overdose compared with the radiation emitted by the Swiss-made RS-25 therapy device, which uses not light but radioactive radium as an energy source. Radiation emitted is so weak, apparently, that it cannot be measured at all: its output is calculated at 1.2 milli-roentgen per hour or less. We would hesitate to mention the RS-25 but for the remarkable claims made for it by doctors in public hospitals in Switzerland and Italy. It sounds even less likely to succeed than the radionic black box: filters are placed between the lead-shielded radiation source and the output, and are soaked in medicine. Treatment consists of no more than beaming radiation from this complex at the patient! Swiss authorities have certified the device as safe, and doctors who use it believe its effects are due to secondary radiations of unknown nature.[21]

Quantum Mechanics and Consciousness

Any attempt to explain the strange effects we have mentioned so far by a single mechanism is doomed to failure. For instance, photochemical reactions may be responsible for cancer induction by strong UV light, but they cannot explain the healing power of the mini-laser or the non-specific effects of the RS-25.

However, changing our frame of reference somewhat, it may be possible to group these widely differing phenomena under the heading of quantum mechanical effects. Quantum mechanics, the "revolution that shook physics," involves the study of very small systems, atoms and their nuclei, with *quantized* energy states as opposed to the continuum in energy distribution of classical Newtonian physics.

The suggested link between quantum mechanics and consciousness has implications so great that we cannot omit a brief discussion of it here. It is complicated and controversial—one scientist with whom we have tried to discuss it became almost apopleptic at the idea—but it is far more exciting than it may sound.

Quantum mechanics generally means the study of "lifeless" atoms, molecules and nuclei and the search for eversmaller particles, sub-particles with their postulated quarks, gluons, strangeness, charm, upness and downness and other occult attributes, none of which has any obvious connection with "real life," the life processes going on inside our bodies.

But atoms and molecules do not exist only in physics laboratories. We are all made of them, about 10^{24} atoms in every cubic centimetre. Is it really so far out to postulate, therefore, that a quantum mechanical process may be directly responsible for life and consciousness? Men and women, like lumps of concrete, are made of atoms and molecules, but in the case of humans, the whole is something far greater than the sum of its identifiable physical parts.

In 1976 Hill had a chance to bring up the question of quantum mechanics in the human brain with a Nobel laureate in neurophysiology. "It has definitely been proved," said the great man, "that neural conduction is purely electrical and chemical in nature." He meant, of course, the kind of electricity and chemistry taught in the nineteenth century: totally mechanistic and deterministic. It is curious that in an age when physicists are moving away from mechanistic models with their either/or, yes/no states, biologists are eagerly embracing models long abandoned by the physicists.

Nobody can dispute the fact that chemical messenger molecules and electrical currents play an important role in synaptic transmission within each neural net and in the brain as a whole, and it is known that the genetic code is determined by the chemical organization of the DNA molecules in their double helix. But what determines *these* processes? What *causes* the right chemical transmitter to cross the synapse at the right time to deliver its electrical message?

The neurophysiologist of today is as unable to discuss the concept of will or consciousness as Aristotle was 2,000 years ago. The ghost of consciousness has continued to elude the electromagnetic machine. Something is missing from the mechanistic-deterministic models, namely the spark that sets life processes in actions at a distance between atoms with a common past which, until death intervenes. What could this missing element be?

David Bohm, a theoretical physicist at Birkbeck College in London, has been interested in consciousness for some time. In 1952, to solve a problem known as the Einstein-Podolsky-Rosen (EPR) paradox, he invented the concept of *hidden variables* in quantum mechanics. These, though hidden from the scientific observer and his instruments, are "really" responsible for the interactions at a distance between atoms with a common past which as in the EPR experiment, became separated but continue to act as if one "knew" what the others were doing.[22]

Bohm did not specify what these variables were, but in 1970 the American physicist Evan H. Walker came out and stated that they represented consciousness itself: the missing link in the mind-body problem that has been the cause of centuries of dispute. Walker's "hidden consciousness variables" can be included in Schrödinger's equation of quantum mechanics in addition to the "dead" variables of time, position and momentum, and Walker has used this approach to tackle problems lying traditionally outside physics, such as the interaction of mind and matter at a distance. He also suggested that quantum mechanical "tunnelling" in the nervous system may be involved in the conduction of nervous impulses, not, as our Nobel Prize neurophysiologist friend might fear, as a replacement for the established chemical and electrical mechanisms, but as an explanation for what *causes* the electrons and molecules involved to function as they do. This would explain the missing link between human will and the neuromuscular interactions that result from it, as for instance when we execute a "psychokinetic phenomenon" by raising a finger or blinking an eyelid.[23]

Walker's formulation, like all quantum mechanics, is mathematical, and far too complex to elaborate here. His theory has yet to make much impact on those scientists most in need of a unifying theory: biologists, psychologists and philosophers, each of whom talks a good deal about life, consciousness and states of being, but who have so far failed to quantify a model beyond vague statements. Walker's, on the other hand, *is* a quantitative theory. It is based on known data, and it has predictive value, without which no model can qualify as a full-blown theory, an essential feature of which is that it can be tested, however wild it may sound, and proved right or wrong. When science denies new theories without testing them, it relapses into dogmatic religion, and many believe this to be happening to an alarming extent today.

This is not to suggest for a moment that there should be no room in science for mystical approaches. As Fritjof Capra, himself a physicist, has written, "science does not need mysticism and mysticism does not need science; but man needs both. Dr. Capra laments the fact that many scientists do not realize the implications of their own theories, and continue to support a mechanistically oriented society. "The world view implied by modern physics is inconsistent with our present society," he says, calling for a "cultural revolution in the true sense of the word."[24]

But let us return, briefly, to the question of how quantum mechanics is involved with interactions between light and the physical body. The absorption and emission of UV radiation normally originates in electronic quantum-transactions between different energy levels of atoms and molecules. UV radiation is absorbed in the course of many important chemical reactions. Proteins and enzymes are excited by it, returning to their ground states by emission of other forms of radiation. Such "photochemical" UV radiation involves an exchange of heat, and is not new. What *is* new is non-thermal emission of UV radiation by biological systems.

According to Canadian scientist J. Bigu, the human body emits a small amount of nuclear radiation: gamma and beta radiation of non-thermal origin. The human body contains several radioactive isotopes, of which one, potassium 40, is biologically significant. Although the amount in the body is small, Bigu reckons that when it decays to a more stable form, it emits either beta particles or gamma rays (some 870 per second of the latter), giving humans in effect a gamma ray aura which could well be strong enough for suitably sensitive persons to perceive it under certain conditions.[25]

At this point, we feel that a brief reference to the bio-engineering of the animal kingdom is relevant, to lead us into further examination of the invisible energy body of man. As we have said, we cannot assume that man has or ever had any specific attributes to be found in non-human living species, yet on the other hand we cannot assume the contrary.

It is well known that certain animal species, notably rays and electric fish, possess highly developed organs for the detection and emission of EM radiation. An electric eel can produce a discharge in an open circuit of up to 600 volts, with a power output of 100 watts. The explorer Humboldt had an arm paralysed for days after he had carelessly put it in water too close to one of these creatures, whose powerful discharge can stun small fishes.

Direct EM communication between organisms is also to be found in the insect world. Ultraviolet light is "visible" to bees, while infrared radiation from female moths' bodies can apparently attract the male of the species. P. S. Callahan, a former U.S. Air Force officer and now an entomologist with the U.S. Department of Agriculture, has constructed an amusing model for biocommunication among moths that goes something like this:

Daylight or darkness programme the night-flying moth for *go* or *no-go* signals, depending on radiation present: *go* for infrared at night, *no-go* for visible and UV light during the day. Shorter UV wavelengths programme the insect with regard to time, while longer infrared wavelengths direct and control its behaviour. As a *go* signal is received, a *camera hold* order is overridden and the IR eye's *camera-on* control takes over. Pitch, yaw and gyro controls come into force, using the temperature readout of background radiation, and the airborne insect wanders on a random course until IR or microwave emissions from a plant or mate lock onto the moth-transducer. Insect mates can be further tracked by IR/microwave emission of scent molecules, enhanced through feedback via movement of molecules towards the insect.[26]

We include this not as a joke, but to make the point that man is still struggling to keep pace with technological advances in the animal world. Here are some more examples:

The common honeybee knows a thing or two about the forces of nature that we never taught it. The honeycomb, with its stacked nest of hexagons, gives maximal structural support against external forces, and can withstand more pressure per weight and surface area than a block of reinforced concrete. (Perhaps the bathyhexagon is already on the drawing boards?) Bees fly in "bee lines" because they can orient themselves by the polarization of UV light from a blue patch of sky even if the Sun is obscured. Ants also have this ability.[27]

As for the ungainly bumblebee, it was once proved conclusively by a mathematician that it cannot possibly leave the ground. (Rather as more serious scientists were proving the same about interplanetary rockets right up to the eve of the launching of Sputnik 1.)

In 1794 an Italian scientist named Spallanzani suggested that bats can navigate in the dark by using what we now call sonar. His colleagues thought this was a huge joke. "If they see with their ears, what do they hear with their eyes?" they asked. Professor Eric Laithwaite, head of the Imperial College (London) department of electrical engineering, reckons that if Spallanzani had been taken seriously we might have had radar long before we did. Indeed, he thinks that living objects might contain all the answers to our technical problems. "All we really have to do," he says, "is to discover the questions." This is an interesting admission from an engineer, of whom more anon.[28]

Bats are still far ahead of us in their use of such "recent" discoveries as echo location and acoustic holography. Boa constrictors may not know that one of the latest U.S. spy satellites can detect a temperature difference of half a degree at a distance of 450 miles. Amazing! The boa has been able for some time to use the receptor pits on its upper lip to distinguish variations of one *thousandth* of a degree. The common sea squid propels itself through the water unaware of the jet engines roaring overhead and using basically the same method of propulsion.

These are just a few examples of ways in which nature has provided us with the new science of bionics: the application of biological principles to electronics and engineering systems; or, in effect, doing our best to imitate nature. Max Delbrück, 1969 Nobel laureate in medicine, has observed that when a physicist or engineer tackles a biological problem, he is usually worried that he does not know enough biology. But, he says, "it invariably turns out that he does not know enough physics or engineering."

Throughout the course of evolution, living beings have had to adapt or die; to find solutions to a number of technical problems concerning moving and sensing in a number of ways, in addition to just keeping alive in a mercilessly hostile environment. Our dumb animal friends, with their "bird brains," have solved many problems, or evolved solutions to them, that still baffle us. How, for instance, do barnacles manage to manufacture a glue that sticks them even to chemically inert substances? The multinational chemical companies would very much like to know. They would also like to be able to manufacture a thread as efficient as a spider's, and to desalinate water as easily as seagulls seem to do with their special glands.[29]

The irony is that the more advanced and complicated our technology becomes, the more often we find that nature has beaten us to it. As Professor Laithwaite has pointed out, carriages used to topple over until some designer decided to put the centre of gravity below the points of suspension—a trick he could have learned many years previously by observing a beetle or spider. Vast sums of money are now being spent on ways to eliminate the wheel, one of man's proudest discoveries, in high-speed transport systems. When we succeed, Laithwaite observes, we shall have taken another step towards nature, not away from it. Nature has managed without the wheel.

As for modern science's greatest triumph, the computer, this is superior to the human brain only in that it can solve single problems faster. In other respects it is inferior to the equipment of the smallest ape. Nor is the human brain necessarily the most marvellous and supreme achievement we tend to assume. Biophysicist John Lilly discovered, when experimenting with dolphins, that the brains of these fascinating mammals are more developed in some areas, notably the acoustic cortex, than ours are. They can communicate with each other by sound at rates of up to ten times faster than the fastest possible human speech. Some dolphins do in fact have larger brains than man. *What are they using them for?*

We do not train dolphins. They train us. For instance, a new "trainer" has to *learn* to hold his hoop properly, before the dolphin will jump through it. Lilly feels that if we are to learn anything useful from the dolphin, we must approach it as an equal rather than as captor and exploiter. (He himself has now dropped out of scientific research involving the catching, caging and dissection of animals.)[30] Indeed, we could well approach them as our superiors, since they have almost totally dominated their environment, which we have not.*

In our final chapter, we turn to a subject that has intrigued and baffled man for centuries: the possible existence in man of a non-physical component. Throughout this book, we have shown that man, like any biosystem, is constantly under the influence of forces beyond his conscious control or comprehension. If we can now show that there is a component associated with our physical bodies that receives such energies and makes use of them, for good or ill, the last piece of the cosmic many-body jigsaw will fall into place, and an entirely new approach to our understanding of ourselves and our environment will become not only possible but essential.

---

* The greatest mystery about the dolphin is its friendliness towards its only natural predator: man, who kills thousands of them every year either deliberately, as in Japan, or accidentally in tuna sieve nets. John Lilly is now campaigning to stop such slaughter, and has set up a fund to save both dolphins and whales.

Chapter 12

THE ENERGY BODY

THROUGHOUT HIS recorded history, man has believed himself to be something more than the sum of his physical and chemical parts. The Egyptians built special chambers in their tombs for the *ka*, or etheric double, shown in their paintings as a winged form joined by a cord to the physical body. Ancient Indian teachings describe two invisible bodies, the *karana sharira* and *sukshma sharira*, or casual and subtle. Tibetan manuscripts give precise instructions on how to separate the *prana*, or vital current, from the dying physical body and ease it into the *bardo*, or clear light of reality.

Since the question of a second body, or energy body, is so closely linked by many to that of our survival of bodily death, the whole subject has become bogged down in theological debate, with members of every religious sect having their own ideas as to how many extra bodies we all have and what they should be called. Scientists, not surprisingly, have been put off by all the nonsense written about the non-physical in general: the behaviourist psychologist B. F. Skinner has even denied the existence of mind, which he calls "a spurious entity." Yet it would be unwise to dismiss all concepts of extraphysical human components, now that it has been shown how many waves, fields and particles, invisible as they may be to our restricted senses, definitely do exist, and just as definitely have some effect on our known physical bodies.

Some of the best anecdotal evidence for the existence of invisible energy bodies comes from Hawaii, where in 1917 an American schoolteacher named Max Freedom Long arrived to find a still surviving system of ancient knowledge and practical magic preserved by the *kahunas*, or keepers of the secret. Christian missionaries of the nineteenth century had done their best to stamp out such knowledge, but Long set about learning all he could about *kahuna* beliefs and practices, which included almost everything known as "psi phenomena" to modern parapsychologists; from firewalking and telepathy to healing by paranormal methods.

The *kahunas* held that man has three non-physical bodies, vehicles of his subconscious, conscious and superconscious selves. Their belief predated Freud's discovery of the subconscious by centuries. Indeed, there are grounds for believing *kahuna* lore to be directly descended from that of ancient Egypt. They regarded the subconscious (*unihipili*) and conscious (*uhane*) as entirely separate entities, with the superconscious or "parental" self (*aumakua*) distant from the others but linked by a vital cord or thread along which two-way communication could take place by thought-form or prayer. The three bodies were also known as the high, middle and low selves, a classification that brings to mind F. W. H. Myers' concept of "subliminal" and "supraliminal" seats of consciousness.[1]

The *kahunas* believed that each of the "shadow" bodies was superimposed on the physical, and contained a *mould* of every cell and tissue of it. Such a concept fits in very well with a number of scientific theories and speculations, especially those of H. S. Burr, to be mentioned shortly. Even Alfred Russel Wallace, whose ideas on natural selection among living beings were presented together with those of Darwin in 1858, insisted firmly that man was a duality, containing an organized non-physical form that evolved with and permeated the physical body.[2]

The Brazilian parapsychologist H. G. Andrade has put forward a detailed hypothesis to explain the interactions between physical and extraphysical bodies, postulating a *biological organizing model* (BOM) that arranges the molecules of matter into a previously determined form; starting with the division of the first cell and recapitulating thereafter the whole evolutionary history both of the individual and of its species. Andrade sees his model as a "hyperform," a pure energy structure rather than any kind of body, and it is tempting to speculate further along the lines of an organizing field as the basis of life, the field of force that would solve the problem mentioned earlier of how the information transmitted by the DNA gets *into* the cell in the first place.[3]

Such a concept is fully compatible with the findings of the late Professor Burr of Yale, mentioned together with those of his pupil Leonard Ravitz in Chapter 9. Burr discovered that all living things are surrounded and permeated by electric fields that order and control them by organizing the physical matter with which they are associated. "Where there is life, there are electrical propertics," Burr says, speaking of nature's "jelly moulds" that *shape* the millions of different forms of life on Earth, from the tiniest seed to the largest tree.

Burr's theory has the advantage of being testable by measurement, and it can also be put to many practical uses, as we shall see. Even so, it attracted little attention from 1935, when he first announced it, until 1972, when a popular book on the subject reached a wide audience;[4] despite the fact that during this time Burr and a number of associates repeatedly tested his "electrodynamic theory of life" on a wide variety of living systems.

Burr's most interesting discovery was that an electric field is set up in the organism at a very early stage, a pure voltage field that seems to originate in the body but to extend a distance from it, and that remains relatively stable though subject to transient changes during pathological processes. From such externally monitored changes, it has been possible to use it to diagnose internal states of the organism; in 1947 Burr and Louis Langman published a paper only a few paragraphs long in *Science* in which they claimed to be able to diagnose cancers of the uterus by using nothing more than a vacuum-tube voltmeter. Simply by placing an electrode in a woman's vagina and comparing the voltage readings of cervix and abdomen, a relation could be found between voltmeter readings and the presence of uterine tumours. This was long before the women themselves, or their doctors using conventional diagnostic methods, were aware of the presence of the tumours.[5]

In the course of this research, in which 860 patients were involved, it was found that voltage readings of the human electric field were useful indicators of many internal states in addition to cancer. Earlier, Burr and Langman had also found that a peak reading in voltage between two fingers of women's hands seemed to occur during ovulation, suggesting that the female fertility period can actually be indicated instrumentally.[6]

It is known that some women have very short fertility periods, of even only a few hours, so that unless timing is very precise, they can be considered effectively sterile. Despite the obvious distress this must cause to childless women, it was more than 30 years after the work of Burr and Langman that an "electronic ovulation detector" appeared on the market. (In the patent application for this, no mention is made of Burr.)[7]

Not only man, but also lower animals and plants have what Burr called "L-fields" (L for Life), which he found to exist long before the organism was fully grown, in fact even before the egg of a salamander was fertilized. Even seeds that were later to bloom into healthy plants showed different readings to those of plants destined to be stunted. If this could be found to apply to human beings, we would virtually have electric precognition of future physical conditions.

There is much evidence to indicate that Burr's L-field is separate from such known electric effects as skin potentials, with which it can be, and often is, very easily confused. The L-field can be picked up at a distance from the body, however, and it remains steady even when the ionic character of the tissue changes. Moreover, when its EM potentials change, they do not do so in a random way, but in response to external influences such as those we mentioned in earlier chapters: ELF wave atmospherics, solar and lunar cycles, also other external disturbances now suspected to affect biological systems at a very basic level. Since man's L-fields respond in similar ways to those of trees and lower life forms, a substratum common to all life is indicated: one far from understood as yet, but one with which the electromagnetic field hypothesis is wholly consistent.

Evidence for an EM factor behind life keeps turning up in the most unexpected places. In 1977 Professor Laithwaite, whom we mentioned in the previous chapter, described an experiment in which he constructed what was in effect a liquid electric motor, consisting of salt water and a substance known as ferrofluid, which is both liquid and magnetic. Laithwaite placed a flat dish of this mixture in alternating magnetic fields, and found that the ferrofluid arranged itself, as expected, according to the lines of the field. But, he found, the shapes were not those of man, but of nature. "There were ferns and tree branches and sunflowers! ... It is an electromagnetic plant, no less."[8]

Now that ferrofluid is available commercially, we trust such experiments will be widely replicated. We cannot help speculating that if a plant-like shape can be "grown" out of inert matter under totally artificial conditions, might not similar forces already be at work behind all forms of life, including, of course, man?

Leaving further speculation until there is more evidence, we must now examine our supposed energy body from another angle. If there is such a body, and if it is electromagnetic in nature, it must be possible to manipulate it artificially. This, of course, is just what acupuncturists claim has been done since long before any kind of modern medicine or surgery came into existence.

The U.S. Patent Office is not in the habit of granting patents to inventions that originate from beyond the borders of established science. However, on 26 July 1976 Patent No. 3,971,366 was granted to Dr. Hiroshi Motoyama, a Japanese physiologist, for his Apparatus for Measuring the Condition of the Meridians and the Corresponding Internal Organs of the Living Body, or AMI Machine for short. This is no less than a computer that gives an automatic diagnosis of the entire human body in about five minutes.

The meridians of traditional acupuncture may be no more real than those on maps of the world used to denote latitude and longitude, yet centuries of practical experience in China and Japan must have gone into the mapping of this complex of lines that cover maps of the human body from top to toe, each line dotted with *loci*, or points, like stations on a railway line. These meridians are considered to form a network wholly or partially independent of the known nervous system, and to be connected in some way to the various organs of the body.

At the end of each meridian there is a terminus, as there is at the end of a railway line. These are known in Japanese as *seiketsu* points, and there are 28 of them in all. To make a diagnosis with his AMI machine, Dr. Motoyama attaches an electrode to each terminus and presses a button, whereupon the computer makes a series of readings of the current flow along each meridian. To oversimplify a very intricate process, what happens is that a very short time later the machine prints out an analysis of the body's electrical energy imbalances and indicates at which points treatment should be given.

Dr. E. Stanton Maxey, the Florida surgeon we mentioned in Chapter 3, has seen the AMI machine in action and has used one himself, probably being the first person outside Japan to do so. "It is quite striking," he says, "to see the computer read-out showing imbalances, hear Dr. Motoyama's diagnosis of what's wrong and what the machine will show upon appropriate acutherapy; and then see the computer reading indicating these improvements have occurred as predicted."[9]

Science fiction readers may recall the late C. M. Kornbluth's story *The Little Black Bag*, in which both diagnosis and treatment have become fully automated and disease is no longer a problem. The AMI machine is a step towards this goal. Dr. Motoyama has published a number of reports indicating that his invention can be very helpful in diagnosis, which can be the most difficult part of a doctor's job. One patient was analysed by the AMI as having particularly severe imbalance in the gall bladder meridian; subsequent x-ray examination revealed more than fifty stones in it.[10]

An AMI machine has now been installed at the Miami Heart Institute, where Dr. Maxey, working with electronics engineer James Beal, has been studying its use in connection with the effects of ions on the human body. He believes that the acupuncture points may play an important part in the mechanism through which electric energy is exchanged between the inside of the body and the world outside. In one intriguing experiment, Maxey recorded data on his AMI machine on nine people flying at an altitude of 39,000 feet, comparing the readings with data obtained at ground level. In each case, there was a difference in the print-out, possibly due to a predominance of positive air ions in the cabin at high altitudes. These, Maxey speculates, may be carriers of modulated electric or magnetic field effects into biological systems which, in extreme cases, could even be sending pilots momentarily to sleep, thus providing a possible explanation for some accidents attributed to pilot "error." The findings of Maxey, himself a pilot, make an interesting comparison with those of Dr. Caymaz in Turkey, mentioned in Chapter 4.[11]

In Europe, scientific research into acupuncture is centred on Vienna, the first city to have a state-sponsored clinic, the Ludwig Boltzmann Institute, devoted to it. Since 1969, when it opened, much progress has been made there in the study of basic problems of skin electrical characteristics, electroconductivity of organisms, veterinary and dental acupuncture, and also in the stimulation of points on the human ear. Vienna is also the headquarters of the Working Group for Radiation Research, under the direction of Peter Kokoschinegg. This has been examining what happens when AC potentials are applied to acupoints and other tissue for both diagnostic and therapeutic purposes. Under investigation by this group are diseases of the urogenital system, cardiac strains, and tumour growth.

Acupuncture, like the sunspots, has been around since long before Western scientists showed much interest in it. A case of dropsy cured by needles was reported in the very first issue of *The Lancet* in 1823, and four years later Dr. John Elliotson announced in the same journal that he had successfully treated more than one hundred cases of chronic rheumatism with what his critics claimed were "hat pins." (Elliotson, who introduced the stethoscope to Britain, also tried to introduce mesmerism into hospital practice, but was forced to give it up.)

Also in 1827, acupuncture was used at the Royal Infirmary in Edinburgh, the city where an experiment to be described took place exactly 150 years later. Yet, like the sunspots in the seventeenth century, acupuncture virtually disappeared from Western eyes soon after it was introduced, and although it managed to survive in France to some extent, it was only very recently that it has enjoyed a sudden revival, largely due to two men—Mao Tse-tung and former U.S. president Richard Nixon. Although Chinese medicine was banned altogether under the Chiang Kai-shek regime in 1929, Mao managed to restore it even before the 1949 proclamation of the Chinese People's Republic, and a clinic was opened in Yenan in 1945. With the thaw in U.S.-Chinese relations that took place during Nixon's presidency, many American doctors went to China, and decided that acupuncture was well worth further study.

One, Dr. Samuel Rosen, declared "I have seen the past, and it works."

Today, acupuncture is well established in Europe and the U.S., although nobody, least of all those who practise it, knows how it works. (Which does not stop them writing popular books about it.) We do not know how it works either, but we have at least tried to round up such new evidence here as might enable an answer to emerge if the evidence holds.

The first question that demands an answer is: do the meridians exist or not? To an embryologist, the idea that an internal organ might have some connection with a certain point on the skin is not altogether unreasonable. While the embryo is developing, all parts of the future body grow from the same tissue (indeed, from the same cell), so that the ectoderm, which later becomes the outer skin, lies very close to the notochord or future spinal cord, and equally close to the other embryonic organs. Could not some kind of neural connection be established early in the development of the embryo between skin and organs, perhaps governed in its development by the overall "blueprint" or "biological organizing model" of the L-field? This seems a hypothesis worth testing.

It is also quite reasonable to suppose that the surface of the skin may be associated with an internal disorder. Doctors with no knowledge of Chinese medicine are familiar with the "McBurney point" that becomes tender, on the skin, during appendicitis. The Danish scientist Wernøe, experimenting with decapitated fish, found that by chemical or electrical stimulation of the rectum, he could produce alterations in the pigment of their skin.

On the traditional meridian maps, which date from at least eleventh century, it is evident that some of the meridians run parallel to the known dermatome innervation system, while others do not. (A dermatome is an area of skin served by sensory fibres from a single spinal segment.) It may be, therefore, as one theory has it, that a nerve signal can jump over several dermatomes at a time in a kind of long-distance reflex process (in medical terms: viscero-cutaneous or intersegmental reflexes). Again, this is a reasonable idea. Long reflexes can be observed in dogs when they scratch themselves, and it has even been suggested that elephant drivers practise a kind of acupuncture when they clout their mounts on certain parts of their hides to make them do what they want them to.

Much interest was aroused in the 1960s by the claims of a North Korean doctor named Kim Bong Han, who announced that he had identified a special "ducting" system through which *ch'i* energy flowed independently of the nervous system and lymph systems. Repeated attempts to support his claims were made in Austria and Germany, but without success, and we have been informed by an East European source that Dr. Bong Han committed suicide in 1976, though we have no official confirmation of this.

A more convincing demonstration of the possible existence of the meridians was provided in Japan by Dr. Yoshio Nagahama of Chiba University Medical School, who was lucky enough to be able to examine a patient who had just been struck by lightning and lived to tell the tale. The man claimed that he could feel a kind of "echo" when a needle was inserted in his skin, so Dr. Nagahama placed needles in each of the fundamental points of the meridians, and found that his patient could actually trace the course through which the "needle echo" was transmitted with his finger. He could also say if the echo was strong or weak. Even more interesting was the fact that the transmission speed of needle echo was much slower than that of an impulse in the spinal or autonomic nerves. It travelled at speeds ranging from 15.2 to 48.1 cm. per second, compared with five to 80 *metres* per second for nervous conduction velocities. Moreover, it seems that the patient, though he knew nothing about acupuncture, indicated the traditional meridians as he pointed to the path followed by the needle echoes.[12]

If this experiment were repeatable, it would virtually settle the matter once and for all, though it seems it can only be repeated with another lightning-struck patient, and these are hard to come by, at least alive. However, any busy doctor of long experience can recall the odd case that had him completely baffled, and the possible knowledge to be gained from such cases makes detailed reporting of them well worth while. Consider, for instance, by way of a brief diversion, what happened in Clinton Prison, Dannemora, N.Y. in February 1920:

One day, tinned salmon was on the menu, and thirty-four

men who ate some were taken ill with botulinus poisoning, which is often fatal. On this occasion they did not die, but instead every one of them suddenly became highly electrically charged—a human magnet. One tried to throw a crumpled piece of paper away, but the paper stuck to his hand. The prison physician, Dr. Julius B. Ransom, had the time and the interest to carry out some simple impromptu experiments while caring for his unusual patients. He found that the men could move compass needles, "electrify" pieces of paper, and cause a suspended steel tape to sway from side to side. Their strange ability, he noted, varied in proportion to the severity of the disease, and when they were fully recovered the unexpected power disappeared.

Commenting on this case in an article that seems to have been overlooked by parapsychologists seeking normal explanations for such "paranormal" events as psychokinesis, Mayne R. Coe Jr. reckons that when the poison injured the men's muscle cells, a low amperage current pushed by about half a million volts was caused to flow through their bodies. He also states that sufferers from botulism glow in the dark with an electrostatic charge—a true aura!

Coe himself made a serious attempt to replicate certain "poltergeist" phenomena by generating currents of bioelectricity in his own body through traditional yoga fasting and breathing exercises aimed at "raising the *kundalini*." He claims he could make a suspended cardboard box swing *towards* him from a distance of eight feet, also that he could train anybody to do the same thing. Once again, it seems that the mystics of the East know more about *Homo electromagneticus* than the scientists of the West.[13]

It would be useful to repeat Dr. Ransom's experiments, but for obvious reasons it will be difficult. Meanwhile, we must be content with more conventional attempts to map out the energy body of man, such as the techniques used by Dr. Motoyama. He is the ideal researcher in this field, being both a skilled scientist and a professor of yoga with more than twenty years of experience. In addition to his AMI machine, he has invented a device to measure energy emitted from the *chakras*, the points of the body where, in yoga tradition, the *prana* energy of the universe interacts with that of the body. As is the case with Dr. Nagahama, Motoyama's research would seem to settle the matter of the existence of the energy body if it could be repeated independently, and two reasons why this has not yet been done may be that he is a man of great modesty who does not advertise himself, and that his machinery is expensive.

Meanwhile, acupuncture seems to be holding up well when submitted to scrutiny by Western scientists, as is the case to some extent with astrology. It may be that the original precepts of the former have survived virtually intact, whereas those of the latter have become somewhat dispersed over the centuries. In any case, we are now in the process of rediscovering the secrets of both, and about time too.

In 1977, Duncan Stewart and two colleagues at Edinburgh University published the results of an experiment set up to find out if acupuncture could produce analgesia (inability to feel pain) under conditions that ruled out suggestion on the part of the acupuncturist. It has often been alleged that acupuncture is no more than a form of suggestion or even hypnotism, though such an allegation is now hard to support.

Stewart took twelve volunteer subjects and did three experiments: one involving no acupuncture at all as control, one using traditional acupoints, and a third with "pseudoacupuncture" or application of the needle at points on the skin deliberately chosen to avoid the traditional loci. In each case, he applied heat to the subjects' skin and told them to push a button when the pain became intolerable. This "pain threshold" is defined as the moment when the heat sensation coalesces to a point accompanied by a sharp pricking feeling.

Results showed that when traditional acupoints were used, subjects took more than twice as long to reach pain threshold as when no acupuncture was used. Their pain tolerance was also significantly increased. However, to confuse matters somewhat, pseudoacupuncture proved also to be quite effective, though less than the genuine method. Stewart concludes that although he feels the practical value of acupuncture has yet to be established, it may indeed have analgesic effects. An interesting detail of his set-up was that real and pseudo acupoints were selected from the same dermatome; it seems that when you needle a dermatome you get some results, but when you hit the traditional acupoint you get even better results. In other words, the Chinese may know exactly what they have been doing all these thousands of years.[14]

In the previous chapter we referred briefly to the work of Canadian researcher Dr. Bruce Pomeranz, who has found that needle stimulation in the right place can lead to the release of a recently discovered *natural* anaesthetic in the human body similar to morphine. What happens, he thinks, is that the needling stimulates deep sensory nerves which in turn cause the pituitary, or midbrain, to release a polypeptide which has been named betaendorphin (endogenous morphine). This then blocks nerve signals on their way from the spinal cord to the higher brain centres, where they would normally have been felt as pain. The process, it seems, takes about 25 minutes to become effective, and the effect itself lasts only about an hour.[15]

Anaesthetists may well ask: why use needles when xylocain is quicker? This is like deriding the existence of telepathy because it is usually easier to use the phone. As far as we are concerned here, acupuncture is of great importance to our discovery of the true nature of the human being, which has been one of man's main concerns for as long as we have historical records. Aside from the fact that xylocain costs money and phones do not always work, it is very interesting to discover a natural substance in our bodies similar to an artificial one with notoriously dire side effects. It leads us to wonder what else remains locked in our bodies awaiting discovery.

It may be thought that talk of energy bodies, auras and the like are inextricably mixed with the unscientific state of mind sometimes (but not always) found in those who believe in spirits, reincarnation, and other subjects not often discussed in scientific circles. This is not the case, for much work into man's nonphysical component is going on in the Soviet Union, where spiritualistic attitudes are not encouraged. Fortunately, however, enough quotes can be extracted from the writings of Marx, Engels and Lenin to justify what is in effect psychical research, which is taken extremely seriously by Soviet authorities—perhaps, we suspect, because it may have military applications.

Since one of us (Hill) has been able to meet with leading Soviet researchers in areas of interest to us here, we close this book with a first-hand account of some of their recent work. We hope in the process to clear up some misunderstandings that have arisen regarding just what they are up to.

In 1967, a group of Soviet scientists put forward a new concept of man's hidden component, which they called the *bioplasmic* body. A couple of years later, word reached the West that a couple named Semyon and Valentina Kirlian had invented a method of photographing some kind of human aura. So much confusion has surrounded both the Kirlian Effect and the bioplasma concept that we must begin with a look at ordinary physical plasma.

Plasma is the fourth state of matter, with higher energy than the others: solid, liquid and gas. When individual atoms are ionized—as electrons, for example, are forced away from their nuclei at high temperatures—an ion gas or high temperature *plasma* is formed containing nuclei, electrons and neutral particles in addition to positive and negative groups of ions. Enormous temperatures are reached (millions of degrees), making it impossible for any conventional container to be used. (Attempts are being made to design a "controlled" H-bomb involving nuclear fusion, with plasma contained by magnetic fields.)

How could such a plasma exist in the human body, with its temperature of a mere 37°C? The Soviets' answer to this question is somewhat complex, but simplified as much as possible it goes something like this:

In solid state electronics, physicists speak of an "electron gas" within semiconductor elements at room temperatures, and they also talk about gases made up of electrons and "holes," i.e. the absence of an electron, and of "excitons" (excited electrons)—such a hole plus an electron. The density of electron-hole plasma in a semiconductor changes according to its temperature, and the number of electrons per square cm. can be increased up to one billion times (to 10^{22}). The Soviets believe that forced vibrations in such a plasma can approach those of visible or UV light, giving rise to plasma radiation.

In a solid, each electron "belongs" to only one atom at a time, whereas in a plasma, each electron or hole has broken free of the crystal lattice of the solid body to manifest itself as part of the totality of its structure. Now, as it happens, there are such "delocalized" electrons to be found in biological processes, and the evidence is strong for the existence of semiconductor properties in a variety of human components.

As we mentioned in Chapter 3, Albert Szent-Györgyi was the first scientist to draw attention to the semiconductor possibilities in biological elements, and in 1960 he suggested investigating such properties experimentally, especially those of membranes.[16] Since then, a number of Western scientists, such as R. O. Becker (Chapter 3), have identified what they consider to be semiconductor properties in living tissues, while others such as Popp (Chapter 11) now feel that the DNA helix can function as a "biolaser" and emit coherent radiation.

Professor Wlodzimierz Sedlak of the Catholic University in Lublin, Poland, was one of the first to produce a comprehensive model of the bioplasma body, which he sees as the ultimate substratum of both chemical and electronic processes, and also as the carrier of *all* information within the system. "Life," he says, "is an electromagnetic wave generated in a medium of protein semiconductors." The biochemical processes familiar to traditional biochemistry take place *within* such a bioelectric medium. "One should think," he says, "of metabolism in terms of transformation of energy rather than of matter," and he goes so far as to state that "the problem of the nature of life can ultimately be reduced to the concepts of plasma and electromagnetic fields."[17]

"Bioplasma is an organized system," says Dr. V. M. Inyushin, a biophysicist at Kazakh State University in Alma-Ata, who, with an engineer named V. S. Grishchenko, first postulated the existence of bioplasma in 1967. It is, he says, no less than the matrix of the biological field or biofield, which he describes as a "frozen-in hologram," every fragment of which possesses a characteristic of the essential properties of the whole organism. The possibilities for speculation on this hypothesis are unlimited. Of particular interest is the fact that bioplasma is a changeable structure, in which several kinds of waves—electromagnetic, acoustic and perhaps gravitational—are distributed, its energy state depending on "the breathing of the cosmos." The bioplasma, says Inyushin, "cannot cease to breathe in one rhythm with the cosmos." Surely there have been few more challenging visions of the unity of all life than this?[18]

Inyushin and his colleagues are particularly interested in cell membrane and mitochondria in connection with the bioplasma theory. A mitochondrion is a tiny body found in cells of practically all organisms, which helps convert food into energy with the enzymes it contains. About one percent of our body weight is made up of mitochondria, some 10^{15} of them in all, and these create energy reserves in the cell due to conversion of energy emitted during biological oxidation. Professor Stefan Manczarski of Warsaw believes that they form "a locus of cold electronic plasma whose density is on average greater than that of the ionosphere."[19] As for cell membranes, the role of which is extremely important in biological processes, Soviet researchers have revealed the existence of semi-conductivity in them, and even regard them as "reservoirs" of exciton plasma.[20]

Solid ice, when heated, becomes liquid water and eventually gaseous steam; and likewise when cooled, steam becomes water and finally ice. So with plasma: just as it is formed by processes of ionization and formation of localized charged particles, so does the reverse process also take place, with the mutual interaction of particles and their return to lower energy states by binding themselves to a lattice of atomic nuclei. Both these processes are accompanied by the emission of absorption of radiation quanta.

The weak bioluminescence of tissues and cells has been a subject of great interest to Soviet researchers ever since their successful replication of the work of Gurvich, which we mentioned in Chapter 3. They now believe that many cells luminesce in the visible part of the spectrum during many electron-exchange reactions, though it is still very difficult to detect light of such weak intensity. However, in an early experiment at Leningrad State University, using a freshly excised frog's heart that was kept beating by artificial stimulation, bioluminescent light from the heart was recorded by a UV-sensitive photodetector, the highest level corresponding to periods of strongest heartbeat. When the heartbeat faltered, the UV light output fell, although the stimulation remained constant, and when the heart finally stopped beating altogether, the output of bioluminescence trailed off to zero.[21] In Chapter 3 (ref. 24) we mentioned the work done in Australia using a yeast culture, in which a good correlation between photon and cell counts was observed, giving some support to Gurvich's long discredited mitogenetic radiation theory.

The Energy Body 311

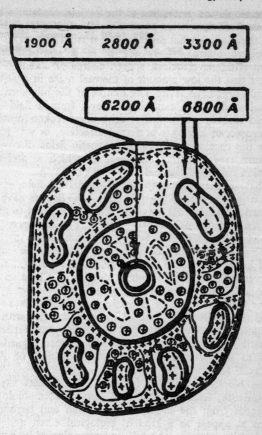

Fig. 12
The human cell as an emitter of electromagnetic radiation, as seen by Soviet biophysicists. According to them, it emits radio waves, hydrodynamic plasma waves from exciton and electron-hole plasma, also visible and invisible light frequency and infrasonic acoustical waves. Different parts of the cell emit different frequencies: as shown here, the nucleus emits invisible UV light between 1,900 and 3,300 Å, while the mitochondria (with their high ion densities) emit visible red light (6,200 and 6,800 Å), so weak that special detection methods are necessary. (Courtesy of D. V. M. Inyushin.)

The bioplasma, according to the Soviets, contains a high degree of order, or what physicists call a low entropy level. This, they say, is what distinguishes it from ordinary semi-conductive plasma. They see it as a non-equilibrium system with, nevertheless, a high degree of stability, and suggest that the low amount of thermal noise in it reflects a highly *organized* state. The matrix of the bioplasma on a whole-body basis, can, they believe, form a field—the biofield already mentioned—which they reckon must be anisotropic, or having a preferred direction, as was found to be the case with the electromagnetic fields discovered by Burr (see previous chapter), in which a preferred direction could be detected at a very early stage.

Whatever they choose to call it—biofield, L-field, energy body or anisotropic bioplasmic body—the Soviets have put forward a concept that may give scientific support to many of man's traditional mystic beliefs. For instance, Hill learned in 1976 that they believe bioplasma to be involved in energy transfer between healer and patient in the "laying on of hands" process, a form of healing far from unknown in several Communist countries despite its Christian and Spiritualist associations. At least one Soviet researcher, Dr. Viktor G. Adamenko of Moscow, has publicly acknowledged the similarity of one of his own theories with the psi-matter hypothesis developed nearly 20 years earlier by Hernani G. Andrade of Brazil, who in turn regards his ideas as wholly consistent with the Spiritism of Allan Kardec.[22] (See p. 297).

The bioplasma is also involved, the Soviets say, in acupuncture, laser therapy, and also in cosmobiological influences from the ionosphere, the Sun and beyond. It begins to appear as a kind of biological ether, and final proof of its existence and full understanding of its properties will surely go far to solve many of the mysteries we have mentioned in this book.

A great many laboratory experiments have been carried out to test the bioplasma hypothesis. One of its most interesting attributes is that it can serve as a medium for both inter- and intra-cellular communication, and during Hill's visit to the biophysics laboratory of Kazakh State University in 1976, he had the privilege of being the first Western observer to witness experiments designed to test this attribute. It was quite the most unusual experiment he has ever seen. As a guest observer, he was not able to impose any controls of his own, but was assured that the basic experiment had been repeated hundreds of times, and a full account of it will eventually be published. For the time being, what follows below is the first and only eye-witness account yet available outside the Soviet Union.

The experiment, supervised by Inyushin, involved a plant, a cat, and a laser. It demonstrated the apparent transfer of energy from one living system to another through the bioplasmic medium, and was basically a variation on the well-known and widely misinterpreted work of Cleve Backster, the New Yorker who claimed in 1968 to have detected a primary perception in plants, thus sparking off the fashion for talking to them, playing music at them, and so fourth. A brief word about his work may be helpful here.

Backster attached electrodes to his plant leaves or stems and hooked them on a normal polygraph. He ran a weak current through the electrodes and measured the resistance of the plant tissue. After various informal observations (as botanists have frequently complained, his work cannot be called an experiment), he theorized that his plants were responding to his actions and even his thoughts. This type of galvanic tissue response measurement is notoriously unstable. The slightest vibration from a slamming door, or even a highly charged electrostatic field carried by the sweater or lacquered hairdo of a passer-by, and the pen can jump off the chart. This does not mean that the plant is jealous.

Backster's own claims were considerably more modest than those of some of his admirers, and he co-operated fully with the only two serious attempts to repeat his original work of which we are aware. Both of these gave totally negative results, the 1977 report of John M. Kmetz of Science Unlimited Research Foundation, San Antonio, seemingly demolishing Backster's hypothesis of a primary perception in plants once and for all.[23]

However, history repeats itself. There are signs that Backster may have made a genuine and important discovery, although he and/or others interpreted it incorrectly. This was the case with Spallanzani and his bats' ears, Gurvich and his onion, the Kirlians (to be discussed) with their electrophotography, and a host of other pioneers.

There has been no stampede of botanists to investigate Backster's theories, but at least one, Dr. Barbara Pickard of the University of St. Louis, has produced evidence (in front of a TV camera*) that plant leaves and shoots show periodic changes in electrical potential that cannot be explained by standard theories of plant physiology. Danish physicist Leif Karlsson has recorded electric potentials from plant tissues and found that such signals appear to be related to the plant's environment. The field is becoming respectable.[24]

In 1975, Hill took part in some experiments at the Research Centre for Psychobiophysics in Stockholm, in which electrical signals were recorded from ordinary hens' eggs. The signals were not completely regular, and they seemed to originate from within the egg. They were not evidently related to outside noise sources, such as power lines. Further investigations are under way to see if such signals can be used, as the Swedes think possible, as a kind of primitive "protocommunication" system. An interesting feature about the signals is that they continue long after the egg has gone completely bad, and only cease when the shell membrane is ruptured. They can also be recorded from unfertilized eggs.

This background gave Hill an opportunity to appreciate Inyushin's experimental set-up, which was as follows:

Inside the laboratory was a double-walled metal room serving as a Faraday cage, allowing no EM radiation of short wavelength inside except the light from the 25 mW helium neon laser (LG-75) used in the experiment. Inside the cage, on a table and intersecting the laser beam stood the subject of the experiment, the plant, while outside it was the other participant, the "agent" or inductor, i.e., the cat. This was placed in a metal box with only its head visible, and connected to an electrical stimulator outside the cage through electrodes attached to its ears.

After a stable baseline of transmitted light had been recorded from the plant leaf, the cat was then put inside the cage with the plant, while the leaf tracing continued. Then, at a signal from Inyushin, a technician applied 50 volts in short pulses across the cats ears, causing it (and Hill) considerable distress. Apparently, this cruel treat-

---

* *The Green Machine.* BBC TV, 5 August 1977.

ment of the cat also affected the plant, because the level of light detected as the amount of circularly polarized light transmitted through the plant leaf increased, indicating less fluid in the leaf and therefore greater transparency and more transmitted light.

In normal finger plethysmography, a small lamp is placed on one side of the finger and a photocell (light detector) on the other. As skin is largely transparent to white light, enough light reaches the cell to be registered on the recorder. If the blood vessels are small and the pulse normal, much light is transmitted; whereas if the pulse is fast and the vessels dilated, more light is blocked by the dark blood and less transmitted, so that the indicator reading falls. The same principle was used by Inyushin to study a plant's "unconscious" system.

Plants, of course, have neither blood nor hearts, but they do have a primitive circulation system whereby a clear liquid is slowly circulated from root to leaf. Changes in this circulation cause fluctuations of the amount of light sent through the leaf that can be recorded instrumentally. Inyushin's experiment seemed well controlled, and has been repeated using another plant as inductor, also with a human being acting as agent. Two different experimental methods were used, indicating the observed effect to be due to some common biological feature rather than an error in measurement technique. For plant/plant tests, both the high voltage discharge method and laser optical registration were used, and though these methods are quite different the results were strikingly similar.

According to Soviet researchers, such experiments in interactions at a distance can be explained by properties of their postulated bioplasma. Its shape, form and extent can be controlled, they say, by an experienced "agent" and used to transfer information to the plant or other receiver. A similar explanation concerns the process of distant healing, PK effects at a distance and other distant interactions.

We have given here the briefest of glimpses of what is a major research programme, not in some eccentric's back yard, but in one of the best equipped laboratories in the Soviet Union, whose director has a decade of research into bioplasma behind him. One of his major works on it is now available in English (co-translated by Hill), and it is to be hoped that with improvements in cooperation among scientists of East and West, the real value, which we think to be considerable, of recent paraphysical research in the East will emerge and become accepted.[25]

Finally, we come to the crucial question: can the energy body be photographed? If it can, then our jig-saw puzzle is complete, and a whole new model of man can be drawn up. As we now show, there is reasonable evidence that it can be.

The Kirlian Effect—The First First-Hand Report

In 1777, long before photography was invented, German physicist Georg Lichtenberg discovered that when an electric battery was discharged, luminous patterns could be seen around the poles. Positive and negative current gave noticeably different "figures," and these are now known as Lichtenberg figures.

In the nineteenth century, at least two researchers, Hippolyte Baraduc and Jacob Jodko-Narkiewicz, showed that a strange aura-like image of a human hand could be projected on to photographic emulsion, and enhanced by passing weak current through the film and the person in a darkened room.

It is not surprising that passing electrical current near or through film will give some visible effect. It is a question of straightforward electrochemistry. An electrical current is a flow of positively and negatively charged particles, and photographic emulsion is no more than a suspension of molecules sensitive to electrical energy. The best-known application of electrochemistry is the xerographic process invented in 1948 by the late Chester Carlson (who, incidentally, gave much of his huge fortune to parapsychological research). This is based on a process in which an electrical potential is developed between two electrodes.

The theory of electron transmission between two *metal* electrodes is well known. Neither biologist nor physicist can tell us very much, however, about electron emission and capture in *biological* electrodes: that is, when we replace one of the metal electrodes with the leaf of a plant or a human finger.

During the 1930s, an electrical technician from Krasnodar in Soviet Caucasia named Semyon Kirlian carried out experiments, with his wife Valentina, in what has come to be known as Kirlian electro-photography.

They did not invent the process, but the name has stuck and we see no reason to unstick it. Kirlian is now retired and his wife deceased, and there is so much confusion about what exactly they did that we would like to set the record straight. Since Hill was the first Western scientist to set foot inside a Kirlian laboratory in the Soviet Union, what is to be described here is, as far as we know, the only authentic account yet to be published outside the U.S.S.R.

Kirlian-type devices used by Western scientists and amateurs alike consist of an apparatus based on a high voltage supply, a positive and negative electrode, and a sheet of film placed on the latter. Fingers are placed directly on the film, and the film is developed in the normal way. The results are visually fascinating, and have always been misinterpreted in the West. The essential difference between commercial Western Kirlian apparatus and the real thing is that in the latter, the film is *not* placed between the electrodes, but at a distance from them, and the light reaching it from the biological object between the electrodes is filtered and analysed. If this is not done, the resulting image is largely "noise" with very little signal.

The Kirlians knew very well that any electrical field over a certain strength applied between two electrodes, metal or biological, will produce what is called a corona discharge, a visible glow caused by the electrons in the electrode (*and* in the gas between the electrodes, unless the experiment is done in a vacuum) being excited to an energetically higher state by the applied electric field. When they return to a lower state, they emit a quantum of light in the process. This, depending on the energies involved, can be in the UV, visible or IR bands, and if the field is strong enough, the electrons will hit the metal with such an impact that secondary radiation (such as "hard" or "soft" x-rays) will be produced.

Since different electrons receive different amounts of energy, a mixture of frequencies can result, making it very difficult to carry out accurate analysis of such a corona flow. However, this is where the Kirlians made a major breakthrough. They developed techniques for analysing bioelectric radiation, by removing the film between the two plates and placing it at a distance from the electrodes, thus eliminating any chance that the electrical current would have any *direct* effect on the film. The light from the discharge was collected by lenses and focused on to the emulsion, the size of the image being enlarged by standard optical techniques. They also brought in a spectrophotometer, so that the exact frequency spectrum of the light involved could be analysed, thus separating the mixture of frequencies involved into distinct lines. In this way, they sorted out the signal from the noise.

However, there was more to it than this. How could they be sure the light was coming from the biological object and not from the electrodes or the gas between the plates? The Kirlians solved this problem by setting up an artificial atmosphere made up of a gas of known composition, such as nitrogen, in which they knew that only a few spectral lines of known frequency would be produced by the excited gas molecules. Since these lines are relatively narrow and of a precise frequency known to spectroscopists, they can be filtered out by imposing the necessary narrow-band filters along with the lenses collecting the light before it reached the film.

In this way, any light reaching the film can only come from one of the electrodes—and one of the Kirlians' first innovations was to make the biological object itself serve as one of the electrodes—so that all that remained to be filtered out were the electrons from the metal plate. By using a metal electrode of known composition and carefully regulating the current, they could eliminate this problem. What is now left, some 15 percent of the original amount, is *only* the light from the biological object: the fingertip, plant or whatever.

This is the Kirlian Effect. Whatever its precise relation to the bioplasma, it is already being put to practical use in Soviet hospitals and research laboratories, for this light, they believe, can give us valuable information about the subtle state of the biological system. Great care has to be taken to exclude changes in light emission due to skin moisture, variations in electrode pressure, and so on. The Kirlians tackled all these problems one by one, earning a number of patents in the process, and the equipment they pioneered has already proved its worth.

At the Moscow Institute of Normal Physiology (under the Ministry of Health), Kirlian techniques are used in the early detection of tumours, comparing favourably with standard x-ray diagnostic methods. High frequency probes are used to obtain photographs from inside the body, and variations from normal in the geometrical characteristics of the high frequency fields can lead to the determination of the dynamic growth and distribution of malignant tumours.[26]

Also in Moscow, Dr. Viktor G. Adamenko has applied the method to the study of arthritic joints. His painstaking work on the Kirlian effect, lasting more than seven years, earned him his technical sciences doctorate in 1976. In that year, he told Hill that he wished Western scientists would appreciate the seriousness and advanced level of research in his field, in which the Soviets are far ahead of anybody else.

"The Kirlian apparatus is far from perfect," says Dr. Inyushin, "but it is one of the few means available for diagnosing the bioplasma. With the aid of the Kirlian effect, it has been possible to penetrate into a new field of knowledge, using the latest achievements of quantum electronics to obtain unique information about the bioenergetic state of the organism, and to prove the existence of the biofield."[27]

We began this book with the ambitious aim of tying up some of the loose ends in many fields of apparently unrelated research. In the process, we have left a good many new loose ends lying about ourselves, and we are fully aware, as a publisher's reader complained of an earlier draft, that it is "all middle with no beginning and no end."

So is life, as far as we can tell. Research into its mysteries never stops, but a book has to stop somewhere, and this one is going to stop here, after a short summary of such loose end-tying as we think we have managed to do:

The whole of observable space is filled with pulsating energies, none of which we fully understand. Ours is a universe of perpetual motion in all directions, with nothing from particle to planet ever in a state of rest. Life on Earth has become adapted to all this activity, and we have been lulled into a false sense of security by the apparent constancy and regularity of our cosmic environment. We take it for granted, which is a mistake.

Our environment is neither constant nor regular. Nor is it predictable beyond certain limits. Much of this uncertainty is due to the fact that our local star, the Sun, is not the perfect and constant deity we assumed it to be until recent observation showed otherwise. We now see it as a turbulent and irregular generator and transformer of the energies we need in order to stay alive, one that operates according to principles beyond our present comprehension. We know that its energy output waxes and wanes in cycles of just over 11 years on average, such cycles being reflected in countless events on Earth. As H. H. Clayton suspected, much human "weal and woe" is entwined with "the surging flames which heave and roar on its surface." Though some Sun/Earth cyclic correlations are so weak that we cannot be sure they exist, others are very strong, the most important being the unmistakable link between solar activity and magnetic conditions on Earth, and the almost equally firm statistical links between these and a wide variety of effects on plants, animals and humans.

Man, like all living creatures, is an electromagnetic system in an EM environment from which he cannot isolate himself. He responds to the tides and currents of this environment in many more ways than he can consciously identify. An event originating from a distant quasar can be connected to an event in our brains. Ours is not a universe of isolated parts; we are made of the same raw material as the rest of it, and we respond to the same forces that drive and shape it all. Star and skin are in regular contact.

The fact that some of our energy inputs are cyclic indicates that some of our future actions are preordained, to the extent that they will result from our predictable reactions, conscious or unconscious, to known environmental stimuli both weak and strong. The effectiveness of the invisible forces is not a function of their strength or weakness. We can battle through a gale that blows a tree down, yet we can be unhinged mentally or physically by a sudden shift in energetic quantity or quality so weak that we can neither perceive it consciously nor measure it accurately. Light we cannot see and sound we cannot hear can do us irreparable harm, though we are at last discovering that they can also be harnessed to do us good.

How safe are we from the invisible forces around us?

Since early in this century, we have been beaming intense electric and magnetic radiation all over the world. We began to do this long before anything was known about possible long-term biological effects of cumulative exposure to such radiation. It is only one or two generations ago that alternating current (AC) was introduced on a wide scale to replace short-range direct current (DC), and AC electrical power is now carried long distances by high-voltage transformers and power networks, so that there can scarcely be anybody in the industrially developed countries who lives or works far from such currents, which are extremely low frequency (ELF) waves.

In Chapter 4 we described some of the biological effects of natural EM and acoustic ELF waves in the environment, and in Chapter 8 we mentioned some of the ways in which artificially produced ELF waves can affect us. We need hardly point out that normal AC power at the common domestic frequencies of 50 or 60 cycles per second is *also* a source of ELF waves, and since the wavelengths concerned are thousands of kilometres long they are virtually impossible to shield out. Glass, wood, plastic and paint give no shielding at all from either electric or magnetic ELF waves. Not even reinforced concrete ten feet thick offers 100 percent shielding from them. We are all exposed to such waves throughout most of our lives, whether we like it or not.

There is no generally accepted international set of rules for maximum levels of human exposure to radiation. Some governments have set standards for exposure to some types of radiation, while others have not. Although much is known about the effects of short-term exposure to such powerful forms of radiation as microwaves, laser beams and ultraviolet light, there is practically no legislation anywhere to govern exposure standards for ELF fields and waves. The initial results of the U.S. Navy's Project Sanguine, mentioned earlier, are contradictory: some reports claim that detrimental effects can be expected from the powerful ELF waves that would be involved in an undersea communications network, while other experts assure us there is nothing to worry about.

Evidence we have presented in this book indicates that "weak is strong" in a number of contexts. The "magic of the minimum dose," to borrow a phrase from homeopathy, may apply only too well to doses of electromagnetic radiation that we all take daily without knowing what we are doing. Animals, fish, birds and insects have adapted themselves to the forces around them, retaining and evolving their sensitivity to them. Man, on the other hand, can hardly adapt himself to something he cannot consciously register at all. Moreover, he has added powerful artificial forces to those of nature without fully understanding the effects of either.

Where indeed, as T. S. Eliot asked, are the wisdom we have lost in knowledge, and the knowledge we have lost in information? This is the age of the information explosion, as we know, with 50,000 scientific periodicals on the library shelves to keep up with. Eight percent of all scientists who have ever lived are alive today. Does this mean that the knowledge explosion and perhaps the wisdom explosion are on the way? It can hardly be said that human progress advances on all fronts in proportion to annual totals of science graduates—indeed it could even be claimed that the less information a man has, the more wisdom his own instincts give him. In this book, we heard at least one scientist (Bob Johannes) declare that an uneducated fisherman taught him more about a subject of great scientific interest than was to be found in the entire literature.

Man's technology is still unable to keep pace with his instincts, despite the technological advances of the Space Age, in which practically all our ideas have changed with regard to both outer and inner space. We continue to listen to our instincts, or we should, for they are often responses to stimuli that actually exist, even if we cannot identify them consciously. If the "psychic" duck of Freiburg, to which we paid our respects in Chapter 5, was able to save lives by responding to a signal inaudible to the human ear (as we think possible), could there not be signals in the environment to which we too could respond, with or without instrumental help?

"Let us learn to dream, gentlemen!" This was Friedrich Kekulé's advice to his fellow scientists after he had literally dreamed the solution to the important problem of the molecular ring structure. "Discovery," says Koestler, commenting on a similar case, "consisted in uncovering what had always been there." But the knowledge, he adds, "had been buried under the rigid crust of a conventional matrix, which made his thoughts turn in a vicious circle."[28]

Kekulé's dreams would, of course, have probably been of less value had he not been a first-rate scientist in his waking state. This can also be said of the great sleepwalkers of earlier times—Copernicus, Kepler, Galileo and Newton —and of the modern sleepwalkers from Chizhevsky, Petersen and Piccardi to Gauquelin, Takata, Nelson,

Inyushin and many more men and women we have mentioned here, each of whom has managed to peep out from under that conventional matrix.

It is barely 150 years since electricity and magnetism were seen to be interrelated, some 20 centuries after they were discovered. Eventually, it must be shown how all the invisible forces of nature are related. Einstein, the greatest of all sleepwalkers, spent much of his later life searching for his "unified field theory," and it has already been suggested that electromagnetism, gravitation, and the strong and weak nuclear forces are merely manifestations of one principle. When this principle is finally uncovered, much human knowledge will become obsolete overnight.

This book will certainly be out of date long before it is published. (We had to make several major revisions in the very week it went to press, and much material became obsolete during early drafts.) However, if in the future Age of Wisdom it is found to be of some historical interest as a record of man's early search for the meaning of the invisible forces, and the principle behind them all, it will not have been written in vain.

LIST OF REFERENCES

Introduction

1 SERGEYEV, G. A., *Biorhythms and the Biosphere*. Moscow: Znaniye, 1976, p. 3 (Russian).
 PICCARDI, G., *Astrophysical Phenomena and Terrestrial Events*. Lecture, Palais de la Découverte, Paris, 1959.

2 GAUQUELIN, M., *Astrology and Science*. (Tr. James Hughes). London: Mayflower, 1972, p. 24.

3 HOLROYD, S., *Psi and the Consciousness Explosion*. London: The Bodley Head, 1977, p. 169.

Chapter One

1 BRADY, J. L., "The effect of a trans-Plutonian planet on Halley's comet." *Publications of the Astronomical Society of the Pacific* 84, 314-322, 1972.
 HILL, S. M., "Evidence for a tenth planet." *Gnostica News* 2 (2), 1972.

2 KOESTLER, A., *The Sleepwalkers*. London: Hutchinson, 1959.

3 DUDLEY, H. C., "Is there an ether?" *Industrial Research* November 15, 1974, pp. 41-46.

4 LANDSCHEIDT, T., *Cosmic Cybernetics*. Aalen: Ebertin-Verlag, 1973.

5 MOLCHANOV, A. N., "The resonant structure of the solar system." *Icarus*, 8, 203-215, 1968. See also *Icarus* 11, 111, 1969.

6 GOLD, T. & SOTER, S., "Atmospheric tides and the resonant rotation of Venus." *Icarus* 11, 356-366, 1969.

7 ALLAIS, M. F. C., "Should the laws of gravitation be reconsidered?" *Aero/Space Engineering* September 1959, 46-52; October 1959, 51-55; and November 1959, 55.
SAXL, E. J. & ALLEN, M., "1970 solar eclipse as 'seen' by a torsion pendulum." *The Physical Review* D 3 (4) 823-825, 1971.

8 GOSLING, J. T. & HUNDHAUSEN, A. J., "Waves in the solar wind." *Scientific American* 236 (6) 36-43, March 1977.

9 DANJON, A., in *Comptes Rendus de l'Academie des Sciences* (B) 249, 2254, 1959; 250, 1399, 1960; 254, 2479 & 3058, 1962.
GRIBBIN, J. & PLAGEMANN, S., "Discontinuous change in Earth's spin rate following great solar storm of August 1972." *Nature* 243, 26-27, 1973.

10 LODGE, O. J., *Ether and Reality*. London: Hodder & Stoughton, 1929.

11 SWENSON, L. S., "The Michelson-Morley-Miller experiments before and after 1905." *Journal for the History of Astronomy* 1, 56-78, 1970.

12 DIRAC, P. A. M., "Is there an Æther?" *Nature* 168, 906-907, 1951.

13 RUDERFER, M., "Parapsychological theory. 1. Boundary conditions imposed by physical laws." *The Journal of Research in Psi Phenomena* 1 (1) 53-71, 1976.

14 DUDLEY, H. C., "Michelson's hunch was right." *Bulletin of the Atomic Scientists* January 1975, 47-48.
JOLLY, W. P., *Sir Oliver Lodge*. London: Constable, 1974. 231-232.

Chapter Two

1 NELSON, J. H., "Shortwave radio propagation correlation with planetary positions." *RCA Review* March 1951, pp. 26-33.
NELSON, J. H., "Planetary position effect on short-wave signal quality." *Electrical Engineering* May 1952, 421-424.
NELSON, J. H., *Cosmic Patterns. Their influence on man and his communication*. Washington: American Federation of Astrologers, 1974. (See also ref. 36 below).

2 EDDY, J. A., in *Scientifically Speaking*, BBC Radio 3, September 22nd 1976.

3 CARRINGTON, R. C., "Description of a singular appearance seen in the Sun on September 1, 1859." *Monthly Notices of the Royal Astronomical Society* 20, 13, 1859.
HODGSON, R., "On a curious appearance seen in the Sun." *ibid.* 20, 15, 1859.

List of References

4 ARNOLDY, R. L., et al. in *Space Physics* by R. S. White. New York: Gordon & Breach, 1970. Chapter 4.

5 SEVERNY, A. B., *ibid.*

6 KOESTLER, A., *The Sleepwalkers*. London: Hutchinson, 1959, pp. 429-432.

7 EDDY, J. A., "The Maunder Minimum." *Science* 192, 1189-1902, 1976.

8 HERSCHEL, W., in *Philosophical Transactions of the Royal Society of London* 265, April 16th 1801.

9 SCHWABE H., in *Astronomische Nachrichten* 20, 495, 1843.

10 WOLF, R., in *Astronomische Mitteilungen* 1 (8) 1856. (R, the Wolf number, $= K(10g + f)$, where $K = 0.6$, g the number of sunspot groups, and f the total number of spots on the visible disc. The data published today in the *Quarterly Bulletin of Solar Activity* is based on observations from about 30 observatories all over the world).

11 BROWN, G. M., "A new solar-terrestrial relationship." *Nature* 251, 592-594, 1974.

12 BRAY, R. J. & LOUGHHEAD, R. E., *Sunspots*. London: Chapman & Hall, 1964.

13 EDDY, J. A., "The Sun from Skylab." *New Scientist* 61 (890) 738-741, 1974. See also *ibid.* 60 (874) 607, 1973.

14 BRAY & LOUGHHEAD, op. cit. above (12). Chapter 8.

15 JOSE, P. D., "Sun's motion and sunspots." *The Astronomical Journal* 70 (3) 193-200, 1965.
COHEN, T. J. & LINTZ, P. R., "Long term periodicities in the sunspot cycle." *Nature* 250, 398-400, 1974.

16 WILLIAMS, D., "Theories of the cause of sunspots." *Cycles*, September 1973, 205-214.

17 NORTON, W. A., *A Treatise on Astronomy*. New York: Wiley, 1867.
BROWN, E. W., "A possible explanation of the sunspot period." *Monthly Notices of the Royal Astronomical Society* 60, 1900.
PORTIG, W. H., in "Position of planets linked to solar flare prediction" by Rex Pay. *Technology Week* May 15, 1967, 35-38.

18 DIGBY, W., *Natural Law in Terrestrial Phenomena*. London: Hutchinson, 1902.

19 MAUNDER, A. S. D., "An apparent influence of the Earth on the numbers and areas of sunspots in the cycle 1889-1901." *Monthly Notices of the Royal Astronomical Society* May 1907.

SANFORD, F., "The influence of planetary configurations upon the frequency of visible sunspots." *Smithsonian Miscellaneous Collections* 95 (11) 1-5, 1936.

20 MELDAHL, K. G., *Tidal Forces in the Sun's Corona Due to Planets.* Copenhagen: Berlingske Forlag, 1938.

21 BIGG, E. K., "Influence of the planet Mercury on sunspots." *Astronomical Journal* 72, 463-466, 1967.

22 BUREAU, R. A. & CRAINE, L. B., "Sunspots and planetary orbits." *Nature* 228, 984, 1970.
WOOD, R. M., "Comparison of sunspot periods with planetary synodic period resonances." *Nature* 255, 312-313, 1975.

23 TOMAN, K., "On the possible existence of a 29-day period in the sunspot number series 1940-1964." *Journal of Geophysical Research* 72 (21) 5570-5571, 1967.

24 WOOD, K. D., "Sunspots and planets." *Nature* 240, 91-93, 1972.
WOOD, R. M. & WOOD, K. D., "Solar motion and sunspot comparison." *Nature* 208, 129-131, 1965.
ÖPIK, E., "Planetary tides and sunspots." *Irish Astronomical Journal* 10, 298-301, 1972.

25 BAGBY, J. P., "Sunspot cycle periodicities." *Nature* 253, 482, 1975.

26 LUBY, W. A., in *Popular Astronomy* 48 (10), 1940.
JOHNSON, M. O., *Correlation of Cycles in Weather, Solar Activity, Geomagnetic Values, and Planetary Configurations.* San Francisco: Phillips & Van Orden, 1946.

27 GRIBBIN, J., *Forecasts, Famines and Freezes.* London: Wildwood House, 1976. Chapter 5.

28 OKAL, E. & ANDERSON, D. L., "On the planetary theory of sunspots." *Nature* 253, 511-513, 1975.

29 WOOD, K. D., op. cit. above (24).

30 in WILLIAMS op. cit. above (16).

31 LANDSCHEIDT, T., *Cosmic Cybernetics.* Aalen: Ebertin-Verlag, 1973, p. 51.

32 SLEEPER, H. P., "Planetary resonances, bi-stable oscillation modes, and solar activity cycles." Washington, D.C.: NASA (Report CR-2035), April 1972.

33 GRIBBIN, J., op. cit. above (27).

34 MEMERY, H., *L'action individuelle des taches solaires sur les phenomènes terrestres.* Observatoire de Talence, 1948.

35 BAILEY, J. V., "Radiation protection and instrumentation," in *Biomedical Results of Apollo.* NASA SP 368, 1975. Chapter 3.

36 NELSON, J. H., Personal communication to Playfair, 1977, in which he promises to publish his correspondence with

NASA in his forthcoming second book. This, he says, "will do much to establish the fact that the planets definitely cause changes in solar radiations." After reading it, he adds, astronomers will either have to accept his theories or ascribe "clairvoyant powers" to him. We await this book with great interest. See also:

NELSON, J. H., "Circuit reliability, frequency utilization and forecasting in the high frequency communication band," in *The Effect of Disturbances of Solar Origin on Communications.* AGARDograph 59, Oxford: Pergamon Press, 1963, pp. 293-301.

37 CAYMAZ, G., "Effect of increased atmospheric electricity on the blood electrolytes of airplane crew," in NATO-AGARD Conference Proceedings 182, B7 1-5, 1976.

38 CAMBEL, A. B., "MHD for spacecraft." *Science Journal* 6(1) 68-73, 1970.

39 WILLIAMS, D., "Predicting power system disturbances." *Cycles,* November 1968.

40 *Nature* 144, 808, 1939.

41 KING, J. W., "Solar radiation changes and the weather." *Nature* 245, 443-446, 1973. (See also *Nature* 247, 131-134, 1974).

42 REITER, R., "Increased influx of stratospheric air into the lower troposphere after solar H-alpha and X-ray flares." *Journal of Geophysical Research* 78 (27) 6167-6172, 1973.

43 PAY, R., "Position of planets linked to solar flare prediction." *Technology Week* May 15, 1967, 35-38.

44 WOOD, C. A. & LOVETT, R. R., "Rainfall, drought and the solar cycle." *Nature* 251, 594-596, 1974.

45 MARSHALL, J. R., "Precipitation patterns of the United States and sunspots." Ph.D. thesis, University of Kansas, 1972.

46 WILCOX, J. R., "A synoptic approach to Sun-weather investigations." *Journal of Atmospheric and Terrestrial Physics* 39, 173-178, 1977.

47 DEAN, G. A., *Recent Advances in Natal Astrology.* Subiaco (Western Australia): Analogic, 1977. Probably the most comprehensive and objective study of the subject yet to appear. (608 pages, over 1,000 references). See especially pp. 492-515 for detailed survey of recent work on solar cycles, including the author's own. Available from Recent Advances, 43 Granville Rd., Cowes, Isle of Wight PO31 7JF, England. Quotation here is from personal communication to Playfair, August 1977. See also:

FOX, L., "Planets and sunspots." Unpublished MS. Department of Geological Sciences, University of British Columbia, 1973.

48 Quoted in *Man, Weather, Sun* by W. F. Petersen. Springfield: Charles C. Thomas, 1947, p. 311.

Chapter Three

1 MAURITIUS, G., "Der verplante Mensch." *Grenzgebiete der Wissenschaft* 24 (4) 197-230, 1975. See also:
ROHRACHER, H. & INANAGA, K., *Die Mikrovibration: Ihre biologische Funktion und ihre klinische Bedeutung*. Berne: Verlag H. Huber, 1969.

2 SPENGLER, L., *Erfaringer angående elektricitets virkinger på sygdomme*. Copenhagen, 1754.

3 PATTIE, F. A., "Mesmer's medical dissertation and its debt to Mead's *De imperio solis ac lunae*." *Journal of the History of Medicine* 11, 275-278, 1956.

4 DINGWALL, E. J., *Abnormal Hypnotic Phenomena*. London: Churchill, 1967. Vol. 1, p. 8.

5 D'ARSONVAL, M. A., "Production des courants de haute fréquence et de grande intensité, leurs effets physiologiques." *Comptes Rendus (Soc. biol.)* 45, 122, 1893.

6 HANSEN, K. M., "Some observations with a view to possible influence of magnetism upon the human organism." *Acta medica scandinavica* 97, 339, 1938.
HANSEN, K. M., "Studies on the influence of magnetism on the oxygen absorption in man." *ibid.* 118, 261, 1944.
LISIN, V. V., in NRCC/TT 1545, p. 122. (For full title see ref. 13 to chapter 6 below).

7 TROMP, S. W., *Psychical Physics*. Amsterdam: Elsevier, 1949, p. 388.
GRAD, B. R., "The biological effects of the 'laying on of hands' on animals and plants: implications for biology," in *Parapsychology: its Relation to Physics, Biology, Psychology and Psychiatry* (ed. G. Schmeidler). Metuchen: Scarecrow Press, 1976, pp. 76-89. Includes list of 5 previous papers by Grad.

8 Details of the Electroson-4T from: V/O MEDEXPORT, Moscow G-200, U.S.S.R. (Telex 247. Cables: Medexport Moscow. Phone 121-0154) and *not* from us.

9 MAXEY, E. S., "Critical aspects of human versus terrestrial electromagnetic symbiosis." Unpublished MS: USNC/URSI-IEEE Meeting, Boulder, Colorado. October 1975.
MAXEY, E. S. Personal communication to Playfair, 1977.

10 TROMP, S. W., op. cit., above (7); 123-125.
BUDYKO, M. I., "The effect of solar radiation variations on the climate of the Earth." *Tellus* 21, 611-619, 1969.

11 BROWN, F. A., Jr., "The 'clocks' timing biological rhythms." *American Scientist* 60 (6) 756-766, 1972.

12 HULL, C. D. et al., "Role of the olfactory system in arousal to x-ray." *Nature* 205, 627-628, 1965.

13 KALMIJN, A. J., "Electro-perception in sharks and rays." *Nature* 212, 1232-1233, 1966.

14 HERRNKIND, W. F. & McLEAN, R., "Field studies of homing, mass emigration and orientation in the spring lobster *Panulirus argus*." *Annals of the New York Academy of Sciences* 188, 359-377, 1971.

15 LISSMANN, H. W. & MACHIN, K. E., "The mechanism of object location in *Gymnarchus niloticus* and similar fish." *Journal of Experimental Biology* 35, 451, 1958.

16 KEETON, W. T., "Magnets interfere with pigeon homing." *Proceedings of the National Academy of Sciences* 68 (1) 102-106, 1971.

17 BOOKMAN, M. A., "Sensitivity of the homing pigeon to an Earth-strength magnetic field." *Nature* 267, 340-342, 1977.
MOORE, F. R., "Geomagnetic disturbance and the orientation of nocturnally migrating birds." *Science* 196, 682-683, 1977.

18 BIGU DEL BLANCO, J. & ROMERO-SIERRA, C., "The properties of bird feathers as piezoelectric transducers and as receptors of microwave radiation." I International Conference on Biomedical Transducers, Paris. November 1975.
TROMP, S. W., *op. cit.* above (7); 389-391.

19 AUDUS, L. J., "Magnetotropism: a new plant-growth response." *Nature* 185, 132, 1960.
TITAYEV, A. A., *The Biology of Large Mushrooms*. Moscow: Nauka, 1976. Chapter 3. (Russian).
KHOLODOV, YU. A. (ed.), *The Influence of Magnetic Fields on Biological Objects*. Moscow: Nauka, 1971. (Russian).

20 PRESSMAN, A. S., *Electromagnetic Fields and Life*. New York: Plenum Press, 1970.

21 GURVICH, A. A., *"Problems of mitogenetic radiation as an aspect of molecular biology."* Leningrad: Meditsina, 1968.

22 KAZNACHAYEV, V. P. et al., "Apparent information transfer between two groups of cells." *Psychoenergetic Systems* 1, 37, 1974.

23 SCHJELDERUP, V. Personal communication to Hill, 1976.

24 QUICKENDEN, T. I. & QUE HEE, S. S., "Weak luminescence from the yeast Saccharomyces cerevisiae and the existence of mitogenetic radiation." *Biochemical & Biophysical Research Communications* 60 (2) 764-769, 1974.

25 OTT, J., *Health and Light*. Old Greenwich: Devin-Adair, 1973.

26 GRATTAN, H., "A successful feat," in *Practical Dowsing* (ed. A. H. Bell). London: Bell, 1965.

27 MABY, J. C. & FRANKLIN, T. B., *The Physics of the Divining Rod*. London: Bell, 1939.

28 TROMP, S. W., *op. cit.* (7) above; 289-378.

29 HARVALIK, Z. V. & DE BOER, W., "Scientific aspects of dowsing." *The American Dowser*, August 1976.

30 JENNY, E., "Das Wünschelrutenproblem." *Schweize Medizinsche Wochenschrift* 66 (21, 22, 24) 1934. See also *ibid.* 65, 1935.

31 POHL, G. von, *Erdstrahlen als Krankheitserreger*. Giessen: Huber, 1932.
RAMBEAU, G., "Krankheiten durch Ausstrahlungen der Erde." *Heilkunde der Gegenwart* 81, 1932.

32 STÄNGLE, J. W. F., "Grundstrahlungsmessungen über geopathischen Reizstreifen." *Wetter, Boden, Mensch* 18, 1146-1153, 1973.

33 PETSCHKE, H., in *Unsichtbare Umwelt* by H. L. König. Munich: Heinz Moos Verlag, 1975, p. 146.

34 TROMP, S. W., "Review of the possible physiological causes of dowsing." *International Journal of Parapsychology* Winter 1968, 363-391.

35 HITCHING, F., *Earth Magic*. London: Cassell, 1976.

36 TROMP, S. W., *op. cit.* (7) above; 310-330.

37 PROKOP, O., *Wünschelrute, Erdstrahlen und Wissenschaft*. Stuttgart: Enke, 1955.

38 ROCARD, Y., *Le signal du sourcier*. Paris: Dunod, 1962.

39 HARVALIK, Z. V., "A biophysical magnetometer-gradiometer." *The Virginia Journal of Science* 21 (2), 59-60, 1970.

40 HITCHING, F., *op. cit.* (35) above; Ch. 6.

41 PUHARICH, A., "What happens when radio waves penetrate the human skin." *Impact of Science on Society* 24 (4), 353-357, 1974.
For information about the TD-100, Contact: Intelectron Corp., 432 W. 45th Street, New York NY 10036.

42 KAZHINSKY, B. B., *Biological Radiocommunication*. Kiev: Nauka, 1962.

43 ANDRADE, H. G., "Psi Matter" in *The Indefinite Boundary* by G. L. Playfair. London: Souvenir Press, 1976, pp. 295-307.

44 BECKER, R. O., "Electromagnetic forces and life processes." *Technology Review* December 1972, pp. 32-38. (Becker's bibliography 1959-1975 contains 84 items).
BECKER, R. O., "Stimulation of partial limb regeneration in rats." *Nature* 235, 109-111, 1972. See also: *Annals of the New York Academy of Sciences* 238, 1974.

45 COHEN, D., "Magnetic fields of the human body." *Physics Today* August 1975, pp. 34-43.
46 FRIEDMAN, H., BECKER, R. O. & BACHMAN, C. H., "Geomagnetic parameters and psychiatric hospital admission." *Nature* 200, 626-628, 1963.
47 DÜLL, T. & DÜLL, B., "Zusammenhänge zwischen Störungen des Erdmagnetismus und Häufungen von Todesfallen." *Deutsche Medizinische Wochenschrift* 61, 95, 1935.
48 TARLING, D. H. & TARLING, M. P., *Continental Drift*. Harmondsworth: Penguin Books, 1975. Chapter 6.

Recommended further reading:
BARNOTHY, M. F., *Biological Effects of Magnetic Fields* (2 vols.) New York: Plenum Press, 1965, 1969.
CHAPMAN, S. & BARTELS, J. *Geomagnetism*. 2 vols. (1,049 pp.) Oxford: Clarendon Press, 1940.

Chapter Four

1 *The Times* June 22nd 1976.
2 WINKLESS, N. & BROWNING, I., *Climate and the Affairs of Men*. London: Peter Davies, 1976. (Quotation from *International Herald Tribune* July 14th 1976).
3 *Sunday Times*, May 16th 1976.
4 GRIBBIN, J., "Man's influence not yet felt by climate." *Nature* 264, 608, 1976.
5 *Daily Mail* October 22nd 1976.
6 PAY, R., "Position of planets linked to solar flare prediction." *Technology Week* May 15th 1967, 35-38.
7 JOHNSEN, S. J., DANSGAARD, W. & CLAUSEN, H. B., "Climatic oscillations 1200-2000 A.D." *Nature* 227, 482, 1970.
8 ROSENBERG, R. L. & COLEMAN, P. J., "27-day cycle in the rainfall at Los Angeles." *Nature* 250, 481-484, 1974.
9 GERETY, E. J., WALLACE, S. M. & ZEREFOS, C. S., "Sunspots, geomagnetic indices and the weather." *Journal of the Atmospheric Sciences* 34 (4) 673-678, 1977.
10 PETERSEN, W. F., *Man, Weather, Sun*. Springfield: Charles C. Thomas, 1947, pp. xi-xii.
For full listing of Petersen's writings on biometeorology, see *International Journal of Biometeorology* 7 (1) 105-109, 1963. See also his monumental 4-volume work *The Patient and the Weather*. Ann Arbor: Edwards Brothers, 1935-37.
11 TROMP, S. W., *Medical Biometeorology*. Amsterdam: Elsevier, 1963.

TROMP, S. W. (ed.), *Progress in Human Biometeorology.* Lisse: Swets & Zeitlinger, 1972—. (Companion volumes on animal and plant biometeorology in preparation).

12 TROMP, S. W., "Effects of weather and climate on mental processes of man," in *Parapsychology and the Sciences.* New York: Parapsychology Foundation, 1972, pp. 135-148. See also:
TROMP, S. W. & BOUMA, J. J., *Study of the possible relationship between atmospheric environment, suicide and suicide attempts in the western part of the Netherlands (period 1954-1969).* Leiden: Biometeorology Research Centre Monograph Series Vol. XII, 1973.

13 STENGEL, E., *Suicide and Attempted Suicide.* Harmondsworth: Penguin Books, 1973.

14 DURKHEIM, E., *Suicide.* London: Routledge & Kegan Paul, 1952.

15 PHILLIPS, D. P., "Motor vehicle fatalities increase just after publicized suicide stories." *Science* 196, 1464-1465, 1977.

16 TENG, H. C. & HEYER, H. E., "The relationship between sudden changes in the weather and the occurrence of acute myocardial infarction." *American Heart Journal* 49, 9-20, 1955.

17 POUMAILLOUX, M. & VIART, R., "Corrélation possible entre l'incidence des infarctus du myocarde et l'augmentation des activitiés solaire et géomagnetique." *Bulletin de l'academie nationale de médecine* 143, 167-170, 1959.
DUBROV, A. P., *The Geomagnetic Field and Life.* Leningrad: Gidrometeoizdat, 1974, pp. 71-86. (English translation in preparation, Plenum Press, New York).

18 SCHULDT, H., "Condensations of field force lines in biological systems, an interpretation of the acupuncture meridians." *American Journal of Acupuncture* 4 (4) 344-348, 1976.

19 SCHUMANN, W. O., "Über elektrische Eigenschwingungen des Hohlraums Erde-Luft-Ionosphäre, erregt durch Blitzenladungen." *Zeitschrift für Angewandte Physik* 9, 373-378, 1957.

20 KÖNIG, H. L. & ANKERMÜLLER, F., "Über den Einfluss besondere niederfrequenter elektrischer vergange in der Atmosphäre auf den Menschen." *Die Naturwissenschaften* 1960, 486-490.

21 ALTMANN, G., in "Research of biological fields of electric environmental factors." *Archiv für Meteorologie, Geophysik und Bioklimatologie* (B) 24, 109-126, 1976.

22 PERSINGER, M. (ed.), *ELF and VLF Electromagnetic Field Effects.* New York: Plenum Press, 1974.
PRESSMAN, A. S., *Electromagnetic Fields and Life.* New York: Plenum Press, 1970.

23. GAVREAU, V., "Infrasound." *Science Journal* 4, 33-37, 1968.

24. COOK, R. K. & YOUNG, J. W., "Strange sounds in the atmosphere." *Sound* 1 (2) 12-16 and 1 (3) 25-32, 1962.

25. PERSINGER, M., *The Paranormal*. New York: MSS Information Corp., 1974. Part 2, Chapter 2.

26. WIKE, E. L. & WIKE, S. S., "Escape conditioning and low-frequency whole-body vibrations." *Psychonomic Science* 27, 161-164, 1972.

27. GREEN, J. E. & DUNN, F., "Correlation of naturally occurring infrasonics and selected human behaviour." *Journal of the Acoustical Society of America* 44 (5) 1456-1457, 1968.

28. STEPHENS, R. W. B., "Infrasonics." *Ultrasonics* 7 (1) 30-35, 1969.

29. FERGUSON, M., *The Brain Revolution*. London: Davis-Poynter, 1974, 240-241.

30. CANTRELL, R. W., "Physiological effects of noise." NATO/AGARD Conference Proceedings CP 171, 1975.

31. TAYLOR, R., *Noise*. Harmondsworth: Penguin Books, 1975, p. 19.

32. KJELLSON, H., *Forsvunden Teknik*. Copenhagen: Nihil, 1974, pp. 52-56.

33. CHIZHEVSKY, A. L., "Action de l'ionisation de l'atmosphère artificielle de l'air sur les organismes sains et les organismes malades," in *Traité de Climatologie Biologique et Médicale* (ed. M. Piéry), Paris: Masson, 1934. Vol. 1, 661-673. (Includes list of 23 papers by Chizhevsky on ionization published between 1925 and 1931).

34. SULMAN, F. G. et al., "Air-ionometry of hot, dry desert winds (Sharav) and treatment with air-ions of weather-sensitive subjects." *International Journal of Biometeorology* 18 (4), 313-318, 1974. See also *ibid.* 14, 45-53, 1970. "Infant asthmatics respond to negative ions." *News from Medion* 5 (1), n.d.

35. *News from Medion* 5 (1), n.d.

36. KRUEGER, A. P. & REED, E. J., "Biological impact of small air ions." *Science* 193, 1209-1213, 1976. (List of 46 references).

37. VASILIEV, L. L. & CHIZHEVSKY, A. L., "An hypothesis of organic electro-exchange." Voronezh: *Transactions of the Central Laboratory of Scientific Research on Ionization* 1, 219, 1933. (Russian).

38. ASSAEL, M., PFEIFER, Y. & SULMAN, F. G., "Influence of artificial air ionization on the human electroencephalogram." *International Journal of Biometeorology* 18 (4), 306-312, 1974. See also: *Elektromedizin* 8 (1), 1963 (Eichmeier) for

earlier work on ions and alpha rhythm, pulse and breathing frequency changes. See also:

KORNBLUEH, I. H., PIERSOL, G. M. & SPEICHER, F. P., "Relief from pollinosis in negatively ionized rooms." *American Journal of Physical Medicine* 37, 18, 1957.

39 KÖNIG, H. L., *Unsichtbare Umwelt*. Munich: Heinz Moos Verlag, 1975.

40 RHEINSTEIN, J., "The influence of artificially generated atmospheric ions on the simple reaction time and on the optical moment." Ph.D. thesis, Technische Hochshule, Munich, 1960.

41 CARSON, R. W., "Anti-fatigue device works by creating electric field." *Product Engineering* February 13 1967.

42 CAYMAZ, G., "Effect of increased atmospheric electricity on the blood electrolytes of airplane crew." Ankara: NATO-AGARD Conference Proceedings 182, B7 1-5, 1976.

43 MAXEY, E. S. Personal communication to Playfair, 1977.

44 PUHARICH, A., *Beyond Telepathy*. New York: Doubleday, 1962. (Appendix A).

45 GADDIS, V. H., *Mysterious Fires and Lights*. New York: Dell, 1968.

HARRISON, M., *Fire from Heaven*. London: Sidgwick & Jackson, 1976.

ARNOLD, L. E., "The flaming fate of Dr. Bentley." *Fate* April 1977, pp. 66-72. Of special interest here is the author's claimed link between spontaneous combustion of people and days of geomagnetic flux. Both this and Harrison (above) include photographs of inexplicably combusted bodies.

Recommended further reading:

SOYKA, F. & EDMONDS, A., *The Ion Effect*. New York: E. P. Dutton, 1977. Based on Soyka's personal sufferings as a *foehn* victim, this lively and persuasive book contains a mass of useful information on "how air electricity rules your life and health."

Chapter Five

1 GRIBBIN, J. & PLAGEMANN, S., *The Jupiter Effect*. London: Macmillan, 1974 and Fontana (revised) 1977. See also: *Icarus* 26, 257-270, 1975 (Meeus, Gribbin & Plagemann); *Icarus* 29, 435-436, 1976 (Ip); *Nature* 265, 4 & 13, 1977, and *Science News* 111 (2) 19, 1977.

2 GRIBBIN, J., "Relation of sunspot and earthquake activity." *Science* 173, 558, 1971.

3 *Earthquake Prediction and Public Policy*. Washington, D.C.: National Academy of Sciences, 1975.

SHAPLEY, D., "Earthquakes: Los Angeles prediction suggests faults in federal policy." *Science* 192, 535-537, 1976.

4 JACKSON, R., *Thirty Seconds at Quetta*. London: Evans, 1960, pp. 131-132.

5 RIKITAKE, T., *Earthquake Prediction*. Amsterdam: Elsevier, 1976. Chapter 2.

6 SHARMA, P. V., "Earthquake prediction—from legends to new methods." Colloquia on Earthquake Prediction, Geophysical Institute, University of Copenhagen, October 1976.

7 MOORE, G. W., "Magnetic disturbance preceding the 1964 Alaska earthquake." *Nature* 203, 508-509, 1964.

8 ROMAN, C., "Earthquakes: after nearly 3,000 years of study we still have much to learn." *The Times*, August 5th 1976.

9 Stanford University Medical Center news release, October 25th 1976.

10 RICHTER, C., "Earthquakes may follow a pattern, but there are still some surprises." *The Times*, August 23rd 1976.

11 CHALLINOR, R. A., "Variations in the rate of rotation of the Earth." *Science* 172, 1022-1024, 1971.

12 KOZYREV, N. A., "On the interaction between tectonic processes of the Earth and the Moon," in *The Moon* (ed. S. K. Runcorn & H. Urey). IAU Symposium 47. Dordrecht: Reidel, 1972.

13 LATHAM, G. et al., "Moonquakes." *Science* 174, 687-692, 1971.
TOKSÖZ, M. N., GOINS, N. R. & CHENG, C. H., "Moonquakes: mechanisms and relation to tidal stresses." *Science* 196, 979-981, 1977.

14 SYTINSKY, A. D., "Effect of solar activity on the Earth's seismicity." *Doklady Akademii Nauk SSSR* 208 (5) 1078-1081, 1973. (English version: *Doklady Earth Sciences* 208, 36-39, 1973).
PUDOVKIN, I. M. et al., "Direct relationship between geomagnetic variations and earthquakes." *ibid.*, pp. 1074-1077.
ZÁTOPEK, A. & KRVISKY, L., "On the correlation between meteorological microseisms and solar activity." *Bulletin of the Astronomical Institutes of Czechoslovakia* 25 (5) 257-262, 1974.

15 FILSON, J., SIMKIN, T. & LEU, L., "Seismicity of a caldera collapse. Galapagos islands, 1968." *Journal of Geophysical Research* 78 (35) 8591-8622, 1973.

16 MAUK, F. J. & JOHNSTON, J. S., "On the triggering of volcanic eruptions by Earth tides." *Journal of Geophysical Research* 78 (17) 3356-3362, 1973.

17 RYALL, A., VAN WORMER, J. D. & JONES, A. E., "Triggering of micro-earthquakes by Earth tides, and other features of

the Truckee, California earthquake sequence of September 1966." *Bulletin of the Seismological Society of America* 58 (1) 215-248, 1968.

18 SIMPSON, J. F., "Earth tides as a triggering mechanism for earthquakes." *Earth & Planetary Science Letters* 2, 473-483, 1967.

19 HUNTER, R. N., "Is there a connection between astrology and earthquakes?" *Earthquake Information Bulletin*, May/June 1971, 18-23.

20 PRIEDITIS, A. A., "Astral dissonances and earthquakes." *Gnostica* 39, 21-24, 1976.

21 TOMASCHEK, R., "Great earthquakes and the astronomical positions of Uranus." *Nature* 184, 177-178, 1959. See also: *Nature* 186, 336-337, 1960.
NEELY, J., in *Phenomena* 1 (5) 1977, p. 9.

22 LANDSCHEIDT, T., "Kosmische Kybernetik: Besteht ein Zusammenhang zwischen der Sonnenaktivität und der Aktivität im Zentrum der Milch-Strasse?" *Jahrbuch der Wittheit zu Bremen* 18, 275-296, 1974.

23 STETSON, H. T., *Sunspots in Action*. New York: Ronald Press, 1947, 140-141.
STETSON, H. T., "The correlation of deep-focus earthquakes with lunar hour angle and declination." *Science* 82, 523-524, 1935.

24 HJORTENBERG, E., "Spectral measurements of *in situ* seismic velocities and their possible use in earthquake prediction." Colloquia on Earthquake Prediction, Geophysical Institute, University of Copenhagen, December 1976.

25 WANG, M. et al., "A preliminary study on the mechanism of the reservoir impounding earthquakes at Hsinfengkiang." *Scientia Sinica* 19 (1) 149-169, 1976.

26 OXBURGH, R., "How Britain's warm substrata could cut down fuel bills." *Telegraph Sunday Magazine*, December 14th 1976. See also:
BERMAN, E. R., *Geothermal Energy*. Park Ridge & London: Noyes Data Corp., 1975.

Chapter Six

1 MEADOWS, A. J., "A hundred years of controversy over sunspots and weather." *Nature* 256, 95-97, 1975.

2 GARCIA-MATA, C. & SHAFFNER, F. I., "Solar and economic relationships." *Quarterly Journal of Economics*, November 1934.

3 Foundation for the Study of Cycles, 124 South Highland Avenue, Pittsburgh, Penn. 15206.

4 DEWEY, E. R. & MANDINO, O., *Cycles. The Mysterious Forces that Trigger Events*. New York: Manor Books, 1973.

5 DAVIES, D. Letter to *The Times*, August 21st 1976. Reprinted by permission of the author.

6 MARGERISON, T., in *New Scientist* 72 (1026) 349-350, 1976.

7 Center for Short-Lived Phenomena, 129 Mount Auburn St., Cambridge, Mass. 02138.

8 PICCARDI, G., *The Chemical Basis of Medical Climatology*. Springfield: Charles C. Thomas, 1962.

9 EICHMEIER, J. & BÜGER, P., "Über den Einfluss elektromagnetischer Strahlung auf die Bismuthchlorid-Fällungsreaktion nach Piccardi." *International Journal of Biometeorology* 13, 239-256, 1969.

10 TAKATA, M., "Über eine neue biologische wirksame Komponente der Sonnenstrahlung. Beitrag zu einer experimentalen Grundlage der Heliobiologie." *Archiv für Meteorologie, Geophysik und Bioklimatologie* Serie B, 11, 486-508, 1951.

11 CHIZHEVSKY, A. L., *The Terrestrial Echo of Solar Storms*. Moscow: Mysl, 1973. (Russian). Second edition, 1976.

12 CHIZHEVSKY, A. L., "L'action de l'activité périodique solaire sur les épidémies," and "L'action de l'activité périodique solaire sur la mortalité générale," in *Traité de Climatologie Biologique et médicale* (ed. M. Piéry). Paris: Masson, 1934, Vol. 2, pp. 1034-1045.

13 GAUQUELIN, M. & GAUQUELIN, F., "Review of studies in the U.S.S.R. on the possible biological effects of solar activity." *Journal of Interdisciplinary Cycle Research* 6 (3) 249-252, 1975.

SHULTS, N. A., "Influence of Solar Activity on the Number of White Blood Corpuscles," in *The Earth in the Universe*, Moscow: Mysl, 1964. (Russian).

YAGODINSKY, V. N., "The cyclic nature of epidemics and the heliomagnetic factors of the external medium." *Materials of the II All-Union Conference on the Study of the Effects of Magnetic Fields on Biological Organisms*. (Moscow, 24-26 September 1969). English version: National Research Council of Canada Technical Translation 1545, Ottawa, 1972, p. 242.

YAGODINSKY, V. N. & SEREBRYANSKY, V. S., "The principle of long-range prediction of infectious disease rates taking into account heliomagnetic variations." *ibid*. p. 245.

14 REINBACHER, L., "Können Höhlen heilen?" *Bild der Wissenschaft* May 1976, pp. 76-80.

15 VAUX, J. E. Personal communication to Hill, 1974.

16 SADEH, D. S. & MEIDAV, M., "Search for sidereal periodicity in earthquake occurrences." *Journal of Geophysical Research* 78 (32) 7709-7716, 1973.

Also recommended:
A worthy and highly qualified successor to Charles Fort is William R. Corliss, a physicist and science writer who founded his "Sourcebook Project" in 1974, and has since published several volumes of "anomalous data" in archaeology, astronomy, biology, geology and geophysics. See, for example, *The Unexplained: A Sourcebook of Strange Phenomena*. New York: Bantam Books, 1976.

Chapter Seven

1 PITTENDRIGH, C. S., "Circadian systems: 1. The driving oscillation and its assay in Drosophila pseudoobscura." *Proceedings of the National Academy of Sciences* 58 (4) 1762-1767, 1967.

2 SOLZHENITSYN, A. I., *The Gulag Archipelago*. London: Collins/Fontana, 1974. Part 1, Chapter 12.

3 LEWIS, P. R. & LOBBAN, M. C., "The effects of prolonged periods of life on abnormal time routines upon excretory rhythms in human subjects living on abnormal time routines. *Quarterly Journal of Experimental Biology and Cognate Medical Sciences* 42, 356-371, 1957.
LEWIS, P. R. & LOBBAN, M. C., "Dissociation of diurnal rhythms in human subjects living on abnormal time routines," *ibid.* 371-386.

4 LUCE, G. G., *Body Time*. St. Albans: Paladin, 1973.

5 GITTELSON, B., *Biorhythm*. London: Futura, 1976.

6 HERSEY, R. B., "Emotional cycles of man." *Journal of Mental Science* 77, 151-169, 1931.

7 RODGERS, C. W., SPRINKLE, R. L., & LINDBERG, F. H., "Biorhythms: three tests of the predictive validity of the "critical days" hypothesis." *International Journal of Chronobiology* 2 (3) 247-252, 1974.

8 RAVITZ, L. J., "History, measurement, and applicability of periodic changes in the electromagnetic field in health and disease." *Annals of the New York Academy of Sciences* 98 (4) 1144-1201, 1962.

9 DEWAN, E. M., "Rhythms." *Science & Technology* January 1969, 20-28.

10 Personal communication to Hill, 1976.

11 LUCE, G. G., op. cit. above (4), chapter 6.

Chapter Eight

1. COLE, F. E. & GRAF, E. R., "Precambrian ELF and abiogenesis," in *ELF and VLF Electromagnetic Field Effects* (ed. M. A. Persinger), London & New York: Plenum Press, 1974.

2. ISKHAKOV, V. P., "Towards the problem of the influence of solar activity in mental illnesses," in *Sun, Electricity, Life.* (Materials from Readings of the Moscow Society of Naturalists (Physics Section) dedicated to the memory of Prof. A. L. Chizhevsky, February 7th, 1968). Moscow, 1969. (Russian).

3. KHOLODOV, YU. M., "Electromagnetic fields and the brain." *Impact of Science on Society* 24 (4) 291-297, 1974.

4. PICCARDI, G., *The Chemical Basis of Medical Climatology.* Springfield: Charles C. Thomas, 1967. Chapter 7.

5. BENTOV, I., "Micro-motion of the body as a factor in the development of the nervous system." Unpublished MS, 1974.

6. See bibliography in LUCE, G. G., *Body Time.* St. Albans: Paladin, 1973 for listing of numerous papers by Pöppel, Aschoff and Wever.

7. BAWIN, S. M., GAVALAS-MEDICI, R. J. & ADEY, W. R., "Effects of modulated very high frequency fields on specific brain rhythms in cats." *Brain Research* 58, 365, 1973.

8. BUSBY, D. E., "Space biomagnetics." *Space Life Sciences* 1 (1) 23, 1968.

9. MAXEY, E. S. & BEAL, J. B., "The electrophysiology of acupuncture: how terrestrial electric and magnetic fields influence aerion energy exchanges through acupuncture points . . ." VII International Biometeorology Congress, University of Maryland, August 1975.

10. DUBROV, A. P., *The Geomagnetic Field and Life.* Leningrad: Gidrometeoizdat, 1974. (Russian).

11. BROWN, F. A., JR. & PARK, Y. H., "Synodic monthly modulation of the diurnal rhythm of hamsters." *Proceedings of the Society for Experimental Biology and Medicine* 125, 712-725, 1967.

12. LUCE, G. G., *Body Time.* St. Albans: Paladin, 1973. Chapter 2.

13. PÖPPEL, E., "Jet travel—body and soul." *New Scientist* August 3rd 1972, 232-235.

14. MAWDSLEY, C., "Epilepsy and television." *The Lancet* 1, 190-191, 1961.

15. *Secrets of a Coral Island* (film) BBC 2, November 29th, 1976.

16 SINCLAIR, A. R. E., "Lunar cycle and timing of mating season in Serengeti wildebeest." *Nature* 267, 832-833, 1977.
NEGUS, N. C. & BERGER, P. J., "Experimental triggering of reproduction in a natural population of *Microtus montanus*." *Science* 196, 1230-1231, 1977.

17 DEWAN, E. M., "On the possibility of a perfect rhythm method of birth control by periodic light stimulation." *American Journal of Obstetrics and Gynaecology* 99 (7) 1016-1019, 1967.
DEWAN, E. M., "Rhythms." *Science & Technology* January 1969, 20-28.

18 JAFAREY, N. A., KHAN, M. Y. & JAFAREY, S. N., "Role of artificial lighting in decreasing the age of menarche." *The Lancet*, August 29th 1970, p. 471. See also *ibid*. September 12th 1970, p. 571 and April 3rd 1971, p. 707.

19 British Patent No. 1,114,787. Information on British patents from The British Library, 25 Southampton Buildings, Chancery Lane, London WC2A 1AW.

Recommended further reading:
WARD, R. R., *The Living Clocks*. London: Collins, 1972.
BÜNNING, E., *The Physiological Clock*. New York: Springer-Verlag, 1967 (revised).
STRUGHOLD, H., *Your Body Clock*. London: Angus & Robertson, 1972.

Chapter Nine

1 TARCHANOFF, J. R., "Voluntary acceleration of the heart beat in man." *Pflügers Archiv* 35, 109-135, 1885.

2 BROWN, B. B., *New Mind, New Body*. New York: Harper & Row, 1974.

3 *ibid*. p. 31.

4 *ibid*. pp. 221-222.

5 KAMIYA, J., "Conscious control of brain waves." *Psychology Today* 1, 56-60, 1968.
BROWN, B. B., "Recognition of aspects of consciousness through association with EEG alpha activity represented by a light signal." *Psychophysiology* 6, 442-452, 1970.

6 BAIR, J. H., "Development of voluntary control." *Psychology Review* 8, 474-510, 1901.

7 BASMAJIAN, J. V., "Electromyography comes of age." *Science* 176, 603-609, 1972.

8 DEWAN, E. M., "Rhythms." *Science & Technology* January 1969, 20-28.

9 TAYLOR, J. D., " 'Radio One'—an unexpected response." *The Lancet* October 16th 1971, 881-882.

List of References

10 SCHWARZ, G. E., "Voluntary control of human cardiovascular integration and differentiation through feedback and reward." *Science* 175, 90-93, 1972.

11 BROWN, B. B., op. cit. above (2), p. 398.

12 HIRAI, T., "Electroencephalographic study on the Zen meditation (Zazen)—EEG changes during the concentrated relaxation." *Psychiatria et neurologia Japonica* 62, 76-105, 1960.

13 WALLACE, R. K. & BENSON, H., "The physiology of meditation." *Scientific American* February 1972, 84-90.

14 DOMASH, L. H., "The transcendental meditation technique and quantum physics." Scientific research on the TM programme; Collected Papers Vol. 1, Maharishi European University Press.

15 GREEN, E., "Biofeedback and voluntary control of internal states," in *The Frontiers of Science and Medicine*. (R. J. Carlson, ed.) London, Wildwood House, 1975, 48-59.

16 PLAYFAIR, G. L., *The Flying Cow*. London: Souvenir Press, 1975. (U.S. title: *The Unknown Power*. New York: Pocket Book, 1976).

17 TRABKA, J., "High frequency components in brain wave activity." *Electroencephalography and Clinical Neurophysiology* 14, 453-464, 1962.
BENTOV, I., "Micro-motion of the body as a factor in the development of the nervous system." Unpublished MS, 1974.

18 PRESSMAN, A. S., *Electromagnetic Fields and Life*. New York: Plenum Press, 1970. Page 240n.

19 FERGUSON, M., *The Brain Revolution*. London: Davis-Poynter, 1974.

20 *Psychic*, June 1975, 48-55.

21 CADE, C. M. & WOOLLEY-HART, A., "The measurement of hypnosis and auto-hypnosis by determination of electrical skin resistance." *Journal of the Society for Psychical Research* 46 (748) 81-102, 1971.

22 BROWN, B. B., op. cit. above (2), 14.

23 BURR, H. S. & NORTHROP, F. S. C., "The electrodynamic theory of life." *Quarterly Review of Biology* 10, 322-333, 1935.
MARKSON, R., "Tree potentials and external factors," in *Blueprint for Immortality* by H. S. Burr. London: Neville Spearman, 1972, 166-184.

24 RAVITZ, L. J., "History, measurement and applicability of periodic changes in the electromagnetic field in health and disease." *Annals of the New York Academy of Sciences* 98 (4) 1144-1201, 1962.

ADAMENKO, V. G., "Voluntary control of the human bio-electric field," *Journal of Psychoenergetic Systems* 1 (1) 1974.

25 VALENSTEIN, E. S., *Brain Control*. New York: John Wiley, 1973.

26 ROUSE, L. et al., "EEG alpha entrainment reaction within the biofeedback setting and some possible effects on epilepsy." (Talk): Annual Meeting, Biofeedback Research Society. Colorado Springs, February 1974.

Chapter Ten

1 BOK, B. J. et al., "Objections to astrology, a statement by 186 leading scientists." *The Humanist* 35, Sept./Oct. 1975, 4-6.

2 FEOLA, J., "The Great Container (II)." *Gnostica* 5 (3) 48-51, 1976.

3 PAY, R., "Position of planets linked to solar flare prediction." *Technology Week*, May 15th 1967, pp. 35-38.

4 TEMPLE, R. K. G., *The Sirius Mystery*. London: Sidgwick & Jackson, 1976.
DEMAREST, K., "The Winged Power," in *Consciousness and Reality*. (Charles Musès & Arthur M. Young, eds.) New York: Discus/Avon, 1974.

5 KOESTLER, A., *The Sleepwalkers*. London: Hutchinson, 1959. Part 3, Chapter 2 (1).

6 GAUQUELIN, M., *The Cosmic Clocks*. Chicago: Henry Regnery, 1967.
GAUQUELIN, M., *Astrology and Science*. London: Peter Davies, 1970.
GAUQUELIN, M., *Cosmic Influences on Human Behaviour*. London: Garnstone Press, 1974.

7 HUNTINGTON, E., *Season of Birth: its Relation to Human Abilities*. New York: Wiley, 1938.

8 COOPER, H. J., "Occupation and season of birth." *Journal of Social Psychology* 89, 109-114, 1973.
COOPER, H. J. & SMITHERS, A., "Do the seasons govern your career?" *New Society* April 5th 1973, 6-8.

9 JUNG, C. G., *Synchronicity. An Acausal Connecting Principle*. London: Routledge & Kegan Paul, 1972. Chapter 2.

10 COOPER, H. J. Personal communication to Playfair, 1976.

11 HARE, E. H., PRICE, J. S. & SLATER, E., "Mental disorder and season of birth." *Nature* 241, 480, 1973.

12 WINDSOR, D. A., "Molecular biologists come of age in Aries." *Nature* 248, 788, 1974.

13 GRIBBIN, J., "Science journalists come of age in Pisces." *Nature* 252, 534, 1974.

14 TROMP, S. W., "Effects of weather and climate on mental processes of man," in *Parapsychology and the Sciences*. New York: Parapsychology Foundation, 1972, 135-148.

15 ADDEY, J. M., *The Discrimination of Birthtypes*. Green Bay & Bournemouth: The Cambridge Circle, 1974.
ADDEY, J. M., *Harmonics in Astrology*. Romford: Fowler, 1976.

16 Biodynamics Research, P.O. Box 2884, Santa Rosa, California 95405. (Europe) Postboks 1168, 1010 Copenhagen K., Denmark.

17 ELWIN, H. V. H., *The Muria and their Ghotul*. Bombay: Oxford University Press, 1947. (730 pages).

18 BILLINGS, J. J., *Natural Family Planning*. Collegeville: The Liturgical Press, 1973.

19 Aquarian Research Foundation, 5620 Morton St., Philadelphia, Pa. 19144. (Publishers of *The Natural Birth Control Book* by A. Rosenblum & L. Jackson, containing extract from (17) above and *Stern* article).
See also: BENSON, M., "Natural Birth Control." *Radionics Quarterly* 23 (7) 21-23, 1976.

20 LIEBER, A. L. & SHERIN, C. R., "Homicides and the lunar cycle." *American Journal of Psychiatry* 129, 101-105, 1972.

21 KRIPPNER, S., "The cycle in deaths of U.S. presidents elected at twenty-year intervals." *International Journal of Parapsycology* 9 (3) 145-153, 1967. See also *ibid*. 10 (1) 107-112, 1968.

22 HUGHES, D. W., "The star of Bethlehem." *Nature* 264, 513-517, 1976.

23 PRIEDITIS, A. A. Personal communication to Playfair, 1977.

24 *Sun, Electricity, Life*. Moscow: Izdatel'stvo MGU, 1969. (Russian).
CHIZHEVSKY, A. L., "Effets de l'activité périodique solaire sur les phénomènes sociaux," in *Traité de Climatologie Biologique et Médicale* (ed. M. Piéry). Paris: Masson, 1934. Vol. 1, 576-586.
CHIZHEVSKY, A. L., "Physical factors of the historical process." *Cycles*, January 1971, 11-23.

25 Zipporah Dobyns, quoted in (26) below.

26 DEAN, G. A., *Recent Advances in Natal Astrology*. Subiaco (W. Australia): Analogic, 1977.

27 MAYO, J., WHITE, O. & EYSENCK, H. J., "An empirical study of the relation between astrological factors and personality." 1977 (in press).

Chapter Eleven

1 YIP, K. et al., in *American Journal of Chinese Medicine* 3, 1976.

2 THEOBALD, G. W., *The Electrical Induction of Labor*. New York: Appleton-Century-Crofts, 1973.

3 POMERANZ, B. Personal communication to Hill, 1976. See also *New Scientist* 73 (1033) 12-13, 1977.

4 For recent developments in acupuncture in the U.S.S.R. see:
HILL, S., "Acupuncture research in the U.S.S.R." *American Journal of Chinese Medicine* 2, 1976.
INYUSHIN, V. M. & CHEKUROV, P. R., *Biostimulation through Laser Radiation and Bioplasma*. Translated by S. Hill & T. D. Ghoshal. Copenhagen: Danish Society for Psychical Research, 1976.
ROMEN, A. S. (ed.), *All-Union Congress on Psychical Self Regulation*. Alma-Ata: Kazakh University Press, 1973, 1974. Vols. 1 and 2. (Russian).

5 "Animal cancers treated by laser energy." *IEEE Spectrum* February 1964, pp. 160-162.

6 *Proceedings of the All-Union Biological Symposium on Psychical Self Regulation*. Alma-Ata: Kazakh State University, 1976. (In press). (Russian).

7 PALEG, L. G. & ASPINALL, D., "Field control of plant growth and development through the laser activation of phytochrome." *Nature* 228, 970-973, 1970.

8 HILL, S., "Heilung durch Licht." *Esotera*, January 1976.

9 NAMENYI, J. et al., "Effect of laser irradiation and immunosuppressive treatment on survival of mouse skin allotransplants." *Acta Chirurgica Academiae Scientiarum Hungaricae* 16 (4) 327-335, 1975. See also *ibid.* 15 (2) 203-208, 1974.

10 *Advantages of the Laser in Acupuncture*. Munich: Messerschmidt-Bölkow-Blohm, 1976.

11 DOMANSKII, A. & TUROVEROV, K. K., *Luminescence of Biopolymers and Cells*. New York, London: Plenum Press, 1969.

12 CZERSKI, P., "Experimental models for the evaluation of microwave biological effects." *Proceedings of the IEEE* 63 (11) 1540-1544, 1975. See also ref. 7 to Chapter 8.

13 HELLER, J. H. & TEIXEIRA PINTO, A. A., "A new physical method of creating chromosomal aberrations." *Nature* 183, 905, 1959.

14 YAO, K. T. S. & JILES, M. M., "Effects of 2450 MHz. microwave radiation on cultivated rat kangaroo cells," in *Biological Effects and Health Implications of Microwave Radiation*, ed. S. F. Cleary. U.S. Dept. of Health, Education & Welfare report BRH/DBE-2/70, 1971, p. 123.

15 HAMEROFF, S., "Ch'i: a neural hologram? Microtubules, bioholography and acupuncture." *American Journal of Chinese Medicine* 12 (2) 163-170, 1974.
HAMEROFF, S., "Light is heavy. Wave mechanics in proteins: a microtubule hologram model of consciousness." *Proceedings, II International Congress on Psychotronic Research* Monte Carlo, 1975, pp. 168-169.

16 AXELROD, J., "The pineal gland—a neurochemical transducer." *Science* 184, 1341-1348, 1974.

17 POPP, F. A., *Biophotonen*. Heidelberg: Ewald Fischer Verlag, 1976. (Contains Popp's theory of DNA helix as biological laser).
POPP, F. A., "Molecular aspects of carcinogenesis," in *Molecular Base of Malignancy* ed. E. Deutsch et al. Stuttgart: Thieme, 1976.
POPP, F. A., & POPP, R. B., "Experimentelle Untersuchungen zur ultraschwachen Photonen-emission biologischer Sisteme." *Zeitschrift für Naturforschung* 31 (C) 741-745, 1976. See also *ibid.* 29 (C) 453, 1974, and *Bild der Wissenschaft* January 1976, pp. 59-62.

18 FINSEN, N. R., "Inflydelse af lyset på sygdomme." Cand. med. dissertation, University of Copenhagen Medical School, 1893. (Revised).

19 URBACH, F. (ed.), "The biologic effects of ultraviolet radiation, with emphasis on the skin." Proceedings, 1 International Conference, International Society of Biometeorologists & Skin and Cancer Hospital, Temple University Health Science Center. London: Pergamon Press, 1969.

20 BLUM, H. F., "Environmental radiation and cancer." *Science* 130, 1545-1547, 1959.

21 Information from Barapio S.A., 159-161 Route d'Hermance, Collonges-Bellerive, 1245 Geneva, Switzerland.

22 BOHM, D. & BUB, J., "A proposed solution of the measurement problem in quantum mechanics by a hidden variable theory." *Review of Modern Physics* 38, 453-469, 1966.

23 WALKER, E. H., "Paraphysical and parapsychological phenomena, foundations of." Proceedings of the Geneva Conference on Quantum Physics and Parapsychology. New York: Parapsychology Foundation, 1975. See also: *Mathematical Biosciences* 7, 131, 1970; *Physics Today* 24, 39, 1971.

24 CAPRA, F., *The Tao of Physics*. London: Wildwood House, 1975. (Epilogue).

25 BIGU, J., "On the biophysical basis of the human aura." *Journal of Research in Psi Phenomena* 1 (2) 8-43, 1976.

26 HSIAO, H. S. & SUSSKIND, C., "Infrared and microwave communication by moths." *IEEE Spectrum* March 1970, 69-76.

27 FRISCH, K. VON, "Die Sonne als Kompass im Leben der Bienen." *Experientia* 6, 210-221, 1950.

28 LAITHWAITE, E. R., "Biological analogues in engineering practice." *Interdisciplinary Science Reviews* 2 (2) 100-108, 1977.

29 GAMOW, R. I. & HARRIS, J. F., "What engineers can learn from nature." *IEEE Spectrum* August 1972, 36-42.

30 LILLY, J. C., *The Mind of the Dolphin*. New York: Doubleday, 1967.
LILLY, J. C., *Programming and Metaprogramming in the Human Biocomputer*. New York: Julian Press, 1973.
KESAREV, V., "The inferior brain of the dolphin." *Soviet Science Review* 2 (1) 52-58, 1971. (Despite its title, offers evidence for superiority in some respects of dolphin brains).

Chapter Twelve

1 LONG, M. F., *The Secret Science Behind Miracles*. Santa Monica: De Vorss, 1954.
MYERS, F. W. H., *Human Personality and its Survival of Bodily Death*. London: Longmans, Green, 1903.

2 WALLACE, A. R., *On Miracles and Modern Spiritualism*. London: James Burns, 1875, p. 213.

3 ANDRADE, H. G., "Psi Matter," in *The Indefinite Boundary* by G. L. Playfair. London: Souvenir Press, 1976, pp. 295-307.

4 BURR, H. S., *Blueprint for Immortality*. London: Neville Spearman, 1972.

5 LANGMAN, L. & BURR, H. S., "Electro-magnetic studies in women with malignancy of cervix uteri." *Science* 105, 209-210, 1947.

6 LANGMAN, L. & BURR, H. S., "Electro-metric timing of human ovulation." *American Journal of Obstetrics & Gynaecology* 44, 223-230, 1942.

7 U.S. Patent 3,920,003. See also ref. 19 to Ch. 10.

8 LAITHWAITE, E. R., "Biological analogues in engineering practice." *Interdisciplinary Science Reviews* 2 (2) 100-108, 1977.

9 MAXEY, E. S. & BEAL, J. B., "The electrophysiology of acupuncture: how terrestrial electric and magnetic fields influence aerion energy exchanges through acupuncture points. Paper read at the VII International Biometeorology Congress, University of Maryland, 1975.

10 MOTOYAMA, H., "The ejection of energy from the chakra of yoga and the meridian points of acupuncture." *Proceedings, II International Congress on Psychotronic Research*, Monte

Carlo, 1975, pp. 375-386. Dr. Motoyama's many publications are available from The Institute for Religious Psychology, 4-11-7 Inokashira, Mitaka-shi, Tokyo.

11 MAXEY & BEAL, op. cit. above (9).

12 NAGAHAMA, Y. & MARUYAMA, M., *Research of Meridians.* Tokyo: Kyorin Shoin, 1950.

13 COE, M. R., Jr., "Does science explain poltergeists?" *Fate* July 1959, pp. 79-90.

14 STEWART, D., THOMSON, J. & OSWALD, I., "Acupuncture analgesia: an experimental investigation." *British Medical Journal* 1, 67-70, 1977.

15 POMERANZ, B. Personal communication to Hill, 1976. See also: *New Scientist* 73 (1033) 12-13, 1977.

16 SZENT-GYÖRGYI, A., *Introduction to a Submolecular Biology.* New York: Academic Press, 1960. (p. 52) See also:
VINOGRADOV, V., "The 'exciton substance' in semiconductors." *Soviet Science Review* 1 (2) 83-86, 1970.

17 SEDLAK, W., "The electromagnetic nature of life." *Proceedings of the II International Congress on Psychotronic Research,* Monte Carlo, 1975, pp. 77-83.

18 INYUSHIN, V. M. & CHEKUROV, P. R., *Biostimulation through Laser Radiation and Bioplasma.* (Translated by Scott Hill & T. D. Ghoshal). Copenhagen: Danish Society for Psychical Research, 1977.

19 MANCZARSKI, S., "Electron plasma in the biological medium." *Proceedings of the II International Congress on Psychotronic Research,* Monte Carlo, 1975, pp. 125-126.

20 PIRUZYAN, L. A. & ARISTARKHOV, V. M., "Some possible energy-bearing mechanisms associated with the development of biopotentials." *Izvestiya Akademii Nauk SSSR (Biol.)* 1, 69-85, 1969. (Russian).

21 GURVICH, A. A., YEREMEYEV, V. F. & KARABCHEVSKY, YU. A., "Registration of mitogenetic radiation of animal heart in *in vivo* experiments with aid of photoelectron multiplier." *Doklady Academii Nauk SSSR* 195 (4) 972-975, 1970. (Russian). (See also *Nature* 206, 20-22, 1965).

22 ADAMENKO, V. G., "Psychoenergetics and extramotoric functions of an organism." *Proceedings of the II International Congress on Psychotronic Research,* Monte Carlo, 1975, pp. 141-143.

23 BACKSTER, C., "Evidence of a primary perception in plant life." *International Journal of Parapsychology* 10, 329-348, 1968.
HOROWITZ, K. A., LEWIS, D. C. & GASTEIGER, E. L., "Plant 'primary perception': Electrical unresponsiveness to brine shrimp killing." *Science* 189, 478-480, 1975.

KMETZ, J. M., "A study of primary perception in plant and animal life." *Journal of the American Society for Psychical Research* 71 (2) 157-169, 1977.

24 PICKARD, B. G., "Action potentials resulting from mechanical stimulation of pea epicotyls." *Planta* 97, 106-115, 1971.

PICKARD, B. G., "Spontaneous electrical activity in shoots of *Ipomea, Pisum*, and *Xanthium*." *Planta* 102, 91-114, 1972.

KARLSSON, L., "Nonrandom bioelectrical signals in plant tissue." *Plant Physiology* 49, 982-986, 1972.

KARLSSON, L., "Instrumentation for measuring bioelectrical signals in plant tissue." *The Review of Scientific Instruments* 43 (3) 458-464, 1972.

25 Op. cit. above (18).

26 GOLVIN, V., "The Kirlian effect in medicine." *Soviet Journal of Medicine* August 11th 1976. (Russian).

27 Op. cit. above (18).

28 KOESTLER, A., *The Act of Creation*. London: Hutchinson, 1964. Chapter 5.

INDEX

Achilles, 278
Acupuncture:
  anaesthesia, 79–80, 277–278, 306–307
  bioplasma and, 312
  current research, 300–308
  labour, 278
  lasers and, 279–82
  Mao Tse-tung and, 302
  meridians, 208, 247–48, 300–301
Adamenko, V. G., 312, 319
Addey, J. M., 258
Adey, W. R., 204, 205–206
Aesop, 24
Air ions, 112, 127–35, 207
Alcock, George, 7
Alfvén, Hannes, 42
Allais, M., 18–20
Allen, Mildred, 19
American Federation of Astrologers, 53, 266
American Society of Dowsers, 89
AMI machine, 300–302, 305–306
Amoco Europe Inc., 146
Ampère, A. M., xi
Andrade, Hernani G., 99, 297, 312

Animal magnetism, 75
Apollo programme, 56–57, 134, 155, 186
Aquarian Research Foundation, 263
Archaeology Institute (Moscow), 267
Argonauts, 247
Arigó, 95
Aristotle, 32, 37
Ashford, Oliver, 104
Asteroids, 5
Astra Birth Control Research Centre, 259
Astrological Association, 258
Astrology, 30, 31–32, 156–158, Ch. 10 *passim*
Ataturk Sanatorium, 57, 134
Atmospherics, 119, 207
Audus, L. J., 83
Auroras, 36
Autolycus, 176

Backster, Cleve, 313–14
Bair, J. H., 225
Balanovski, Eduardo, 92
Baldwin, S., 270
Baraduc, H., 316
Barratt, Michael, 116

351

Index

Basmajian, J., 226
Basov, N. G., 280
Bats and sonar, 293
BBC Radio One, 227
Beal, James B., 207, 301
Becker, R. O., 96–99, 100, 203, 309
Bees, 293
Behaviourism, 223
Benner, Samuel, 170–71, 173
Benson, Herbert, 232
Bentov, Itzhak, 203, 234
Bentze, G., 198
Berger, Hans, 70
Bernhardt, Clarissa, 144n
Bernheim, H-M., 76
Berzelius, J., 129
Beta-endorphin, 278, 307
Bethlehem, Star of, 266
Bigg, E. K., 46–47
Bigu (del Blanco), J., 292
Bikur Holim Hospital, 131
Billings, J., 262
Bio-dynamics Research, 261
Bio-electronics, Inc., 217
Bioentrainment, 121, 200, Ch. 8
Biofeedback:
  blood pressure, 228
  brain, 224–25, 232–34, 239–42
  cell control, 226
  heart, 226–27, 232–33
  hypertension, 228
  metabolism, 236–37
  muscles, 225–26
  music, 226–28
  self-healing, 229–31
  skin, 228–29
Bio-Feedback Research Society, 223, 224
Biological Organizing Model (BOM), 99, 297

Biometeorology, 110ff
Biometeorological Research Centre, 79, 110
Bioplasma, 308ff
Bioresonance, 284–87
Biorhythm, 187, Chs. 8 & 9
Birkbeck College, 290
Birth control, 259–63
Bishop, Jim, 265
Blizard, Jane, 48, 49–50
Blum, H. F., 288
Bode, J., 11
Bogart, Humphrey, 248n
Bohm, David, 290
Bohr, Aage, 22
Bohr, Niels, 22
Bok, Bart, 245, 248–49, 275
Bong Han, Kim, 304
Bookman, M., 82
Brady, J. L., 5–6
Braid, J., 76
Brain and brain waves, 94, 112–13, 132, Ch. 8 *passim*, 224–25, 239–42. See also EEG
Brain Research Institute (UCLA), 204
Brasília, 271
Brekhman, I., 219
Breuer, J., 76
Brown, Barbara B., 223, 225, 229, 231, 235–36
Brown, E. W., 44
Brown, F. A., Jr., 81, 203, 208–10
Browning, Iben, 104, 271–272
Bruno, Giordano, 273
Bryson, Reid, 105
Buckle, H. T., 111
Budyko, M. I., 81
Burckhardt, G., 240
Burr, Harold S., 196, 238, 263, 298–99

Calandra granaria, 214
Callahan, P. S., 292
Cambel, A. B., 58
Camerata Hospital, 180
Capra, Fritjof, 291
Carlson, Chester, 316
Carrington, R. C., 34, 38, 43–44, 170
Carson, Rachel, 132
Caymaz, G., 57–58, 134–35, 219, 302
Center for Short Lived Phenomena, 161, 176–78
Central Intelligence Agency, 105
Cerletti, Ugo, 240
Chacornac, M., 249
Chakras, 305
Chandler, S. C., 7
Chandler Wobble, 7, 149, 156
Charcot, J. M., 76
Ch'i, 198, 286
Chiang Kai-shek, 269, 302
Chizhevsky, A. L., xv, 129–130, 132, 183–85, 267–273, 287–88, 322
Christian Science, 75, 76
Chronopharmacology, 197–199
Churchill, W. S., 123
Circadian rhythms, 188ff
Clark, J. H., 52
Clarke, Arthur C., 50, 107
Clayton, H. H., 65, 320
Climate, 38, 101
Clinton Prison, 304–305
Coe, Mayne, R., Jr., 237, 305
Cohen, David, 70, 99
Comité Para, 252
Concorde, 148
Consciousness:
 altered states, 234–35
 quantum mechanics and, 289–90
Cookworthy, W., 94
Cooper, H. J., 253–55
Copernicus, N., 8–9, 248, 322–23
Corliss, W. R., 340
Cosmobiology, 273
Coulomb, C. A. de, 128
Crab nebula, 23
Cristofv, C., 133, 135
Cycles (Ch. 6):
 characteristics of, 171–72
 circadian, 188–89
 commodity prices, 170–171
 emigration, 269
 fertility, 213–16, 259, 298–99
 historiometric, 267ff
 industrial production, 172
 infradian, 188, 190
 in living systems, Ch. 7, 8 *passim*
 mass movements, 267ff
 presidential deaths, 264–267
 sunspots, 37–51, 169–70 178–85. (See also Sunspots)
 ultradian, 188, 190
 voting patterns, 269, 271–72
 weather, 107–108
Czerski, P., 283–84

Danjon, A., 25
Dansgaard, W., 108
D'Arsonval, M. A., 77, 78, 99
Darwin, C., 297
Davies, David, 175–76
Day, length of, 25, 149
Dean, G. A., 65, 274, 275

354   Index

de Broglie, L. V., 27
Delbrück, Max, 294
Delgado, José, 240
Dermatome, 303
Dewan, Edmond, 214–15, 226, 259
Dewey, E. R., 171, 173–74, 178, 265n, 270
Digby, W., 44
Dingwall, E. J., 76
Dirac, P. A. M., 27
DNA, 202, 284, 285, 287, 290, 298, 309
Dobrovolsky, G. T., 57
Dogon, 247
Dolphins, 295
Domash, L., 232
Dowsing, 76, 87–94
Du Bois-Reymond, P. D. G., 70
Dubrov, A. P., 208
Dudley, H. C., 27
Durkheim, E., 114

Earthquakes:
  general, xii, 102, 137, Ch. 5, 175
  animal behaviour, 143–47
  causes of, 142
  in 1960, 158
  in 1976, 149
  Moon and, 149ff, 161–64
  oil fields and, 146
  planets and, 140, 157ff
  power plants and, 164–65
  precursors of, 142–47
  prediction of, 144–47, 156ff
  reservoirs and, 165
  Sun and, 149ff
  tides (Earth) and, 149ff
  Individual earthquakes:
    Agadir, 1960, 158, 159
    Alaska, 1964, 145
    Bucharest, 1977, 158
    California, 139–40, 144n, 147, 159
    China, 148–49, 158, 165
    Concepción, 1835, 144
    Denver, 165
    Hollywood, 1977, 158
    Kuril, 1965, 161
    Lisbon, 1755, 122
    New Hebrides, 1976, 158
    Quetta, 1935, 141, 142–143, 156, 162
    Quetta, 1955, 163
    Sofiades, 1954, 144
    Stoke-on-Trent, 1976, 163
    Truckee, 1966, 155
    Tyup, 1970, 154
Ebony, 240
Eddy, J. A., 32, 37, 140–41
Eddy, Mary Baker, 76
Eggs, signals from, 314
Eichmann, A., 254
Einstein, A., xi, 26, 27, 323
Electroauragraph, 71
Electrocardiograph (EKG), 71, 93, 213, 221, 226, 227
Electroencephalograph (EEG), 70, 80, 205–206, 217, 221, 224–25, 231, 233, 235
Electromagnetic field, 74. (See also Geomagnetic field)
Electromagnetic pollution, 71–72, 219
Electromagnetic spectrum, 20, 118
Electromagnetism (EM), 20–22, 69, Ch. 3 *passim*
Electromyograph (EMG), 221

Electroshock, 218
Electrosleep, 79, 216ff
Eleutherococcus senticosus max., 219
ELF fields and waves, 83, 111, 118–25, 209
Eliot, T. S., xvi, 322
Elizabeth I, 74
Elliotson, J., 76, 302
Elster & Geitel, 128–29
Elwin, H. V. H., 262
Energy body, xiii, Ch. 12 *passim*
Engels, F., 307
Environmental Health and Light Inst., 86
Epilepsy, 212, 241
EPR paradox, 290
Erskine, D. R., 188
Esdaile, J., 76
Eskdailemuir, 61
ESP, 83
Ether, 25–27
Exciton, 308
Eysenck, H. J., 253, 275

Fabricius, J., 37
Faraday cage, 21–22, 133, 211, 219, 314
Fate, 237
Federal Bureau of Investigation, 137
Feola, J., 245
Ferguson, Marilyn, 234
Ferrofluid, 299–300
Finsen, N. R., 288
Finsen Institute, 288
Fischer, V., 234
Fitzroy, R., 144
Fliess, W., 189, 192ff
Foehn, 114, 130
Forbes, 173
Forbes, G., 5

Forbush decrease, 35
Fort, Charles H., 176–77
Foundation for the Study of Cycles, 56, 110, 173–74, 185
Franco, F., 157
Franklin, Benjamin, 128, 143
Freud, S., 76, 193
Fritz, H., 43

Gable, Clark, 194
Gage, Phineas, 239–40
Galileo G., 8–9, 27–28, 37, 248n, 268, 273, 322–23
Galvani, L., 69, 71
Gamow, G., 22
Garfield, J. A., 264
Garland, Judy, 194
Garrett, Eileen J., 136
Gassner, J. J., 75
Gauquelin, Françoise, 249
Gauquelin, Michel, xiv, xv, 181–82, 191, 249–53, 256, 258, 273, 274, 275, 322
Gavreau, Vladimir, 120–22
Geller, Uri, 95
Geobiology, 72
Geomagnetic field (GMF):
  general, 22, 81
  animal behaviour, 81–83
  auroras, 36
  blood, 116
  growth of children, 116
  heart beat rate, 116, 202
  human activity, 208
  menstruation, 117
  mushrooms, 83
  nervous system, 202
  pigeons, 81–83
  plants, 83
  pole reversal, 100

Geomagnetic field (*cont.*)
  psychiatric disturbance, 100
  timing of birth, 246, 251–53
Geopathic zones, 72, 90–91
Geothermal energy, 165
Gilbert, W., 74
Gnostica, 157
Grad, Bernard, 79
Grattan, H., 88
Gravity, 14–21
Green, Alyce, 223, 235
Green, Elmer, 223, 235
Greenhouse effect, 106
Gribbin, John, 25, 48, 56, 139, 140, 149, 159, 160, 256
Grishchenko, V. S., 309
Grunion, 213–14
Guliayev, P. I., 71
Gulliver's Travels (Swift), 7
Gurvich, Anna A., 84
Gurvich, A. G., 84–86, 283, 310, 313
Gutenberg, B., 148n
Gymnarchus niloticus, 81

Halberg, F., 199
Hale, G. E., 38–39
Hall, Asaph, 7
Halley, E., 9
Halley's comet, 6
Hameroff, S., 285
Hansen, K. M., 78
Harding, W. G., 264
Harmonics, 12
Harriot, T., 37
Harrison, W. H., 264
Harvalik, Zaboj V., 89–90, 93–94
Head, R., 48–49
Heath, R., 241

Hell, M., 75
Helmholtz, H. L. F. von, 70
Helmont, J. B. van, 74
Hermes Trismegistos, 248
Herschel, J., 38
Herschel, W., 6–7, 9, 38, 64, 170
Hersey, R. B., 195, 196–97
Hill, Graham, 230
Hill, Scott, 93–94, 161, 208, 226–27, 236–37, 277, 287, 289, 307–19
Hillary, W., 183
Hippocrates, 112
Hodgson, R., 34
Holiday, E. R., 131
Holmes, Sherlock, 248n
Holroyd, Stuart, xv
Home Office, 283
Homer, 32
Hopkins, F. G., 131
Hoskins, Chapin, 173
Hosty, Graham, 7
Humboldt, A. von, 129
Hunderfossen, 164–65
Hunter, R., 157
Huntington, Ellsworth, 173, 254
Hurkos, Peter, 235
Huxley, Julian, 174
Huygens, C., 210
Hyfreeg, 234
Hypothalamus, 113

IMP-3, 36
IMP-7, 24–25
Imperial College (London), 18, 293
Infrared radiation (IR), 21, 282, 292
Infrasound, 120–24, 147
Institute for Arctic and Antarctic Research (Leningrad), 151

Institute for Biologically Active Substances (Vladivostok), 219
Institute for Brain Research (Leningrad), 85
Institute of Earth Physics (Moscow), 208
Institute of Normal Physiology (Moscow), 318
International Geophysical Union, 159
International Institute for Interdisciplinary Cycle Research, 174, 176
International Society for Biometeorology, 110
Inyushin, V. M., xv, 309–311, 314, 319, 322–23
Irwin, J. B., 134

Jenny, E., 90
Jet-lag, 210–11
Jevons, W. S., 170
Jodko-Narkiewicz, J., 316
Johannes, Bob, 214, 322
Johnson, L. B., 265
Johnson, M. O., 47
Jonas, Eugen, 259–63
Jose, P. D., 42–43, 48
Jung, C. G., 32, 229, 254–55
Jupiter, xii, 5, 23, 43, 44, 48, 160, 247, 250
Jupiter Effect, The (Gribbin & Plagemann), 140–41, 266
Jupiter/Saturn conjunction, 140–41, 266

Ka, 296
Kahuna, 297
Kaibyshev, M. S., 116
Kamiya, Joe, 223, 225
Karajan, H. von, 228

Kardec, Allan, 312
Karlsson, Leif, 314
Kazhinsky, B. B., 96
Kaznachayev, V. P., 85–86
Keeton, W. T., 82, 148
Kekulé, F., 322
Kennedy, J. F., 264–65, 266
Kepler, J., 7, 8–9, 10–11, 34, 37, 52, 160, 248–49, 250, 252, 266, 273–74, 322–23
Keratin, 83, 95, 143
Kesey, Ken, 218
Khamsin, 130
Kholodov, Yu. A., 202–203, 206–207
Kinch, Brite, 226–27
King, J. W., 60–63
Kircher, A., 74
Kirlian, S. D. & V. Kh., 308, 313
Kirlian effect, 316–19
Kjellson, Henry, 125–27
Kmetz, J. M., 313
Koestler, A., 8–9, 322
Kokoschinegg, P., 302
Komarov, V. M., 57
König, H. L., 119
Kornblueh, Igho, 132
Kornbluth, C. M., 301
Kosygin, A., 254
Kraemer, Helena, 147
Krippner, Stanley, 265, 266
Kritzinger, H. H., 6
Krueger, A. P., 131
Kulikovsky, P., 7n
Kundalini, 305

Lacey, Louise, 215
Laithwaite, E. R., 18, 293, 294, 299–300
Landscheidt, T., 12, 14, 49, 160, 161
Lane, Frank W., 144

Langman, Louis, 298
Laser:
  in acupuncture, 281–82
  bioplasma and, 314–15
  eye surgery, 279
  facial paralysis, 281, 288
  vegetable growth, 280–81
Lauda, Niki, 230–31, 238
Lawrence, J. L., 95
Leacock, Stephen, 3, 4
Lenin, V. I., 254, 272, 307
Levavasseur, R., 121
Lewis, W. A., 92
L-field, 196, 298–299, 312
Lichtenberg, G., 316
Liébeault, A., 76
Lilly, J. C., 295
Lincoln, A., 264, 265
Lobban, Mary, 191
Lockyer, N., 171
Lodge, Oliver J., 26–27
Long, Max Freedom, 297
Louis XIV, 37–38
Lovell, B., 7n.
Luby, W. A., 47
Luce, G. G., 188n, 192, 199
Ludwig Boltzmann Institute, 302
Lycée Charlemagne, 249

Mabaan, 124
Maby, J. Cecil, 88, 91
McBurney point, 303
McDonnell Douglas Astronautics, 47
Machu Picchu, 127
McKinley, W., 264
Magnetic storm, 35, 36
Magnetism, 74ff, 78–79, 80, 98–99. See also Geomagnetic field
Magnetocardiograph (MKG), 71
Magnetoencephalograph (MEG), 70
Magnetohydrodynamics (MHD), 42
Magneto-meteorology, 62
Magnetotropism, 83
Maharishi International University, 232
Maharishi Mahesh Yogi, 232
Malinowski, B., 262
Mammoth Cave, 189
Manczarski, S., 310
Many-body problem, 7–8, 14
Mao Tse-tung, 302
Marino, A., 96
Mars, 5, 7, 9, 250
  moons of, 7
Marxism, 280, 307
Mason, B. J., 104, 106
Massachusetts Institute of Technology, 82, 232
Mattuck, Richard D., 7
Maunder, A. S. D., 44
Maunder, E. W., 42
Maunder minimum, 42, 65
Maxey, E. Stanton, 80, 135–36, 207, 301
Max-Planck Institute (Physiology), 204–205, 221
Mayakovsky, V. V., 267
Mayo, Jeff, 275
Meidav, M., 186
Meldahl, K. G., 45–46
Mémery, H., 56
Menarche, 216
Menninger Foundation, 232
Menses, 213
Mercalli scale, 148n
Mercury, 5, 6, 10, 37, 46, 48, 176
Mesmer, F. A., 75–78, 79, 80

Mesmerism, 75–78, 79, 129, 302
Messerschmidt-Bölkow-Blohm, 282
Miami Heart Institute, 301
Michelson-Morley experiments, 26
Microseism, 154
Microtubules, 285
Middlesex Hospital, 74
Miller, Dayton C., 27
Miller, Stanley, 201
Mishima, Y., 115
Mistral, 130
Mitogenetic radiation, 84, 283
Molchanov, A. M., 12–13, 50
Molotov, V. M., 254
Moniz, Egas, 240
Monroe, Marilyn, 194
Montgomery, Betty, 143
Montgomery of Alamein, 143
Months, 17–18, 196, 213
Moon:
  animal sensitivity to, 208–210, 213–14
  and earthquakes, 149ff
  and homicide, 263–64
  and human fertility, 213ff
  motion of, 17
  red spots on, 150–51
  tides, 15–17
Moore, F. R., 82
Moore, Patrick, 7n
Moss, Stirling, 230
Motoyama, H., 300–301, 305–306
Mt. Olympus, 26
Mt. Wilson, 27, 38
Mozart, W. A., 75
Mulholland, T., 223
Muria, 262

Museum of Natural History (Washington), 154
Music and the Brain, 228
Music and heartbeat, 226–228
Mussolini, B., 269–70
Myers, F. W. H., 297

Nagahama, Y., 304
NASA, 9, 36, 54, 55, 56–57, 106, 150–51, 328
National Science Foundation, 140
National Cancer Institute, 288
Nationwide, 103
NATO, 135
Nature, xi, 59–60, 159, 160, 175, 256, 280
Neely, J., 160
Nelson, J. H., 29–32, 34, 48, 51–57, 65, 158, 159, 161, 274, 275, 322, 328–329
Neptune, 47
New Scientist, 18n, 175–76
Newton, I., 8–9, 14–15, 18, 30, 42, 44, 248, 322–23
Nixon, R. M., 302
NORSAR, 164–65
North Middlesex Hospital, 227
Northrop, F. S. C., 238
Northrop Services Inc., 49
Norton, W. A., 44
Nova Sagitta 1977, 7
Nudel, B. M., 234
Nut, 32

Obrig laboratories, 86–87
Ohmi Railway Co., 193–94
Öpik, Ernst, 7n, 47
Orgel, L., 202

Ørsted, H. C., xi, 69
Ott, J., 86
Owen, Dr, 78
Oxburgh, Ron, 165–66

Padma-lax, 198
Palau islands, 214
Panax ginseng, 219
Panek, S., 116–17
Paracelsus, 74
Patsayev, V. I., 57
Pauli, W., 27
Pavlov, I. P., 218, 229
Pearce, A. J., 157
Pendulum, 19
Perkins, Elisha, 78
Persinger, M., 122–23
Petersen, W. F., xv, 109–10, 322–23
Petschke, H., 91
Phase locking, 210ff
Photon code, 85
Photoperiodism, 189
Piccardi, Giorgio, xv, 178–181, 190, 202–203, 251, 322–23
Pickard, Barbara, 314
Piezoelectricity, 82–83, 96
Pigeons, 82–83
Pineal gland, 216
Pioneer 5, 36
Pioneer 10, 24–25
Pioneer 11, 24–25
Pittendrigh, C. S., 189
Plagemann, S., 25, 139, 149, 159, 160
Planck, Max, 20
Planets:
  alignments, 140–41
  aspects, 31–32, 254
  discovery of, 5
  distance from Sun, 11
  geomagnetic field and, 246
  periods, 9–10
  radio reception and, 31, 51–57
  tidal effects of, 16–18
  timing of birth and, 249–256 (See also individual planets)
Planet X, 6, 11
Plasma, 308
Plate tectonics, 140
Playfair, G. L., 50–51, 124, 141–42, 148, 233
Pledge, H. T., 258
Pluto, 5, 6, 15, 47
Pogson, N. R., 171
Pohl, Baron von, 91
Point Foundation, 261
Poltergeist, 305
Pomeranz, Bruce, 307
Popp, F. A., 286–87, 309
Pöppel, E., 205, 211
Portig, W. H., 44
Poumailloux, M., 116
Prana, 26
Press Wireless Inc., 52
Pressman, A. S., 83
Prieditis, A. A., 157–58, 159
Project Mohole, 142
Project Sanguine, 117–18, 321
Prokhorov, A. M., 280
Proxima Centauri, 5
Psychical self-regulation, 236–237
Psychosurgery, 240
Puharich, Andrija, 94–95, 136
Pulkovo observatory, 150
Pulsar, 22–23, 186
Pythagoras, 10, 52

Quantum mechanics, 289–292

Quasar, 22–23, 119
Quimby, P., 76

Radio One therapy, 227
Radiosleep, 217
Rambeau, G., 91
Ransom, J. B., 305
Ravitz, L. J., Jr., 196, 238–239, 298
RCA Communications Inc., 29–32, 51ff
Regenerative growth, 98–99
Reiter, R., 62
Rentsch, Dr, 217
Resonance, 12–14, 121, 203
Richter, C., 148n.
Richter scale, 148n.
Rigshospital (Copenhagen), 78
Rocard, Yves, 93
Rock, J., 215, 259
Rock Reproductive Clinic, 215
Rocky Mountain Arsenal, 165
Roman, C., 146
Romen, A. S., 236–37
Roosevelt, F. D., 264
Rosen, S., 302–303
Rostand, J., 252–53
Rutty, I. A., 183
Ryan Aeronautical Co., 106, 246

Sadeh, Dror, 186
St Clair, David, 144n.
Salazar, A., 269
San Andreas fault, 139, 140, 146, 149
Sanford, Fernando, 45
Santa Ana (wind), 130
Satori, 233–34
Saturn, 5, 44, 247, 249–50
Scheiner, C., 37

Schizophrenics, 112–13, 257
Schjelderup, V., 85–86
Schrödinger, E., 290–91
Schuldt, H., 117
Schultz, J., 237
Schumann, W. O., 119
Schumann resonance, 119, 207
Schwabe, H., 38
Schwarz, Jack, 232, 233, 235
Science, 298
Science Museum (London), 18–19
Science since 1500 (Pledge), 258
Science Unlimited Foundation, 313
Season of birth and:
  brain disorder, 257
  extraversion, 275
  intelligence, 257
  menarche, 257
  mental disease, 255–56
  neuroticism, 275
  physique, 257
  profession, 249ff, 253ff
Sedlak, W., 309
Semiconduction, 97, 281, 308
Serotonin, 131
Shakespeare, W., 16, 176
Sharav, 130
Shults, N. A., 184
Siddhi, 237
Simkin, T., 154
Simpson, J., 155
Sirius, 247
Skinner, B. F., 223, 296
Skylab, 41
Sleeper, H. P., Jr., 49–51, 246
Smith, Bruce, 147

Index

Smithers, Alan, 253
Solar cycle, see Sunspots
Solar flare, 34–37
Solar induced currents, 59
Solar system centre of mass, 33, 140
Solar wind, 24
Soletta, 106
South African Air Force, 133–34
Soyuz I, 57
Soyuz XI, xii, 57, 134, 204
Spallanzani, L., 293, 313
Spannagel, K., 185
Speleotherapy, 185
Spin-orbit coupling, 50
Spiritism, 312
Spiritualism, 75, 76
Spontaneous combustion (human), 137–38
Sputnik I, xi, 9, 24–25, 293
Stalin, J. V., 269–70
Stängle, J. W. F., 91
Starfighter, 133–34
Stengel, Erwin, 114
Stern, 261
Stetson, Harlan T., 164
Stewart, Duncan, 306
Stimoceiver, 241
Stonehenge, 261
Suicide, xii, 113–15, 130
Sulman, F. G., 130
Sun:
    general, Ch. 1, 2 *passim*
    corona, 33–34
    statistics, 33
    tides, 15–16
    weather and, 107–108
    See also Sunspots
Sun, Electricity, Life, 273
Sunspots:
    blood flocculation and, 181–82

    cycle, 50–51, 169, 170, 178–84
    description of, 38–39
    discovery of, 37ff
    earthquakes and, 139–41
    epidemics and, 183–85
    growing season and, 61
    heart failure and, 115–16
    mortality rate and, 184
    planets and, 43–50, 65
    radio conditions and, 51–57
    rainfall and, 170, 171
    tree rings and, 170
    voting trends and, 270–272
Superhigh frequency (SHF), 24
Swami Rama, 232–34, 237
Swift, J., 7
Swoboda, H., 192–93
Sytinsky, A. D., 151–53
Szent-Györgyi, A., 96–97, 309

Takata, Maki, xv, 116, 181–82, 251, 322–23
Tamrazyan, G. P., 151
Taylor, J. D., 227
Taylor, J. G., 92
Teltscher, A., 193
Teng Hsiao-ping, 149
Tetens, Johan, 111
Theophrastus, 37
Thermovision, 282–83
Tibet, 125–27
Tides, x, 15–17, 44, 45
Times, The, 60
Titayev, A. A., 83
Titius, J., 11
Titius-Bode series, 11
Tomaschek, R., 155, 159–160, 161

Tombaugh, Clyde, 6
Tomorrow's World, 72
Townes, C. H., 280
Trabka, Jan, 234
Transcendental meditation (TM), 231–32, 235
Transdermal Instrument (TD 100), 95–96
Tramontana, 130
Tromp, S. W., xv, 79, 80, 89, 90, 92, 93, 110–14, 176, 190, 207
Turkish Air Force, 134

Ultraviolet radiation (UV), 21, 84–87, 111–12, 282–287, 291–93
UNESCO, 110, 174
United Airlines, 193
University:
  Akron State, 155
  Bradford, 253
  California, 131
  Cambridge, 124, 191
  Chiba, 304
  Clemson, 82
  Copenhagen, 8
  Cornell, 82
  Edinburgh, 306
  Graz, 218
  Harvard, 164
  Hebrew (Jerusalem), 130
  Illinois, 123
  Kazakh State, 280, 309, 312–15
  Leningrad State, 310
  London, 275
  Lublin, 309
  Marburg, 286
  Michigan, 241
  Munich, 119
  Northwestern, 81, 203
  Pennsylvania, 195
  Pittsburgh, 174
  St Louis, 314
  Semmelweiss, 281
  Sorbonne, 249
  Stanford, 64, 189
  Toho, 181
  Tulane, 241
  UCLA, 204
  Utah, 214
  Vienna, 75
  Wayne State, 58
  Wyoming, 195
  Wisconsin, 105
Uranus, xii, 5, 6, 7, 47, 159–60, 161–63
U.S. Air Force, 133
U.S. Department of:
  Agriculture, 292
  Commerce, 173
  Defense, 207, 216–17
  Health, Education, Welfare, 284
U.S. National Ocean Survey, 157
U.S. Navy, 118, 321

Valenstein, E. S., 241
Van Allen belts, 23–24
Van de Kamp, P., 7n
Vasiliev, L. L., 85
Vaux, J. E., 185
Venus, 5, 13, 23, 44, 45, 48–50
Verein für Geobiologie, 72
Viart, R., 116
Volcanoes:
  Galapagos, 154
  Krakatoa, 122
  Mt Pelée, 158
Volkov, V. N., 57
Volta, A., 128
Volta cross, 77, 78
Vulcan, 6, 7, 37

Walker, E. H., 290–91
Wallace, A. R., 297
Wallace, R. K., 232
Wallendas, 80
Wall Street crash, 173, 270
Weather, 59ff, 101, Ch. 4 passim. See also Biometeorology
Weber, Joe, 21, 96
Wernøe, 303
Wesley, J., 75
Westinghouse Electric Co., 173
Wilcox, J. M., 64
Williams, David, 59, 266
Wilson, A., 38
Wilson, C. T. R., 128–29
Wisden's Cricketers' Almanac, 61

Wolf, Christian, 11
Wolf, Rudolf, 43
Wolf number, 43, 327
Wood, K. D., 47, 48
Wood, R. M., 47
World Meteorological Organization (WMO), 104

Xerography, 316

Yoga, Ch. 9 *passim*

Zeitgeber, 189, 203
Zen, 231–32, 239
Zerbino, Gustavo, 238
Zodiac, see Season of birth
Zugspitze, 62

MONTH AFTER MONTH ON BESTSELLER LISTS ALL OVER AMERICA!

THE SECRET LIFE OF PLANTS

PETER TOMPKINS and CHRISTOPHER BIRD

The fascinating, astonishing account of the physical, emotional, and spiritual relations between plants and man.

OVER 100,000 HARDCOVERS IN PRINT!

"FANTASTIC REVELATIONS"

The New York Times

"DON'T MISS THIS ONE"

Library Journal

"I BEGAN A SKEPTIC, ENDED BY CROONING A LULLABY TO MY PHILODENDRON."
Harriet Van Horne, *New York Post*

19901/$1.95

PLANT 4-77

SIX MONTHS ON NATIONAL
BESTSELLER LISTS!

DR. WAYNE W. DYER

goes beyond
his record-breaking bestseller

YOUR ERRONEOUS ZONES

in his remarkable new book

PULLING YOUR OWN STRINGS

Now Dr. Dyer zeroes in on those people
and institutions that manipulate you and tells
you how to get them off your back forever!

Each chapter is organized around a major
principle for not being victimized, with concrete
tactics for dealing from a position of strength with
co-workers, clerks, bureaucrats, relatives, lovers
—and all too often, yourself.

SELECTED BY FOUR MAJOR BOOK CLUBS

 Avon/44388/$2.75

PYS 8-79